ROUTLEDGE HANDBOOK OF MEDIA GEOGRAPHIES

This Handbook offers a comprehensive overview of media geography, focusing on a range of different media viewed through the lenses of human geography and media theory. It addresses the spatial practices and processes associated with both old and new media, considering "media" not just as technologies and infrastructures, but also as networks, systems and assemblages of things that come together to enable communication in the real world.

With contributions from academics specializing in geography and media studies, the *Routledge Handbook of Media Geographies* summarizes the recent developments in the field and explores key questions and challenges affecting various groups, such as women, minorities and persons with visual impairment. It considers geographical aspects of disruptive media uses such as hacking, fake news and racism. Written in an approachable style, chapters consider geographies of users, norms, rules, laws, values, attitudes, routines, customs, markets and power relations. They shed light on how mobile media make users vulnerable to tracking and surveillance but also facilitate innovative forms of mobility, space perception and placemaking. Structured in four distinct sections centered around "control and access to digital media," "mass media," "mobile media and surveillance" and "media and the politics of knowledge," the Handbook explores digital divides and other manifestations of the uneven geographies of power. It also includes an overview of the alternative social media universe created by the Chinese government.

Media geography is a burgeoning field of study that lies at the intersections of various social sciences, including human geography, political science, sociology, anthropology, communication/media studies, urban studies, and women and gender studies. Academics and students across these fields will greatly benefit from this Handbook.

Paul C. Adams is Professor of Geography at the University of Texas at Austin. His research is situated at the intersection of media studies, communication theory and human geography. His work considers how socio-spatial perceptions, representations, actions and infrastructures are intertwined through mediated communications.

Barney Warf is a Professor of Geography at the University of Kansas. His research and teaching interests lie within the broad domain of human geography. His research includes telecommunications and political geography viewed through the lens of political economy and social theory. He edits *Geojournal* and co-edits *Growth and Change*.

ROUTLEDGE HANDBOOK OF MEDIA GEOGRAPHIES

Edited by Paul C. Adams and Barney Warf

LONDON AND NEW YORK

First published 2022
by Routledge
2 Park Square, Milton Park, Abingdon, Oxon OX14 4RN

and by Routledge
605 Third Avenue, New York, NY 10158

Routledge is an imprint of the Taylor & Francis Group, an informa business

British Library Cataloguing-in-Publication Data
A catalogue record for this book is available from the British Library

Library of Congress Cataloging-in-Publication Data
A catalog record has been requested for this book

ISBN: 978-0-367-48285-5 (hbk)
ISBN: 978-1-032-11916-8 (pbk)
ISBN: 978-1-003-03906-8 (ebk)

DOI: 10.4324/9781003039068

Typeset in Bembo
by Taylor & Francis Books

CONTENTS

List of illustrations   *viii*
List of contributors   *x*

1   Media geographies: An introduction   1
    Paul C. Adams and Barney Warf

PART I
Control and access to digital media   **17**

2   Internet censorship: Shaping the world's access to cyberspace   19
    Barney Warf

3   Digital divides   29
    James B. Pick and Avijit Sarkar

4   Hacking in digital environments   49
    Mareile Kaufmann

5   The internet media in China   60
    Xiang Zhang

6   Digital media and persons with visual impairment or blindness   74
    Susanne Zimmerman-Janschitz

PART II
Mass media   **93**

7   Newspapers: Geographic research approaches and future prospects   95
    Paul C. Adams

8   Fake news: Mapping the social relations of journalism's legitimation crisis   106
    James Compton

9   Film geography   118
    Elisabeth Sommerlad

10 Approaches to the geographies of television   132
    James Craine

11 Geographical analysis of streaming video's power to unite and divide   145
    Irina Kopteva

PART III
Mobile media and surveillance   **159**

12 Evolving geographies of mobile communication   161
    Ragan Glover-Rijkse and Adriana de Souza e Silva

13 Moving: Mediated mobility and placemaking   172
    Roger Norum and Erika Polson

14 Geographies of locative apps   183
    Peta Mitchell, Marcus Foth and Irina Anastasiu

15 Digital surveillance and place   196
    Ellen van Holstein

PART IV
Media and the politics of knowledge   **207**

16 Race, ethnicity and the media: Absence, presence and socio-spatial
    reverberations   209
    Douglas L. Allen and Derek H. Alderman

17 Nationalism, popular culture and the media   220
    Daniel Bos

18 Eurocentrism/Orientalism in news media   232
    Virginie Mamadouh

Contents

19 Sex, gender and media   245
    Marcia R. England

20 Media, biomes and environmental issues   256
    Hunter Vaughan

Index   268

ILLUSTRATIONS

Figures

2.1 Map of internet censorship scores 23
3.1 Number of individuals using the internet by development status,
2005–2019 30
3.2 Worldwide subscriptions and use of technologies, 2005–2019 31
3.3 Networked readiness index framework 33
3.4 van Dijk model of divides of digital media use 34
3.5 SATUM model of correlates of level of technology 35
3.6 Mobile broadband subscriptions and penetration worldwide,
2008–2019 36
3.7 Internet use to watch videos online, United States, 2017 41
3.8 Internet use for social networking, United States, 2017 42
3.9 Internet use for social networking, United States counties, 2015 43
3.10 Facebook penetration, Latin America and the Caribbean, 2013–2015 44
5.1 Number of internet users in China, December 2013 – March 2020 61
5.2 Number of mobile internet users in China, December 2013 – March
2020 61
5.3 The hierarchical structure of state media in China 62
5.4 Governance structure of the CPC Publicity Department 64
5.5 A comparison across tradition media 65
6.1 Essential research topics in the context of orientation and navigation for
persons with VIB 79
6.2 Detailed turn-by-turn directions in ways2see for people with VIB 84
9.1 Relational perspectives on film geography: Overview 120
9.2 Types of screen-tourism 123
9.3 An example of the integration of filmic practice into a university curriculum
in the MA program "Human Geography: Globalisation, Media, and Culture"
at Johannes Gutenberg University Mainz 126
11.1 Countries with the top 26 independent YouTube influencers and their
combined subscriber base 146
11.2 Top 10 countries with the most YouTube unique users and internet users,
2016 148

Tables

2.1   Global population and internet users by severity of internet censorship, 2020   22
5.1   Top news and information accounts on the WeChat platform (June 2020)   68
5.2   Top news and information accounts on the Weibo platform (June 2020)   69

CONTRIBUTORS

Paul C. Adams is Professor of Geography at the University of Texas at Austin. He is the author of *Geographies of Media and Communication* (Wiley-Blackwell, 2009), co-author of *Communications/Media/Geographies* (Routledge, 2016), and co-editor of the *Research Companion to Media Geography* (Ashgate, 2014). His research bridges media studies, communication theory and human geography.

Derek H. Alderman is Professor of Geography at the University of Tennessee and Past President of the American Association of Geographers (2017–2018). He is the (co)author of over 140 articles, chapters and other essays, many focused on the intersection of race, memory and place in popular culture, tourism and place-naming.

Douglas L. Allen is an Assistant Professor of Geography in the Department of Social Sciences at Emporia State University. His recent work on placemaking, music/festival performance and Black geographies has been published in *Progress in Human Geography, Antipode, Cultural Geographies* and *Geoforum*.

Irina Anastasiu is conducting postdoctoral research at the QUT Design Lab at Queensland University of Technology, Australia, contributing to the ARC-funded study "Digital Media, Location Awareness, and the Politics of Geodata." Her interests include (geo)privacy, and data and technological sovereignty in smart cities.

Daniel Bos is a Senior Lecturer in Human Geography at the University of Chester, UK. Daniel's research explores the relationship between visual culture, geopolitics and militarization. He is the co-author of *Popular Culture, Geopolitics, and Identity* (2nd edn, Rowman & Littlefield, 2019).

James Compton is Associate Professor in the Faculty of Information & Media Studies at the University of Western Ontario. He is author of *The Integrated News Spectacle: A Political Economy of Cultural Performance* (Peter Lang, 2004), and co-editor of *Converging Media, Diverging Politics* (Lexington Books, 2005).

James Craine is Professor of Geography at California State University, Northridge. He is co-editor of the *Research Guide to Media Geography* (Ashgate, 2014), co-editor of *The Fight to Stay Put* (Franz Steiner, 2013), and has authored numerous articles on media geography. His research focuses on advanced cartographic design and visual methodologies.

Adriana de Souza e Silva is Professor of Communication at North Carolina State University (USA). She is the co-editor and co-author of several peer-reviewed articles and five books, including *The Routledge Companion to Mobile Media Art* (Routledge, 2020, with Larissa Hjorth and Klare Lanson) and *Hybrid Play* (Routledge, 2020, with Ragan Glover-Rijkse).

Marcia R. England is Professor and Chair in the Department of Geography at Miami University. Her research focuses on geographies in pop culture, of mental health, and on representations and spaces of the body. Recent work includes a monograph entitled *Public Privates: Feminist Geographies of Mediated Spaces* (University of Nebraska Press, 2018).

Marcus Foth is Professor of Urban Informatics in the QUT Design Lab at Queensland University of Technology, Australia. He is also an Honorary Professor in the School of Communication and Culture at Aarhus University, Denmark. He is a chief investigator of the ARC-funded study "Digital Media, Location Awareness, and the Politics of Geodata" and tweets @sunday9pm.

Ragan Glover-Rijkse is a PhD candidate at NC State University. Her research examines the intersections between mobile media, infrastructures and space/place. She has a book, titled *Hybrid Play* (co-edited with Adriana de Souza e Silva, Routledge, 2020), and her work has appeared in *Mobile Media & Communication* and *Communication Education*.

Mareile Kaufmann is a Postdoctoral Researcher at the Department of Criminology and Sociology of Law, University of Oslo, and the Peace Research Institute Oslo. Mareile studies data governance and surveillance technologies, and how people engage with these from within. She heads the ERC-project "Digital DNA: The Changing Relationships between Digital Technologies, DNA and Evidence."

Irina Kopteva is an Assistant Professor—Research at the University of Colorado in Colorado Springs. She has taught geography in the United States, Europe and Russia. Her research interests include geographical education, human geography and environmental sustainability. Her article on human geography online education received an award from the National Council for Geographic Education in 2019.

Virginie Mamadouh is Associate Professor of Political and Cultural Geography at the Department of Geography, Planning and International Development Studies of the University of Amsterdam, The Netherlands. Her research interests pertain to (critical) geopolitics, political culture, media, European integration, (urban) social movements, transnationalism and multilingualism. She recently co-edited the *Handbook of the Geographies of Globalization* (Edward Elgar, 2018).

Peta Mitchell is Associate Professor in the Digital Media Research Centre and School of Communication at Queensland University of Technology, Australia. Her research focuses on digital geographies, locative media and geoprivacy in everyday digital media use. She is a chief investigator of the ARC-funded study "Digital Media, Location Awareness, and the Politics of Geodata."

Roger Norum is a Social Anthropologist whose research focuses on linkages between environment, infrastructure and mobility, primarily among transient communities in the Arctic and Asia. He holds degrees from Cornell and Oxford, and is Founding Editor of Palgrave's Arctic Encounters book series.

James B. Pick is Professor of Business at University of Redlands. He has authored or co-authored 13 books on GIS, IS and environment including *The Global Digital Divides* (Springer, 2015) and *Exploring the Urban Community: A GIS Approach* (Pearson, 2011). His current research concerns the US digital divides, GIS strategies, the sharing economy and locational privacy.

Erika Polson is an Associate Professor in the Department of Media, Film and Journalism Studies at the University of Denver, USA. Her research involves critical cultural studies of digital media and mobility in global contexts, and specifically on new ways that status is accrued or projected through mobilities. She is author of *Privileged Mobilities: Professional Migration, Geo-social Media, and a New Global Middle Class* (Peter Lang, 2016) and co-editor of the *Routledge Companion to Media and Class* (Routledge, 2020).

Avijit Sarkar is Professor of Business Analytics at the University of Redlands, School of Business. His research interests include technology adoption and diffusion and private sector use of location-based technologies. He has co-authored the book *Global Digital Divides: Explaining Change*. His research has appeared in journals in IT and telecommunications.

Elisabeth Sommerlad is a Postdoc at the Institute of Geography, JGU Mainz. She studied Geography, Communication and Sociology. Her dissertation on Intercultural Encounters in Feature Films (approved 2019) investigates the cinematic staging of interculturality in US movies. Her research focuses on interrelations between media, place, identity and belonging. She is Managing Editor of the book series Media Geography at Mainz.

Ellen van Holstein is Research Fellow in Urban Geography at the University of Melbourne. Her research focuses on technologies, policies and everyday practices that shape opportunities for citizen participation and inclusion in cities. She is currently focusing on the accessibility of urban spaces for people with intellectual disability.

Hunter Vaughan is the Environmental Media Scholar-in-Residence at the University of Colorado Boulder. He is the author of numerous books and articles, including *Hollywood's Dirtiest Secret: The Hidden Environmental Costs of the Movies* (Columbia University Press, 2019), and is a Founding Editor of the *Journal of Environmental Media*.

Barney Warf is a Professor of Geography at the University of Kansas. His research and teaching interests lie within the broad domain of human geography. His research includes telecommunications and political geography viewed through the lens of political economy and social theory. He edits *Geojournal* and co-edits *Growth and Change*.

Xiang Zhang is a Lecturer in International Development at Nottingham Trent University, UK. His research focuses on the economic geography of the internet, social and power relations in cyberspace, and China's role in globalization. He holds a PhD in Geography from the University of Kansas.

Susanne Zimmermann-Janschitz is Associate Professor at the Department of Geography and Regional Science, University of Graz, Austria. She is past Chair of the Disability Specialty Group of the American Association of Geographers. Her current research is focused on GIS in the context of sustainability and persons with disabilities.

1

MEDIA GEOGRAPHIES

An introduction

Paul C. Adams and Barney Warf

Media geographies are all around us. Networks of world-spanning infrastructure keep us in touch with family, friends, business colleagues, customers and clients. News companies headquartered in major cities form our sense of historical and contemporary reality by providing "the news," a shared here and now that circulates through geographical and social space, on newspaper and magazine pages, large and small screens, radio and podcasts. These are media geographies, as are the maps and GPS that guide people's movements through space, facilitating movements from shopping trips to vacations to intercontinental migration. Media geographies also include invented and fictionalized places created for the purposes of entertainment and escapism, including immersive games and virtual reality experiences. Media geographies surround us, guiding us, following us and taking us places.

Our understanding of the term "media" involves more than mere technology. Technologies alone do not communicate. Rather, communication depends on social processes of encoding and decoding (Hall 2001) that inevitably produce a shift in meaning, a kind of translation, between sender(s) and receiver(s), a process that reflects their positionality relative to social power and ideology. Producing and sharing meaning involves complex processes that conjoin the social, political, psychological, linguistic and geographical into a seamless whole. Media are sociotechnical processes (Marwick 2018) moving through all of the following: technologies (communication infrastructures and devices), matter (book pages, ink, wires, satellites, fiber optics, screens), codes (ways of making sense of marks and tracings in matter), information (data, facts, rumors, stories, discourses, narratives), actors (people, bookworms, companies, governments, algorithms, switching devices), and systems (economies, legislative apparatuses, bureaucratic organizations, libraries). The list could be expanded indefinitely, but the point is that a medium is not a technical object, a thing, but rather a network or assemblage, a hybrid mix of the tangible and intangible. Not all of the authors contributing to this volume would espouse actor network theory or assemblage theory, and our wish is not to promote a single approach to media geography; the issues our authors address treat media as neither wholly social nor wholly technical, but rather as *sociotechnical* phenomena and processes.

When we were making our first forays into media geography, there were particular media that interested us. Adams was interested in television (1992) and Warf in telecommunication networks supporting financial transactions (1989). Geographers interested in media usually

DOI: 10.4324/9781003039068-1

focused on particular media, such as paintings (Cosgrove 1985), newspapers (Goheen 1990), writing (Barnes & Duncan 1992) and films (Aitken & Zonn 1994). This interest in particular media did not deny that various media "texts" interrelated to each other intertextually (Kristeva 1980), but it was assumed that literature, news media, telephones, television, music recordings and film were all distinct and separate media that could be studied on their own. Of course, no one in the 1970s or 1980s picked up a telephone to watch a movie, read a newspaper, play games, or monitor their bank account. Now these things are all common-place, indicating a process of media convergence made possible by the digital revolution (Cupples 2015). One can still use a phone to call someone, but one can also check email, only to be interrupted by one's calendar, or a reminder to order a prescription, or a ping indicating a friend has posted on social media. The phone is no longer just a phone, but also a mailbox, bank teller, pharmacist, calendar, movie theater, television, newspaper, toy, puzzle, game, notepad and tape recorder; we can also use a laptop or desktop computer for these very same things (Koeze & Popper 2020). When we speak in terms of "apps" rather than particular devices, we are speaking of families of software that can be hosted on various devices in a plethora of places. It becomes such a tangled mess that "media" has become a singular noun because it is hard to identify a particular medium. The media is a muddle of things from which most of us cannot escape.

Media geographies are, nonetheless, accessed to different degrees and intensities depending on one's social position and geographic location. There are high and low degrees of accessibility whether one compares countries, regions or parts of a city. The typical American household includes seven screens, and two-thirds of the world's population now has access to at least one cell phone (ReportLinker 2017; Taylor & Silver 2019). There are still many places where digital media are scarce and where traditional face-to-face media hold their own. Roughly a third of the world's population does not have access to the internet, and for many people cyberspace is some distant, foreign continent. The digital divide remains a serious issue. Within any of these more or less connected places, people build their own distinct media environments, personalized by their contacts in their physical or digital phone books, their friendship networks and their social media. Media geographies are not just "out there"; they are also in us, in the languages we know, our highly internalized systems of meaning, the nonverbal communications we understand, the way our fingers find the right keys on keyboards and keypads, the way our speech automatically adjusts to various apps and interfaces, the way we sense our phone's vibration even when it is silenced.

In addition, our lives are increasingly *mediatized*: daily routines, activities and social inter-actions evolve in response to configurations of media encountered, used and appropriated (Lundby 2009; Hjarvard 2013). Mediatization is the incorporation of media into our various projects, allowing things to be done across space rather than in place, letting people come together around shared interests, tasks and objectives without physically converging (Kwan 2000). Mediatization reflects, among other things, the increasingly information-intensive nature of capitalism, both in the domains of production and social reproduction, the pro-liferation of new technologies, and the growth of the demand for information of all kinds. It is the ongoing reworking of our "life paths" (Hägerstrand 1970) to incorporate media rou-tines, cycling over spans of days, months and years. The substitution of physical gathering by mediated forms of gathering creates a risk of *context collapse*; as "social media environments become a place where person-to-person conversations take place around user-generated content amidst potentially large audiences" (Marwick & Boyd 2011, 129–130) the *situational integrity* of life breaks down. The boundaries between situations (such as home, work, school, the bank and the doctor's office) that help us maintain control over aspects of our self-iden-tity have become steadily eroded (Nissenbaum 2009).

Geographers can contribute to understanding the challenges of mediatization, the risks of context collapse and the threats to situational integrity, by returning to fundamental geographical concepts of space and place, and reconsidering them in light of sociotechnical transformations associated with new media. Geographers can "ground" media in the reality of lived experience and the highly variegated places in which people live. They can map, literally and figuratively, the flows of information that media create, linking the producers and consumers of information in complex webs of interaction.

Unfortunately, the geography of communication is still plagued by "invisibility," as Ken Hillis noted more than two decades ago (1998). The reason for this lack of visibility strikes us as a truncated understanding of "geography." Outsiders still view geography as studying a "space of places" and are unaware that the discipline has grappled with the "space of flows" (Castells 1989) for decades. To be sure, how the media relates to places is still a topic of significant concern (Halegoua 2019). Media geography as a subdiscipline addresses both patterns and processes, stasis and flows, communication infrastructure and "traffic" on that infrastructure. Nonetheless, we are optimistic that media geography is on the verge of broader recognition, in part because of the longevity of relevant concerns within the discipline, and in part because of the convergence of pertinent questions from all segments of society.

History of media geography

The earliest work in media geography can be traced to the 1970s. Geography's "quantitative revolution" brought attention to human movement and information flows, which led in turn to the key observation that the speeding up of transportation and communication brings locations closer together in time–distance, progressively shortening the time required to move or communicate between points. Costs to move or communicate among places also tend to decrease over time, leading to cost-space convergence (Brunn & Leinbach 1991; Janelle & Hodge 2000). For this reason, media geographies are an integral part of the successive rounds of time-space compression that have swept the world repeatedly since the Industrial Revolution (Warf 2008). Viewed comprehensively, this process means that the spaces in which people live and act can be understood as shrinking, collapsing, compressing or converging.

Media have long played a central role in the reconstruction of relational space. By bridging space effortlessly, by bringing ever larger audiences into reach, telecommunications changed the scale of the community in which people imagined themselves. The telegraph, whose invention is often credited to Samuel Morse in 1844, was the first form of telecommunications and was essential to the expansion of the United States and the formation of a national economy (Pred 1977). Shortly thereafter telegraph lines crossed the Atlantic Ocean to link North America and Europe (Hugill 1999). The telephone, unveiled in 1876, was originally the preserve of the wealthy, only to become a household tool over time. Its use greatly expanded the spatial range of interpersonal networks, the opportunities for interaction, undermined longstanding boundaries between public and private spaces (de Sola Pool 1977) and helped in the formation of "communities without propinquity," or people tied by common interests rather than physical proximity (Marvin 1988). The mass-produced camera, the creation of George Eastman in 1888, led to an explosion in photography and its acceptance as an accurate, unbiased and objective mirror of the world (Sontag 1977). Few innovations had such a power to make distant places seem near. The evolution of photography into the cinema represents one of the most powerful extensions of visual experience in the history of modernity. Cinema allowed people to "get a sense of the world without moving

3

very far at all" (Allen & Hamnett 1995, 3). The radio, an outgrowth of wireless telegraphy, for the first time brought news and entertainment directly into the homes of the masses. Finally, of course, television stitched together the world as a collage of simultaneous sights and sounds divorced from their historical or geographical context. Television exceeds at entertaining (Postman 1985), and forces all other discourses to imitate it: education, religion and politics must be entertaining to be successful. In drawing the multitudes indoors, television helped to eclipse the public agora, deepening the bourgeois process of individualization and commodification. The world's foremost source of entertainment and news, television has shortened attention spans, engendered immediate gratification, led to unrealistic stereotypes, desensitized viewers to violence and endlessly promoted commodities in a never-ending series of advertisements.

In the late 20th century, two major technologies—satellites and fiber optics—became the backbones of the global communications and media network. From their Cold War origins, satellites were deployed by telecommunications companies to provide services for financial firms, the media and transnational corporations. Starting in the 1980s, a global skein of fiber optics cables became the most preferred mode of telecommunications in the world, greatly altering global flows of financial funds and information, and laying the basis for the internet (Warf 2006). Such systems well illustrate the notion of "power-geometries" (Massey 1993) that ground the space of flows within concrete material and spatial contexts. In providing a largely homogenous diet of Western television and video programs around the world, these technologies have had important repercussions for local and national forms of consciousness and subjectivity. Appadurai (1990) views such phenomena as part of a global "mediascape" that interacts with other "scapes" to redefine the cultural geographies of global postmodernism. Thus, from the most intimate spaces of the body to the rarified domain of the global economy, media geographies inform, teach, enlighten, entertain, amuse and at times mislead people, producing subjects and reproducing and changing social relations.

Claims regarding the transformation of space, such as compression or convergence, depend on a particular view of geographic space as *relational*, rather than absolute (Murdoch 2006); space is relevant to human life because of *how it shapes relations between things*, and these relations are stretched and compressed rather than absolute metrics like Euclidean space (Sack 1980). In the current global capitalist system, capitalism drives the compression of space according to its inherent logics; spatial compression is a temporary, provisional resolution of the contradictions inherent to capitalism (Harvey 1990), giving rise to a series of shifting "spatial fixes" over time. The transformation of space through mediated communication and transportation is a broad-brush interpretation and does not foreclose the possibility of local trends in the opposite direction, such as traffic jams and declines in communication connectivity in neglected and marginalized places (Janelle 1969; Massey 1993). Communications of various sorts move through various kinds of spaces, each with its own properties, many of which change dynamically over short and long timespans, including daily and weekly cycles of expansion and contraction (Gould 1991). The people living in any given place are always positioned unequally in their ability to benefit from time-space compression/convergence. Media and place are thus intimately intertwined (Halegoua 2019). This means that social power imbalances are expressed and perpetuated through mediated communications leading to "power geometries" wherein one's race, ethnicity, national origins, class, sex, age, education and other social criteria affect the quality and quantity of one's ability to access, and benefit from, distant places (Massey 1993). In general, the implications and conditions for the transformation of relational space occur within a general process of compression or convergence with regard to time and cost, but in a way that differently impacts differently situated actors, through unevenness, asymmetry and bias in communication flows. Media geography therefore constitutes a particular lens on the issue of globalization.

In addition to these transformations of space and time, media geography has also returned frequently to questions of place, as captured in mediated representations. This approach was developed in the 1980s with humanistic geographic studies of prose and poetry, by authors such as Silk (1984), Porteous (1985), and Pocock (1988). Around the same time, geographers examined visual media, including television, film and advertising (Gould et al. 1984; Burgess & Gold 1985; Adams 1992; Aitken & Zonn 1994). Research on verbal and visual representations of place demonstrated a sharpened recognition of the "cultural" in cultural geography, drawing on ideas of the Frankfurt School, British cultural studies and French social theory, all of which prompted a recognition of culture as a contested terrain. "High" and "low" culture are deeply implicated in material struggles (Bourdieu 1984), and one cannot speak of a place's culture in monolithic terms but only as a particular place-based struggle over culture (Mitchell 1995) among participants who define culture differently. This idea breaks with earlier cultural geography dominated by ideas of areal differentiation (Hartshorne 1968) and the cultural landscape (Sauer 1969), by refusing to accept culture as a patchwork where similarity is a function of distance, and instead positing multiple layers and countercurrents of culture in any given place.

In addition, communication is central to the process of knowing the world, and knowledge inevitably is sutured to power. As Foucault (1993) stressed, discourses—constellations of meanings, narratives and ideologies—do not simply mirror the world, but enter into its making. The implication for media geography is that a place represented in the media is not one thing but multiple perspectives aligned with axes of social power, offering an ideologically vested way of knowing the world. Bringing related ideas "home" to geography's own communications, authors such as Brian Harley (1988; 1989) and Denis Wood (1992) subjected cartography to new kinds of critical reflection, revealing maps not simply as technical objects but as social constructions imbued with power. On this account, cartography is central to geographic ways of knowing, rather than serving merely as a tool. It is a discourse that plays instrumental roles in social and political life, a tactic or strategy to promote various agendas, a projection of social as well as geodetic relations.

In yet another shift, a more critical approach evolved with regard to "landscape," questioning landscape's ontological status; no longer "out there," landscape became seen as a way of representing, seeing and interacting with the world (Cosgrove & Daniels 1988; Duncan 1990; Barnes & Duncan 1992; Duncan & Ley 1993). This work highlighted ways in which power relations embodied in discourses and images of landscape worked to naturalize social inequality. Feminist geographers brought a more complicated and nuanced understanding of social inequality, reminding other geographers that this was not a matter of class but also a complex intersection of gender, race and ethnicity running through modes of representation, discourse and power (Rose 1993; Gibson-Graham 1994; Kobayashi & Peake 1994; Nash 1996). Critical studies of landscape and feminist geographic research both fostered geographic interest in discourse, and the latter situated discourses about space and place within an intersectional, multidimensional model of social power.

Throughout this period, a particularly important figure was Yi-Fu Tuan, who employed a unique, humanistic approach to bear, exploring representations of space and place in media as varied as language, literature, mythology, photography, motion pictures and dance (Tuan 1978; 1991; 2004), while also disclosing how landscape could function as a medium to send disciplinary messages, maintaining social hierarchies and power relations (Tuan 1979; 1984). His work epitomized a humanistic, phenomenological approach, and while no one directly addressing media geography could replicate his style, he nonetheless contributed to the complex interplay between critical and interpretive approaches to media geography.

In short, by the mid-1990s, geographers saw media of all sorts as means of projecting order onto the world through representations, shaping and organizing how people see the world and their place in it, solidifying perceptions and expectations while buttressing material relations, actions and interactions. Few geographers self-identified as media geographers, but many studied discourses of one kind or another, which brought attention to an array of particular media. Most such work shared certain assumptions, foremost among them the idea that a way of showing is also a way of seeing (Berger 1972), and a description is also a script (Ó Tuathail 1992). Stated less obliquely: representations do not just re-present, they also present ways of perceiving, they guide action and they offer people positions and identities. In all of these ways, media and communications are deeply implicated in the dialectics between self and world, here and there, Us and Them. The historical foundations of media geography sketched here have been explored in greater depth elsewhere (e.g. Adams 2009; Adams & Jansson 2012; Adams et al. 2014; Mains et al. 2015; Adams et al. 2017). The core ideas to take away are that media geography is now more than 35 years old, it has engaged with diverse media as sources of place representations and spatial systems, and the area of inquiry has benefited from humanistic, critical and analytical approaches.

Recent trends in media geography

The emergence of "non-representational" geography (Lorimer 2005; Thrift 2008) emphasized communications as perceptions and actions, many of which precede language, more than representation, directing attention to less easily translated communications such as dance, sports, and the everyday geographies of the body. Here people are communicating and their communications saturate spaces and places, but it makes little sense to describe such communications in terms of representation. Non-representational and more-than-representational approaches to media geography can deepen our understanding of images, rhythms, emotions, embodiment and multisensory experiences that contribute to a person's sense of self and sense of place (Lorimer 2005; Latham & McCormack 2009). They can also shed light on how nonverbal communications flow between people in the form of affect and emotion (Pile 2010). Such work demonstrates a preference for writing about flows, fluidity, movement and mobility and avoids nitty gritty details about the infrastructure and images carrying such flows, but this is not universally the case (Carter & McCormack 2006). The intersection between media geography and the discussion of affect remains a promising and largely unexplored area.

Another area of intense current interest in media geography has its roots in the 1990s. Initiated by Gearóid Ó Tuathail and John Agnew, critical geopolitics reconceptualizes geopolitics as a discursive practice (1992, 192), a move that led to fruitful investigations by Sharp (2000), Dodds and Atkinson (2002), Dittmer (2010), and many others. Such work owes a huge debt to Benedict Anderson's (1983) famous and well received idea of imagined community, the shared sense of national identity that characterizes everyday patriotism with its accompanying self-identities, media practices and worldviews. Anderson's "print capitalism" is a social formation linking a particular form of state power to the practice of publicly defining historical time and geographical space through commercial journalism. Not just news media, strictly defined, but all media are in the business of condensing and "interpreting" the nation, including popular magazines and comics, and they help fuse personal identities to particular ways of inhabiting and performing national(ist) identities (Dittmer 2012).

In this reading, popular and formal politics intersect and are hopelessly intertwined. Critical geopolitics constitutes a significant influence on media geography, with somewhat different emphases but shared interests, objectives and assumptions.

A wave of impressive recent work in geography has exposed the peculiarities of digital media, including the following: a compendium of efforts to map cyberspace (Dodge & Kitchin 2000), the geography of the internet industry (Zook 2005), the origins and growth of cyberspace, its uneven social and spatial diffusion, and its innumerable impacts (Malecki & Moriset 2008; Warf 2012; Kellerman 2016), digital code's relation to places and spaces (Kitchin & Dodge 2011), the interface between the human and the machine (Ash 2015), how online interaction incorporates aspects of ritual and fetishism into contemporary life (Hillis 2009), and how uneven geographical coverage in map-based online services perpetuates biases (Zook & Graham 2007). "Old" media like television and film continue to attract attention (Lukinbeal & Zimmermann 2008; Christophers 2009) but the focus of the discipline has shifted to issues relating to new media, and particularly digital communications. Never have so many people been able to contact one another so easily, obtain news, file complaints, pay bills, be entertained and save time than today. For large numbers of users, the real and virtual worlds have become inextricably intertwined; for them, the internet is a necessity, not a luxury. Seen this way, the dichotomy offline/online does not do justice to the diverse ways in which the "real" and virtual worlds are interpenetrating. However, for those without access to the information highway, the internet may represent a new source of inequality.

We sympathize with the effort to promote the study of "digital geographies" (Zook et al. 2004; Ford & Graham 2016; Ash et al. 2018), and the effort to understand the interfaces between humans and technologies as objects in their own right (Ash 2015). However, it may be avoiding difficult questions that come up when speaking of digital media *as media*, that is to say, as sociotechnical communication systems. It is important to frame "the digital" as communication, even if that communication has unfamiliar powers. Alternatively, work in digital geographies is at times putting old wine in new bottles. While the novelty of digital media deserves attention, understanding many aspects of new media requires a return to fundamental questions about communication flows and processes in space and place and about space and place. Fundamental questions about relationships between representation, subjectivity and the world arise whether one is examining digital media or earlier media. Digital media like earlier media can be used to represent places, enhance the functioning of places and connect through physical and social spaces (Adams 2009). The study of digital media benefits from a historical geographic perspective that attends to longstanding questions about geographical ontology, epistemology and methodology. So, rather than encouraging digital geographies, per se, we would encourage geographical attention to digital media within media geography.

Media geography and current issues

The COVID-19 crisis has accelerated key facets of mediatization whereby elements of life, including face-to-face communications and embodied mobility, are hybridized with digital communications and virtual gathering. We cycle between our online workplaces, leisure places, places of learning, marketplaces and information places. "Going to work" increasingly means a particular way of logging onto a particular app or database, with entry to the workplace controlled by a username, a password, and site-specific ways of uploading, downloading, networking and collaborating. Going to school is often just a different way of logging onto apps and databases, a different username and password, and different site-specific ways of uploading, downloading, networking and collaborating. Shopping and hanging out with friends also involve these elements, where each generally constitutes a bounded activity space defined by different ways of "entering" and interacting.

One result of the COVID-based onlining of lives, particularly in the economically developed world, is rapid transformation of social norms and expectations. A blurring of social boundaries follows from the blurring of spatial boundaries. Social media norms spill into business meetings, as kids, cats and dogs poke their noses in and disrupt consultations and boardrooms. The norms of social media also spill into classrooms as professors ask students for a thumbs-up if they understand a particular concept from the lecture. With so much of life migrating online, we are afforded daily demonstrations of how media geographies support construction of disembodied identities (e.g. faces on a screen), but also how media geographies are embodied, emotional and material. In this light, the divisions between the public and private spheres become porous. Video links allow us to peek into one another's homes; Facebook allows people to post the most intimate details of their lives. The boundaries between the public and private realms have become porous indeed.

Despite this plunge into the latest bout of time-space compression, accompanied by context collapse and multiple risks to situational integrity, there are still quite a few of the "old" pre-digital communications around the corners of our worlds. Many people still read books printed on paper, and some read newspapers and magazines that way. We continue to have face-to-face conversations safely distanced from our neighbors as we pass on the street; hours are spent leafing through the encrusted pages of forgotten cookbooks for ways to spice up a homebound life; mail fresh from the mailbox waits on the corner of the table. These things remind us that our lives still depend on old media. Media geography is suited to reflection on the current situation since it considers old and new media, high-tech and the low-tech infrastructure, mobile and stationary uses of media, entertaining and utilitarian media, simple and complex media, and all in light of spatial activity routines and the evolving meaning of place.

During the COVID-19 crisis, media-geography questions have become a general preoccupation. Why is food ordered from a restaurant only half as good as the same food eaten in the restaurant? How do intimate social relations like dates and family get-togethers survive transplantation to the relatively (though not entirely) disembodied space of online togetherness? Why is seeing people in 2D inferior to talking with them in 3D? How can we teach students technical skills, sports, dance, art, social sciences or even social skills when each learning environment is a numbingly similar array of boxes on a Zoom screen? The mass interest and publicity in this moment around what we recognize as media geography will have major implications for the study of media geography, and perhaps by the time "COVID-19" has faded from the scene through widespread vaccinations, "media geography" will be more familiar.

Outline of chapters

The first section of the book engages with issues around the control of, and access to, digital media. These topics include state-backed censorship of media and digital divides, both of which create forms of exclusion from mediated communication flows. The next chapter in this section deals with efforts to overcome controls built into digital media; in a word, hacking. This is followed by chapters on the Chinese internet, which demonstrates a particularly severe form of state control, and a chapter exploring how digital media both exclude and include people with visual impairment and blindness.

Many governments around the world fear the emancipatory power of the internet, which can undermine monopolies over the control of information. In Chapter 2, Barney Warf examines internet censorship, which varies widely across the globe. He summarizes the major forms and levels of severity that censorship assume, then displays their geographies using data from Reporters Without Borders. Next he focuses on the world's most egregious practitioners of censorship, such as China, where the "Great Firewall" is notorious, as well as Vietnam, Iran, Russia, and Central Asian republics such as Turkmenistan. The chapter concludes with a warning that early utopian expectations of the internet have given way to more sobering but politically realistic assessments.

Social and spatial inequalities in access to cyberspace, better known as the digital divide, have long been central to understanding who uses the internet and who does not. In Chapter 3, James B. Pick and Avijit Sarkar provide a comprehensive overview of digital divides among the world's major regions. Although divides in most cases have narrowed, particularly with the growth of cell phones and the mobile internet, significant discrepancies remain among countries, and often within them as well. Pick and Sarkar point out the growing multidimensional complexity of divides, which now include technical literacy, affordability, technophobia, broadband access and social capital. The growth of information technologies the world over has led to new types of divides with varied geographies, which they illustrate with a wealth of examples.

Mareile Kaufmann, in Chapter 4, offers a comprehensive overview of hackers and hacking that departs from conventional representations that portray hackers as criminals. Rather, she emphasizes, hackers are motivated by a range of economic, political, affective and philosophical inclinations. After summarizing discourses about hacking, she focuses on an empirical case study of hacking dataveillance—surveillance using digital data—in three European countries, which redefines the contours of data flows and governance. Her chapter fruitfully depicts hackers in terms that emphasize the multiplicity of views surrounding the practice, their embodiment and the techno-political dimensions of governmentality and resistance.

China has by far the world's largest single population of netizens, more than 850 million in 2020. Xiang Zhang, in Chapter 5, explores the rise of social media there and how it differs from conventional media platforms. Inspired by the theoretical perspectives of Michel Foucault, in which knowledge and power are seamlessly fused, he turns to the social impacts of internet media there, such as news apps on smartphones. In contrast to the rigidly hierarchical structure of state media, a plethora of new media companies such as Sina and Tencent have unleashed enormous changes in the Chinese media landscape. In response, the Communist Party, adamant to retain its authoritarian control over the country, has amplified its surveillance and censorship. How long this status is retained in light of growing internet penetration rates and citizen activism remains to be seen. Chinese usage of video sharing platforms, microblogs, and services similar to Twitter has markedly altered how they obtain and share information, with uncertain long-term consequences.

For people with visual impairment or blindness (VIB), media offer particular challenges and opportunities. In Chapter 6, Susanne Zimmerman-Janschitz explores geographical aspects of VIB and how they intersect with media. Ranging from the large-scale geography of legal and technological conditions structuring VIB experiences in different countries, to the question of how to facilitate access for VIB to the built environment, she shows multiple geographies of media that are encountered by those with visual impairment and blindness. The overwhelmingly visual quality of contemporary digital media effectively shuts out access to much information (including spatial information) when users are limited to aural-haptic interfaces. In contrast, automated aural and tactile navigation assistance is being developed to facilitate navigation by the VIB and help them avoid environmental obstacles and hazards.

The second section is dedicated to geographies of mass media. Chapters range from the newspaper and (fake) news content, to audiovisual media: film, television and video. Running through these chapters are thematic interests in the evolution of the democratic polity, public discourse and the audience segments constituted as publics. An intersecting theme is the ontological status of representations of the world, in particular their ability to falsify or misrepresent, an ability that requires a renewed encounter in the wake of postmodern critiques that destabilized notions of the true and the real.

Chapter 7, by Paul C. Adams, considers the newspaper from various geographical viewpoints: as a venue for disseminating geographical findings, a source of geographical data, a means to identify and critique public discourse, and an institutional actor that plays an important role in social processes. Studies roughly aligned with these various approaches have established the newspaper as one of the most important media for geographers to understand, and have drawn attention to important issues such as the role of newspapers in defining national culture and worldviews, as well as contextualizing social contestation. The chapter closes with a cautionary section addressing the ways in which digital newspapers deviate from the longstanding assumptions about newspapers, requiring geographers to adopt new research approaches.

One of the more distressing trends in global media today is the explosion of fake news, or false stories that masquerade as real. James Compton, in Chapter 8, points out that this phenomenon has created a legitimation crisis for journalism. He traces the origins of fake news, with its deep roots in yellow journalism, and its utilization by demagogues such as Donald Trump. As the variety of news outlets has proliferated with abandon, the opportunities to manufacture fake news have grown accordingly. Coupled with a growing crisis of traditional media, this trend has led to large numbers of misinformed people who are gullible enough to swallow conspiracy theories. He concludes by noting that the right-wing mediasphere—Fox News, Breitbart, Infowars, and the like—have seized on fake news with a vengeance, sowing enormous distrust of the media among large swaths of the public.

Within media geography, a special position is held by studies of film and cinema. Many questions later directed toward other media were originally posed in relation to film. In Chapter 9, Elisabeth Sommerlad reviews the rich history of film geography, with particular attention to geographies in and of film, screen tourism, cinematic cartography, the didactic potential of film critique in geography education, and finally filmmaking as a research methodology and a venue for geographical findings. Sommerlad's chapter concludes with a prospective glance at how these multifarious perspectives on film may become more integrated.

In Chapter 10, James Craine turns to the geographies of television, still arguably the world's most important media outlet. Revolutionary technological changes such as digitization, virtuality and streaming have unleashed new televisual landscapes. Craine analyzes these trends within the context of feminist thought, affect and the literature on spaces of difference. The multidimensional semiotics of televised spaces and places reveal how the virtual and the real have become interpenetrated in complex, often unpredictable ways.

Among the most popular applications of digital media today is streaming video. Irina Kopteva, in Chapter 11, delivers an in-depth profile of the largest such service, YouTube, which has given millions of people a chance to express themselves visually to large audiences. The results include YouTube stars and influencers with millions of subscribers, videos of dangerous stunts, and a surge of material in vernacular languages. Even countries such as China, where YouTube is banned, have seen imitators emerge. As YouTube has grown, so too have debates about the legality of its content, marketing opportunities, advertising, intellectual property rights and repercussions for education and entertainment. YouTube thus unites and divides people in diverse geographic contexts around the globe.

The third section of the book addresses a spectrum of issues ranging from mobile media to surveillance. The functioning of mobile media implies the collection of locational data from users, opening up the possibility of a sort of microscopic, multidimensional surveillance that is unprecedented in human history. While this potential is exposed in the final chapter of the section, prior chapters indicate the ways in which mobile media have facilitated various forms of mobility. A complicated nexus of issues bringing together mobility, digital media, media convergence, loss or erosion of privacy, surveillance, and the commodification of mobility data preoccupies the authors contributing to this section.

Just as mobile communication technologies have been undergoing rapid evolution, so are the geographical insights that can be obtained from the study of these technologies, as revealed in Chapter 12, by Ragan Glover-Rijkse and Adriana de Souza e Silva. The chapter reviews the short but complicated history of mobile media, from the late 1990s to the present, with particular attention to the reworking of human mobility through incorporation of mobile media into everyday spaces and practices. The chapter introduces corresponding transformations of relations between public and private, near and far, present and absent, space and place. The use of mobile media permitted the development of hybrid spaces that are simultaneously digital and physical, but as the authors insist, this process was as much a social as a technological transformation.

Roger Norum and Erika Polson offer a deeper dive into the co-construction of media and mobilities in Chapter 13. Applying the concept of "connective media" to the present period, they question how connective media support both digital placemaking as well as movements to, in and through these places. They delve into the complex intertwinement of mobility and mediation, showing that media are now deeply involved in spatial connections whether one looks at flows of goods, services or people. Moving beyond the more familiar elements of this story, they explore more theoretically challenging aspects of the media-mobility nexus, considering how media function as intermediaries, coming between yet connecting, making what is distant immediate, and thereby altering conceptions of reality.

Chapter 14 continues this dive into theoretical complexity as Peta Mitchell, Marcus Foth and Irina Anastasiu examine geographies of locative apps. Here the focus narrows to the "location-aware" applications running on digital devices. The chapter traces the historical emergence of mobile geolocation and offers a way of theorizing the new, "hybrid" forms of spatiality being generated in its wake. The authors move on to discuss the spatial affordances associated with location-based apps and services. They next offer a sobering reflection on how mobile geolocation has contributed to an emerging economic sector driven by the collection, collation and processing of personal locational data from the users of these apps. Their review of the literature demonstrates that locative apps present simultaneously an unwanted intrusion into personal privacy, a means of enhancing safety and security, and a way of engaging with hybrid spatiality.

In Chapter 15, Ellen van Holstein writes of digital surveillance and place, notably the "culture of watching and being watched." This set of practices is increasingly central to questions of privacy, fear and risk. Moving beyond conventional understandings of the panopticon, she portrays digital surveillance in terms of networks and assemblages, in which new geographies of power and resistance are continually produced and reproduced. She concludes by calling on geographers to come to terms with their own complicity in this phenomenon.

The fourth and final section of the book deals with media and the politics of knowledge. Geographies of media involve intersectionality defined by race and ethnicity, sex and gender, nationality, regional identity and anthropocentric understandings of the natural.

In Chapter 16, Douglas L. Allen and Derek H. Alderman explore the politics of race and ethnicity in the media, bringing to bear geographic theories regarding presence and absence, socio-spatial representations and racialized landscapes. They demonstrate ways that media portrayals of racial and ethnic minorities stereotype, essentialize and marginalize, but also show that media can be used to affirm presence in the face of these processes, contest dominant narratives and images, and subvert oppressive systems. Cell phone videos, for example, have brought racist police brutality to the public eye, while also precipitating an alternative, re-envisioned, aspirational sense of place.

Recent years have brought right-wing movements fusing racist and nationalist ideologies, in the US and throughout the world. In Chapter 17, Daniel Bos explores how media weave nationalist ideas in popular culture. His chapter explores how nationalism is experienced, embodied and performed through everyday mediated encounters. The nation is not merely represented in the media; in many ways, mediated communication is a key process through which the nation is created in an ongoing fashion. This wide-ranging chapter considers old and new research that bears on this question of how the nation comes to be and the part media play in this process. It reflects on the audience and various modes of contributing to, and engaging with, the circulation of nationalist imagery.

Nationalist media content depends in part on representing outside people and places as significantly different, in other words as inferior, alien, bizarre, primitive, threatening, failed and so on. This Othering process is examined in Chapter 18 by Virginie Mamadouh, focusing on how news media perpetuate Eurocentric and Orientalist worldviews. These mediated realities situate negative stereotypes of non-Western places and people within an ethnocentric worldview prejudiced toward Western people and places. Through globalization and digitalization, these Eurocentric and Orientalizing perspectives have diffused outside of their Western source regions. However the same processes of globalization and digitalization have supported alternative perspectives, for example Al Jazeera and the Chinese broadcaster CCTV, as well as local perspectives posted by amateurs on social media platforms.

If nationalism is obviously a spatial expression of power relations in the media, more subtly spatialized aspects of power involve sex and gender. These aspects are treated by Marcia R. England in Chapter 19 through a feminist approach to media geography. She outlines how notions of masculinity and femininity have been naturalized by media, as women's agency is pushed to the margins. She considers how media both affirm or challenge social norms and stereotypes governing men's and women's spatial behaviors. She also reveals how geographical concerns with embodiment inherently invoke dynamics of mediated sex and gender.

In the final chapter, Hunter Vaughan troubles the idea of nature as something we come to know through media representations by showing that media impact the environment and extend associated social justice violations in ways that are often disregarded.

No single volume can hope to address all of the issues that swirl around the theme of media geographies. Obviously there are omissions in this volume: there are no chapters on radio, or Facebook, or the dark web. But we hope that the work presented here is useful for those studying the complex intersections of media and place, the ways in which spatiality and information are wrapped up in one another, and how the continuous, ongoing transformation of both shape our societies, politics, cultures and lives.

References

Adams, P. C. 1992. Television as gathering place. *Annals of the Association of American Geographers*, 82 (1), 117–135.
Adams, P. C. 2009. *Geographies of media and communication*. Malden, MA: Wiley-Blackwell.

Adams, P. C. 2018. Geographies of media and communication II: Arcs of communication. *Progress in Human Geography*, 42 (4), 590–599.

Adams, P. C., and Jansson, A. 2012. Communication geography: A bridge between disciplines. *Communication Theory*, 22 (3), 299–318.

Adams, P. C., Cupples, J., Glynn, K., Jansson, A., and Moores, S. 2017. *Communications/media/geographies*. London and New York: Routledge.

Aitken, S. C., and Zonn, L. (eds.) 1994. *Place, power, situation and spectacle: A geography of film*. Lanham, MD: Rowman & Littlefield Publishers.

Allen, J., and Hamnett, C. 1995. Introduction. In J. Allen and C. Hamnett (eds.) *A shrinking world? Global unevenness and inequality*, pp. 1–10. Oxford: Oxford University Press.

Anderson, B. 1983. *Imagined communities: Reflections on the origin and spread of nationalism*. London: Verso.

Appadurai, A. 1996. *Modernity at large: Cultural dimensions of globalization*. Minneapolis, MN: University of Minnesota Press.

Ash, J. 2015. *The interface envelope: Gaming, technology, power*. London: Bloomsbury Publishing.

Ash, J., Kitchin, R., and Leszczynski, A. 2018. Digital turn, digital geographies? *Progress in Human Geography*, 42 (1), 25–43.

Atkinson, D., and Dodds, K. (eds.) 2002. *Geopolitical traditions: Critical histories of a century of geopolitical thought*. London and New York: Routledge.

Barnes, T., and Duncan, J. (eds.) 1992. *Writing worlds: Discourse, text and metaphor in the representation of landscape*. New York: Routledge.

Bourdieu, P. 1984. *Distinction: A social critique of the judgement of taste*. Cambridge, MA: Harvard University Press.

Brunn, S. D., and Leinbach, T. (eds.) 1991. *Collapsing space and time: Geographic aspects of communications and information*. London and New York: Routledge.

Burgess, J., and Gold, J. R. (eds.) 1985. *Geography, the media and popular culture*. Beckenham: Croom Helm.

Carter, S., and McCormack, D. 2006. Film, geopolitics and the affective logics of intervention. *Political Geography*, 25 (2), 228–245.

Castells, M. 1989. *The informational city: Information technology, economic restructuring, and the urban-regional process*. Oxford and Cambridge, MA: Basil Blackwell.

Christophers, B. 2009. *Envisioning media power: On capital and geographies of television*. Lanham, MD: Lexington Books.

Cupples, J. 2015. Development communication, popular pleasure and media convergence. In S. Mains, J. Cupples and C. Lukinbeal (eds.) *Mediated geographies and geographies of media*, pp. 351–366. Dordrecht and Heidelberg: Springer.

de Sola Pool, I. (ed.) 1977. *The social impact of the telephone*. Cambridge, MA: MIT Press.

Dittmer, J. 2010. *Popular culture, geopolitics, and identity*. Lanham, MD: Rowman & Littlefield Publishers.

Dittmer, J. 2012. *Captain America and the nationalist superhero: Metaphors, narratives, and geopolitics*. Philadelphia, PA: Temple University Press.

Dodge, M., and Kitchin, R. 2000. *Mapping cyberspace*. London and New York: Routledge.

Ford, H., and Graham, M. 2016. Provenance, power and place: Linked data and opaque digital geographies. *Environment and Planning D: Society and Space*, 34 (6), 957–970.

Foucault, M. 1980. *Power/knowledge: Selected interviews and other writings, 1972–1977*. C. Gordon (ed. and trans.). New York: Pantheon Books.

Gibson-Graham, J. K. 1994. "Stuffed if I know!": Reflections on post-modern feminist social research. *Gender, Place and Culture*, 1 (2), 205–224.

Gould, P. 1991. Dynamic structures of geographic space. In S. Brunn and T. Leinbach (eds.) *Collapsing space and time: Geographic aspects of communication and information*, pp. 3–30. London: Harper Collins.

Gould, P., Johnson, J., and Chapman, G. 1984. *The structure of television*. New York: Taylor & Francis.

Hägerstrand, T. 1970. What about people in regional science? *Papers of the Regional Science Association*, 24, 7–21.

Halegoua, G. 2019. *The digital city: Media and the social production of place*. New York: New York University Press.

Hall, S. 2001. Encoding/decoding. In M. Durham and D. Kellner (eds.) *Media and Culture Studies: Keyworks*, pp. 163–173. Malden, MA: Blackwell.

Harley, B. 1988. Maps, knowledge, and power. In D. Cosgrove and S. Daniels (eds.) *The iconography of landscape*, pp. 277–312. Cambridge: Cambridge University Press.

Harley, B. 1989. Deconstructing the map. *Cartographica*, 26(2), 1–20.

Hartshorne, R. 1968. *The nature of geography: A critical survey of current thought in the light of the past.* Lancaster, PA: Association of American Geographers.

Harvey, D. 1990. Between space and time: Reflections on the geographical imagination. *Annals of the Association of American Geographers*, 80 (3), 418–434.

Hillis, K. 1998. On the margins: The invisibility of communications in geography. *Progress in Human Geography*, 22 (4), 543–566.

Hillis, K. 2009. *Online a lot of the time: Ritual, fetish, sign.* Durham NC: Duke University Press.

Hjarvard, S. 2013. *The mediatization of culture and society.* London and New York: Routledge.

Hugill, P. 1999. *Global communications since 1844: Geopolitics and technology.* Baltimore, MD: Johns Hopkins University Press.

Janelle, D. 1969. Spatial reorganization: A model and concept. *Annals of the Association of American Geographers*, 59 (2), 348–364.

Janelle, D., and Hodge, D. (eds.) 2000. *Information, place and cyberspace: Advances in spatial science.* Berlin and Heidelberg: Springer.

Kellerman, A. 2016. *Geographic interpretations of the internet.* Dordrecht: Springer.

Kitchin, R., and Dodge, M. 2011. *Code/space: Software and everyday life.* Cambridge, MA: MIT Press.

Kobayashi, A., and Peake, L. 1994. Unnatural discourse: "Race" and gender in geography. *Gender, Place and Culture*, 1 (2), 225–243.

Koeze, E., and Popper, N. 2020. The virus changed the way we internet. *New York Times* (April 7). www.nytimes.com/interactive/2020/04/07/technology/coronavirus-internet-use.html.

Kristeva, J. 1980. *Desire in language: A semiotic approach to literature and art.* New York: Columbia University Press.

Kwan, M.-P. 2000. Human extensibility and individual hybrid-accessibility in space-time: A multi-scale representation using GIS. In D. Janelle and D. Hodge (eds.) *Information, place, and cyberspace: Advances in spatial science.* Berlin and Heidelberg: Springer.

Latham, A., and McCormack, D. 2009. Thinking with images in non-representational cities: Vignettes from Berlin. *Area*, 41 (3), 252–262.

Lorimer, H. 2005. Cultural geography: The busyness of being "more-than-representational." *Progress in Human Geography*, 29 (1), 83–94.

Lukinbeal, C., and Zimmermann, S. (eds.) 2008. *The geography of cinema: A cinematic world.* Stuttgart: Franz Steiner Verlag.

Lundby, K. (ed.) 2009. *Mediatization: Concept, changes, consequences.* New York: Peter Lang Publishing.

Mains, S., Cupples, J., and Lukinbeal, C. (eds.) 2015. *Mediated geographies and geographies of media.* Dordrecht and Heidelberg: Springer.

Malecki, E., and Moriset, B. 2008. *The digital economy: Business organisation, production processes, and regional developments.* London: Routledge.

Marvin, C. 1988. *When old technologies were new: Thinking about electric communication in the late nineteenth century.* New York: Oxford University Press.

Marwick, A. 2018. Why do people share fake news? A sociotechnical model of media effects. *Georgetown Law Technology Review*, 2 (2), 474–512.

Marwick, A., and Boyd, D. 2011. I tweet honestly, I tweet passionately: Twitter users, context collapse, and the imagined audience. *New Media & Society*, 13 (1), 114–133.

Massey, D. 1993. Power-geometry and a progressive sense of place. In J. Bird, B. Curtis, T. Putnam, G. Robertson and L. Tickner (eds.) *Mapping the futures: Local cultures, global change.* London and New York: Routledge.

Mitchell, D. 1995. There's no such thing as culture: Towards a reconceptualization of the idea of culture in geography. *Transactions of the Institute of British Geographers*, 20 (1), 102–116.

Murdoch, J. 2006. *Post-structuralist geography: A guide to relational space.* London: Sage.

Nash, C. 1996. Reclaiming vision: Looking at landscape and the body. *Gender, Place and Culture*, 3 (2), 149–170.

Nissenbaum, H. 2009. *Privacy in context: Technology, policy, and the integrity of social life.* Stanford, CA: Stanford University Press.

Ó Tuathail, G. 1992. Foreign policy and the hyperreal: The Reagan administration and the scripting of "South Africa." In J. Duncan and T. Barnes (eds.) *Writing worlds: Discourse, text and metaphor in the representation of landscape*, pp. 155–175. London and New York: Routledge.

Ó Tuathail, G., and Agnew, J. 1992. Geopolitics and discourse: Practical geopolitical reasoning in American foreign policy. *Political Geography*, 11, 190–204.

Pile, S. 2010. Emotions and affect in recent human geography. *Transactions of the Institute of British Geographers*, 35 (1), 5–20.

Pocock, D. 1988. Geography and literature. *Progress in Human Geography* 12 (1), 87–102.

Porteous, J. 1985. Literature and humanist geography. *Area*, 17 (2), 117–122.

Postman, N. 1985. *Amusing ourselves to death: Public discourse in the age of show business*. New York: Viking.

Pred, A. 1977. *City-systems in advanced economies*. London: Hutchinson.

ReportLinker. 2017. *When you give a kid a tablet, he'll ask for more time*. www.reportlinker.com/insight/give-kid-tablet-ask-time.html.

Rose, G. 1993. *Feminism and geography: The limits of geographical knowledge*. Cambridge: Polity Press.

Sack, R. D. 1980. *Conceptions of space in social thought*. London: Palgrave Macmillan.

Sharp, J. 2000. *Condensing the Cold War Reader's Digest and American identity*. Minneapolis, MN: University of Minnesota Press.

Silk, J. 1984. Beyond geography and literature. *Environment and Planning D: Society and Space*, 2 (2), 151–178.

Sontag, S. 1977. *On photography*. New York: Farrar, Straus, and Giroux.

Taylor, K., and Silver, L. 2019. *Smartphone ownership is growing rapidly around the world, but not always equally*. Washington, DC: Pew Research Center.

Tuan, Y. F. 1978. Literature and geography: Implications for geographical research. In D. Ley and M. Samuels (eds.) *Humanistic geography: Prospects and problems*. Chicago, IL: Maaroufa Press.

Tuan, Y. F. 1979. *Landscapes of fear*. Minneapolis, MN: University of Minnesota Press.

Tuan, Y. F. 1984. *Dominance and affection: The making of pets*. New Haven, CT: Yale University Press.

Tuan, Y. F. 1991. Language and the making of place: A narrative-descriptive approach. *Annals of the Association of American Geographers*, 81 (4), 684–696.

Tuan, Y. F. 2004. *Place, art, and self*. Chicago, IL: Center for American Places.

Warf, B. 1989. Telecommunications and the globalization of financial services. *Professional Geographer*, 41 (3), 257–271.

Warf, B. 2006. International competition between satellite and fiber optic carriers: A geographic perspective. *The Professional Geographer*, 58, 1–11.

Warf, B. 2008. *Time-space compression: Historical geographies*. London: Routledge.

Warf, B. 2012. *Global geographies of the internet*. Dordrecht: Springer.

Wood, D., and Fels, J. 1992. *The power of maps*. New York: Guilford Press.

Zook, M. 2005. *The geography of the internet industry: Venture capital, dot-coms, and local knowledge*. Malden, MA: Blackwell Publishing.

Zook, M., and Graham, M. 2007. Mapping digiPlace: Geocoded internet data and the representation of place. *Environment and Planning B: Planning and Design*, 34 (3), 466–482.

Zook, M., Dodge, M., Aoyama, Y., and Townsend, A. 2004. New digital geographies: Information, communication, and place. In S. Brunn, S. Cutter and J. W.Harrington, Jr. (eds.) *Geography and technology*, pp. 155–176. Dordrecht: Kluwer Academic Publishers.

PART I

Control and access to digital media

2

INTERNET CENSORSHIP

Shaping the world's access to cyberspace

Barney Warf

In mid-2021, more than 5.1 billion people used the internet, making it a tool of communications, entertainment and other applications accessed by roughly 65 percent of the world's population (www.internetworldstats.com/stats.htm). For countless numbers of people, cyberspace has become indispensable for entertainment, communications, shopping, bill payments and other uses. Increasingly the boundaries between the real and virtual worlds are evaporating.

Numerous geographers have written about the nature and growth of cyberspace, its uneven social and spatial diffusion, and its multiple impacts, ranging from cybercommunities to digital divides to electronic commerce (Warf 2012; Kellerman 2016). This literature offers a valuable means for spatializing the internet, rooting it in the concrete social and material circumstances that vary across the world, which serves to demolish utopian notions that it is somehow placeless or leads to a "flat earth."

One of the most insidious notions that swirls around cyberspace is that it is an inherently and inevitably emancipatory tool, and thus always serves to undermine authoritarian governments. Ronald Reagan once asserted that "The Goliath of totalitarianism will be brought down by the David of the microchip" (quoted in Kalathil & Boas 2003, 1), and the chair of Citicorp, Walter Wriston (1997, 174), argued that "the virus of freedom... is spread by electronic networks to the four corners of the earth." At times this notion is wedded to libertarian interpretations in which the global community of netizens is seen as a self-governing community in which the state has become largely irrelevant (Goldsmith & Wu 2006). Such views conveniently overlook how the internet can be used against people as well as for them.

In contrast, more realistic assessments take seriously the ability of governments to limit access to the internet, regulate what people can see, and how they wield power against cyberdissidents (Diebert et al. 2008; Diebert 2009; Morozov 2011). Most of the world's governments regulate internet access and contents to one degree or another, although the nature and extent vary widely. Indeed, opposition to censorship and political activism is typically confined to small groups of educated individuals, often diasporas, and has relatively little impact among the masses of their respective states (Kalathil & Boas 2003).

While the geographic literature has delved into issues of geosurveillance and govern-mentality, it has been largely silent about how governments erect obstacles to internet access as a form of political control (but see Warf 2010). Warf (2009a; 2009b) addressed the geographies of internet censorship in Latin America and the states that comprised the former

DOI: 10.4324/9781003039068-3

Soviet Union, and Warf and Vincent (2007) addressed the marked government restrictions found in the Arab world.

This chapter explores internet censorship in several steps. It opens with an overview of the dimensions of censorship and the various forms that it can take. Second, it offers a broad overview of the geography of global internet censorship. The third part highlights some of the world's most egregious offenders in this regard, while the conclusion summarizes major themes.

Dimensions of internet censorship

Internet accessibility reflects, among other things, incomes, the cost of access, the prevalence of computers (at home, work, and libraries), literacy rates, gender roles, and how willing governments are to allow their populations to utilize cyberspace. Repressive governments typically fear the potential of the internet to allow people to circumvent their monopoly over information. Several geographers have drawn upon Foucauldian conceptions of power to analyze geosurveillance, invasions of privacy and digital panopticons (Dobson & Fisher 2007). These works illustrate that clearly the internet can be made to work against people as well as for them. Rather than being inherently emancipatory in nature, the internet can sustain dominant authorities and be used to track and harass political opponents of the state.

Governments have varying motivations for internet censorship, including: the political repression of dissidents, suppression of civil rights groups, and the prevention of exposure of corruption or publication of comments insulting to the state (e.g. in China, Iran, Myanmar); religious controls to silence ideas deemed heretical or sacrilegious (as found in many Muslim countries); or cultural restrictions that exist as part of the oppression of ethnic minorities (e.g. refusal to allow government websites in certain languages) or sexual minorities. Often internet censorship is done on the ostensible grounds of protecting public morality from pornography or gambling. Preventing terrorism is another favorite rationale, often backed by vague notions of national security.

Governments may limit the *scope* (or range of topics) permitted on the internet and engage in different degrees of intervention, ranging from permitting information to flow utterly unfettered (e.g. Scandinavian states) to essentially prohibiting access to the internet altogether (e.g. North Korea) (Warf 2015). States with highly centralized power structures tend to be the worst internet censors, particularly those run by a single political party (e.g. China, North Korea). Often their policies earn them great enmity not only domestically but internationally as well, and severe censorship can discourage tourism, foreign investment and innovation (Villeneuve 2006).

Internet censorship centers on control over access, functionality and contents (Eriksson & Giacomello 2009). A broad range of methods may be deployed: discriminatory ISP licenses, content filtering based on keywords, redirection of users to proxy servers, rerouting packets destined for a specific IP address to a blacklist, website blocking of a list of IP addresses, tapping and surveillance, chat room monitoring, discriminatory or prohibitive pricing policies, hardware and software manipulation, hacking into opposition websites and spreading viruses, denial-of-service (DOS) attacks that overload servers or network connections using "bot herders," temporary just-in-time blocking at moments when political information is critical, such as elections, and harassment of bloggers (e.g. via libel laws or invoking national security). Content filtering often relies on algorithms that identify target words or phrases, and can adapt as a new lexicon emerges over time. Filtering may occur at the levels of the individual service provider, the domain name, a particular IP address or a specific URL. Most forms of censorship are difficult to detect technically: users may not even know that censorship is in effect. Sometimes governments use foreign (usually American) software for this purpose. For example, the governments of Iran, Yemen, the UAE and Sudan use Secure Computing's SmartFilter, a program produced in the US (Lee 2001; Villeneuve 2006). Once formal censorship begins, there is the inevitable temptation to expand the list of prohibited topics and websites, or what Villeneuve (2006) calls "mission creep."

By far the most common and insidious form of censorship is self-censorship. Users in countries with authoritarian regimes typically know the boundaries of politically acceptable use and rarely cross them. Most are understandably intimidated by the threat of arrest or harassment, or less commonly, fines, for visiting prohibited websites. From the perspective of government authorities, self-censorship is more efficient and effective than brute force. Since the vast majority of internet usage is not for political purposes, only a minority of users are affected in this way.

The degree and type of internet censorship obviously varies widely and reflects how democratic and open to criticism different political systems are. In Scandinavia, censorship is non-existent. In North Korea, internet access is illegal except for a small cadre of Communist Party elites and hackers trained in cyberwarfare (Warf 2015). In between these extremes lies a vast array of states with modest to moderate forms of censorship. These variations reflect the complex geographies of democracy, civil society and governance systems, as well as resistance to authoritarianism in the defense of freedom of speech.

Viewed this way, internet censorship is a contested terrain of social relations. Opposition to censorship often involves a diverse array of social movements, political parties, activists, hackers, bloggers, journalists, labor unions, human rights groups, religious figures, women's rights campaigns and others. Resistance may be successful at times, such as when cyber-activists use anonymizing proxy servers in other countries that encrypt users' data and cloak their identities. Because cyberspace and physical space have become inseparable, the internet simultaneously reflects and in turn shapes the contours of politics. Moreover, this arena is constantly changing in size and scope, reflecting, among other things, rising penetration rates and growing computer literacy, fluctuations in government policies, and variations in popular sentiment. Internet censors and their subjects play a cat-and-mouse game that results in path-dependent, contingent and unpredictable results.

The empirics of global internet censorship

Reporters Without Borders, an NGO headquartered in Paris and one of the world's preeminent judges of censorship, ranks governments across the planet in terms of the severity of their internet censorship (Quirk 2006). Their index of internet censorship is generated from surveys of 50 questions sent to legal experts, reporters and scholars in each country. Thus, countries in northern Europe, the US and Canada, Australia and New Zealand, and Japan exhibit minimal or no censorship (scores less than 10). Conversely, a rogue's list of the world's worst offenders, including China, Vietnam, Myanmar, Iran and Turkmenistan, exhibit the planet's most severe and extensive restrictions (scores greater than 80). Table 2.1 summarizes the distribution of the world's population and internet users according to the level of severity of censorship. Thus, only 13% of the world's people, but a third of internet users, live in countries with minimal censorship; conversely, roughly one-quarter of the world's people and internet users live under governments that engage in very heavy censorship (the vast bulk of whom are located in China).

Most of the world lives under governments that censor the internet. Using the Reporters Without Borders scores, only 2.9% of the planet, and only 4.7% of netizens, lives in countries

Table 2.1 Global population and internet users by severity of internet censorship, 2020

Internet RWB score	Population (millions)	%	Users (millions)	%
<10.0	26.8	0.3	25.9	0.6
10.0–19.9	204.2	2.6	186.8	4.1
20.0–29.9	1,270.6	16.3	1,008.5	22.2
30.0–39.9	1,350.6	17.4	759.7	16.7
40.0–49.9	2,691.3	34.6	1,298.6	28.5
50.0–59.9	410.1	5.3	221.2	4.9
60.0–69.9	139.1	1.8	108.8	2.4
70.0–79.9	1,555.2	20.0	930.7	20.5
>80.0	124.9	1.6	9.1	0.2
Total	7,772.7	100.0	4,549.2	100.0

Source: Calculated by author

[a] Reporters Without Borders

with zero or minimal government intervention. These include Scandinavian states, Canada and New Zealand. The majority of the world's people and internet users (68%) live under regimes with moderate levels of censorship (RWB scores 20–50), including Russia, most of the Arab world and parts of Africa. About one-fifth of the world, notably including China, lives under governments that practice extreme censorship.

These broad categories fail to reveal important geographical variations in censorship levels (Figure 2.1), which may be mapped using Reporters Without Borders scores. Most of the world's worst internet censors are located in Asia, including China, Turkmenistan, Vietnam and Iran. The majority are run by Communist Parties, including Cuba, the worst offender in the Western Hemisphere. Censorship is a common tool of totalitarian governments, which tolerate little dissent and fear unfettered lines of communication. The Middle East and Arab world do not fare well, nor does most of Africa. Most Latin American countries are moderate internet censors as well. In contrast, prosperous, stable democracies, including northern Europe, the US, Canada, Japan, Australia and New Zealand, exhibit very low levels of censorship, as does South Africa.

Egregious examples of censorship

While the types and severity of censorship vary widely across the globe, it is worth noting a few of the most extreme examples as a means of understanding the lengths to which governments can go to regulate internet access, the strategies they deploy, and their effects.

Chinese internet users, who numbered roughly 989 million people in mid-2021, face some of the world's greatest restrictions (Roberts 2018; Qiang 2019). The Chinese government has been blunt in its justification for censorship, asserting its necessity to maintain a "harmonious society." It deploys a vast array of measures commonly known as the "Great Firewall," which includes publicly employed internet monitors and citizen volunteers, and screens blogs and email messages for potential threats to the party's hegemony. International internet connections to China operate solely through a selected group of state-controlled backbone fiber networks. Access to common Web services, such as Google and Yahoo!, is heavily restricted (Paltemaa & Vuori 2009). The national government hires commentators,

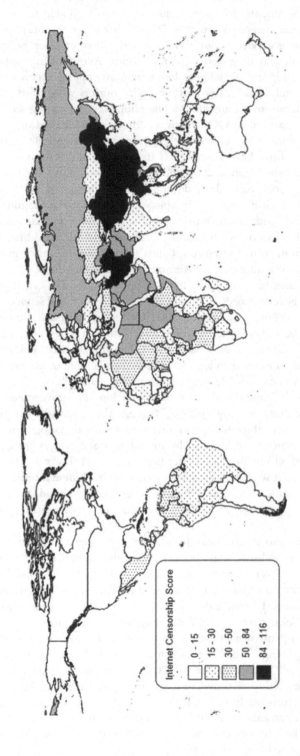

Figure 2.1 Map of internet censorship scores

commonly referred to by the derogatory term the "five-mao party," to monitor blogs and chat rooms, inserting comments that "spin" issues in a way favorable to the Chinese state. Internet service providers censor themselves (Zhang 2020) by monitoring monitor chat rooms, blogs, networking services, search engines and video sites for politically sensitive material in order to conform to government restrictions. Anonymizing websites that help users circumvent censorship are prohibited. Users who attempt to access blocked sites are confronted by Jingjing and Chacha, two cartoon police officers who inform them that they are being monitored. Instant messaging and mobile phone text messaging services are heavily filtered, including a program called QQ, which is automatically installed on users' computers to monitor communications. Blogs critical of the government are typically dismantled within days. Notably, American firms have assisted the Chinese state in this regard (Simonite 2019).

The Great Firewall system began in 2006 under an initiative known as "Golden Shield," a national surveillance network that China developed with the aid of US firms Nortel and Cisco Systems (Lake 2009; Griffiths 2019). It rapidly extended beyond the internet to include digital identification cards with microchips containing personal data that allow the state to recognize the faces and voices of its 1.4 billion inhabitants. Golden Shield has been exported to Cuba, Iran and Belarus. In many respects, China's state-led program of internet development serves as a model for authoritarian governments around the globe.

The Chinese government has periodically initiated shutdowns of data centers housing servers for websites and online bulletin boards, disrupting use for millions. Email services like Gmail and Hotmail are frequently jammed; before the 2008 Olympics, Facebook sites of critics were blocked. In 2007, the State Administration of Radio, Film and Television mandated that all video sharing sites must be state-owned. Police frequently patrol internet cafes, where users must supply personal information in order to log on, while website administrators are legally required to hire censors popularly known as "cleaning ladies" or "big mamas."

China has clashed with foreign parties over its censorship. The government long blocked access to the *New York Times* (Hachigian 2002). Google, the world's largest provider of free internet services, famously established a politically correct website, Google.cn, which censors itself to comply with restrictions demanded by the Chinese state, arguing that the provision of incomplete, censored information was better than none at all (Dann & Haddow 2008). In 2010, Google announced it would no longer cooperate with Chinese internet authorities and withdrew from China altogether. The Chinese government responded by promoting its home-grown search engines such as Baidu, Sohu and Sina.com, which present few such difficulties.

The Chinese state has also arrested and detained internet users and activists who ventured into politically sensitive websites. The state pursues the intimidation strategy popularly known as "killing the chicken to scare the monkeys" (Harwit & Clark 2001). China has incarcerated cyberdissidents and bloggers, and waged intensive campaigns against activists for democracy in Hong Kong, Tibetan independence, Taiwanese separatism, those who investigate the Tiananmen Square massacre, and the religious-political group Falun Gong. However, given the polymorphous nature of the web, such restrictions are inevitably met with resistance. By accessing foreign proxy servers, a few intrepid Chinese netizens engage in *fanqiang*, or "scaling the wall" (Stone & Barboza 2010). Using programmers in the US, Falun Gong has developed censorship-circumventing software called Freegate, which it has offered to dissidents elsewhere, particularly in Iran (Lake 2009). The Chinese state and its opponents are thus engaged in a cat-and-mouse game common across the globe. As one Chinese blogger put it, "It is like a water flow—if you block one direction, it flows to other directions, or overflows" (quoted in James 2009).

Vietnam is another serious internet censor. Only one service provider, Vietnam Data Communications, is licensed for international connections, and it is a subsidiary of the government telecommunications monopoly. Domestic content providers must obtain special licenses from the Ministry of the Interior and lease connections from the state-owned Vietnam Post and Telecommunications Corporation. Like China, the Vietnamese government uses firewalls and encourages self-censorship. Government censors routinely search email for keywords. The government has also imprisoned advocates for internet freedom (International Censorship Explorer 2006). Owners of cybercafés who permit searches of internet sites considered to be "offensive to Vietnamese culture" face stiff fines. Vietnamese bloggers have been routinely harassed and imprisoned. However, recently, in an attempt to curry favor with foreign investors, Vietnam has begun to ease its censorship (Sicurelli 2017).

Likewise, Myanmar operates a highly restrictive censorship regime (Ochwat 2020; Sinpeng 2020). The government uses software purchased from the US company Fortinet to block access to selected websites. At times it has shut down the internet altogether to silence protestors. The state has also tolerated, if not promoted, incendiary attacks on Facebook against the Muslim Rohingya minority (Caryl 2015).

Iran runs a brutal regime that closely monitors internet traffic through its Ministry of Information and Communication Technology (Michaelsen 2018; Yalcintas & Alizadeh 2020). The state uses software purchased from Nokia and Siemens to engage in deep packet inspection of web traffic. The state has blocked access to millions of websites, ostensibly on the grounds of combatting pornography and limiting immoral behavior or that deemed insulting to Islam, reasons that are commonly found throughout the Muslim world. It has arrested numerous bloggers. However, using software developed by Falun Gong, dissidents have found ways to circumvent state controls.

In the Arab world, where censorship of different types has long been practiced, the fascistic regime of Saudi Arabia stands as a particularly offensive case (Pan & Siegel 2020). It has created formidable firewalls to regulate flows of information and banned access to millions of webpages. All internet service providers with international connections must go through the government-owned King Abdul Aziz City for Science and Technology, which uses Smart-Filter software developed by the US company Secure Computing. The government has imprisoned and tortured activists and blocked sites associated with Shia rights and the Muslim Brotherhood (Pan & Siegel 2020).

Vladmir Putin's Russia is another heavy-handed censor of the internet (Ognyanova 2015; Maréchal 2017; Soldatov 2017). Putin long regarded the internet as an avenue for the promotion of American interests inside his country, and upheld censorship as a path toward "information sovereignty" (Nocetti 2015). The state's internet surveillance law, the System for Operational-Investigative Activities, allows security services unfettered physical access to ISPs. The government also promotes websites that offer views supportive of its policies and fosters networks of nationalist bloggers.

Belarus, whose government Reporters Without Borders called one of the world's "bitterest enemies of the Internet," likewise rules the net with an iron fist (European Federation of Journalists 2018). In 2010, President Alexander Lukashenka officially imposed censorship to combat "anarchy on the internet," a move the European Union called, in a classic bit of understatement, a "step in the wrong direction." All service providers are required to connect through Belpak, a subsidiary of the state-controlled ISP Beltelecom. The government occasionally stations troops at cybercafés, where users must register their names, and launches denial-of-service attacks against opposition party websites. Newspapers critical of the state have had their websites blocked by the Ministry of Information.

Another former Soviet republic, Turkmenistan, is also a severe internet censor; Reporters Without Borders lists it as an "internet enemy." For years private internet cafes were illegal, although the government monopoly Turkmen Telecom operated a handful of them, with troops stationed at the front (Eurasianet 2007; Reporters Without Borders 2009). The state-owned monopoly, Turkmentelecom, keeps a list of blacklisted internet undesirables, and regularly blocks their IP addresses. The state's use of deep packet inspection goes without saying.

In the Americas, Cuba stands out as the worst internet censor. For years, individual access to the internet was essentially prohibited (Kalathil & Boas 2003), and the state only grudgingly began to allow access starting in 2006. A limited infrastructure and high prices remain a serious access problem for many. The state has used Avila Link software to monitor users, which it may have obtained from China. It has also harassed dissidents, such as the famous blogger Yoani Sánchez.

Concluding thoughts

Early accounts of the internet celebrated its emancipatory potential, particularly the ways in which it allowed billions of people unfettered access to information and permitted them to bypass government monopolies. The reality has been much more sobering. Many governments have become adept at monitoring and controlling digital data flows (Mozorov 2011). Relatively few countries, and only a small minority of the world's netizens, permit unfettered access to the web. Most states control their residents' access to the internet and its contents. Restrictions can vary from invisible filters to the imprisonment of dissidents and bloggers. American and European companies have played tragic but important roles in this process: there is profit to be made in selling software to unsavory regimes (China plays this game too). And of course, self-censorship is likely the most pervasive form of internet regulation of all. The notion of the "dictator's dilemma," which posits that totalitarian states must choose between censorship and economic stagnation, is thus false; many authoritarian regimes can have it both ways.

These comments serve as a sobering reminder that new technologies rarely, if ever, live up to early utopian expectations. The class mistake is to herald such events as uniformly positive. Yet the reality of internet censorship testifies that it is a serious error to underestimate the flexibility of governments in regulating cyberspace. Many repressive regimes, such as China's, are impervious to international criticism on this account. Hopes for overcoming and reducing censorship, therefore, often lie in the networks of rhizomic resistance that invariably form when the state curtails freedoms of online expression. Dissident groups, Falun Gong, human rights activists, religious parties, expatriate communities and others have long played major roles in combating censorship.

References

Caryl, C. 2015. Burma gives a big thumbs-up to Facebook. *Foreign Policy Online* (November 13). https://foreignpolicy.com/2015/11/13/burma-gives-a-big-thumbs-up-tofacebook/.

Dann, D., and Haddow, N. 2008. Just doing business or doing just business? Google, Microsoft, Yahoo! and the business of censoring China's internet. *Journal of Business Ethics*, 79 (3), 219–234.

Deibert, R. 2009. The geopolitics of internet control: Censorship, sovereignty, and cyberspace. In H. Andrew and P. Chadwick (eds.) *The Routledge handbook of internet politics*. pp. 212–226. London: Routledge.

Deibert, R., Palfrey, J., Rohozinski, R., and Zittrain, J. (eds.) 2008. *Access denied: The practice and policy of global internet filtering*. Cambridge, MA: MIT Press.

Dobson, J., and Fisher, P. 2007. The panopticon's changing geography. *Geographical Review*, 97 (3), 307–323.

Eriksson, J., and Giacomello, G. 2009. Who controls what, and under what conditions? *International Studies Review*, 11, 206–210.

Eurasianet.org. 2007. *In Turkmenistan, internet access comes with soldiers.* www.eurasianet.org/departments/insight/articles/eav030807.shtml.

European Federation of Journalists. 2018. *Belarus: More media censorship and control with new amendments of the Media Law.* https://europeanjournalists.org/blog/2018/06/24/belarus-more-media-censorship-and-control-with-new-amendments-of-the-media-law/.

Goldsmith, J., and Wu, T. 2006. *Who controls the internet? Illusion of a borderless world.* New York: Oxford University Press.

Griffiths, J. 2019. *The Great Firewall of China.* London: ZED.

Hachigian, N. 2001. China's cyber-strategy. *Foreign Affairs*, 80 (2), 118–133.

Harwit, E., and Clark, D. 2001. Shaping the internet in China: Evolution of political control over network infrastructure and political content. *Asian Survey*, 41 (3), 377–408.

James, R. 2009. A brief history of Chinese internet censorship. *Time* (March 18). www.time.com/time/world/article/0,8599,1885961,00.html.

Kalathil, S., and Boas, T. 2003. *Open networks, closed regimes: The impact of the internet on authoritarian rule.* Washington, DC: Carnegie Endowment for International Peace.

Kellerman, A. 2016. *Geographic interpretations of the internet.* Dordrecht: Springer.

Lake, E. 2009. Hacking the regime. *The New Republic* (September 3), www.tnr.com/article/politics/hacking-the-regime.

Lee, J. 2001. Companies compete to provide Saudi internet veil. *New York Times* (November 19), p. A1.

Maréchal, N. 2017. Networked authoritarianism and the geopolitics of information: Understanding Russian internet policy. *Media and Communication*, 5 (1), 29–41.

Michaelsen, M. 2018. Transforming threats to power: The international politics of authoritarian internet control in Iran. *International Journal of Communication*, 12, 3856–3876.

Morozov, E. 2011. *The net delusion: The dark side of internet freedom.* New York: PublicAffairs.

Nocetti, J. 2015. Russia's' "dictatorship-of-the-law" approach to internet policy. *Internet Policy Review*, 4 (4), 1–19.

Ochwat, M. 2020. Myanmar media: Legacy and challenges. *The Age of Human Rights Journal*, 14, 245–271.

Ognyanova, K. 2015. In Putin's Russia, information has you: Media control and internet censorship in the Russian Federation. In M. M. Merviö (ed.) *Management and participation in the public sphere* (pp. 62–79). Hershey, PA: IGI Global.

Paltemaa, V., and Vuori, J. 2009. Regime transition and the Chinese politics of technology: From mass science to the controlled internet. *Asian Journal of Political Science*, 17 (1), 1–23.

Pan, J., and A. Siegel. 2020. How Saudi crackdowns fail to silence online dissent. *American Political Science Review*, 114 (1), 109–125.

Qiang, X. 2019. The road to digital unfreedom: President Xi's surveillance state. *Journal of Democracy*, 30 (1), 53–67.

Reporters Without Borders. 2009. *Internet enemies.* www.rsf.org/en-ennemi26106-Turkmenistan.html.

Roberts, M. 2018. *Censored: Distraction and diversion inside China's Great Firewall.* Princeton, NJ: Princeton University Press.

Sicurelli, D. 2017. The conditions for effectiveness of EU human rights promotion in non-democratic states: A case study of Vietnam. *Journal of European Integration*, 39 (6), 739–753.

Simonite, T. 2019. *US companies help censor the internet in China, too.* www.wired.com/story/us-companies-help-censor-internet-china/.

Sinpeng, A. 2020. Digital media, political authoritarianism, and internet controls in Southeast Asia. *Media, Culture & Society*, 42 (1), 25–39.

Soldatov, A. 2017. The taming of the internet. *Russian Social Science Review*, 58 (1), 39–59.

Stone, B., and Barboza, D. 2010. Scaling the digital wall in China. *New York Times* (January 16), p. B1.

Villeneuve, N. 2006. The filtering matrix: Integrated mechanisms of information control and the demarcation of borders in cyberspace. *First Monday*, 11 (1–2). https://journals.uic.edu/ojs/index.php/fm/article/download/1307/1227. Warf, B. 2009a. Diverse spatialities of the Latin American and Caribbean internet. *Journal of Latin American Geography*, 8 (2), 125–146.

Warf, B. 2009b. The rapidly evolving geographies of the Eurasian internet. *Eurasian Geography and Economics*, 50 (5), 564–580.

Warf, B. 2010. Geographies of global internet censorship. *Geojournal*, 76 (1), 1–23.

Warf, B. 2012. *Global geographies of the internet*. Dordrecht: Springer.

Warf, B. 2015. The hermit kingdom in cyberspace: Unveiling the North Korean internet. *Information, Communication and Society*, 18 (1), 109–120.

Warf, B., and Vincent, P. 2007. Multiple geographies of the Arab internet. *Area*, 39, 83–96.

Wriston, W. 1997. Bits, bytes, and diplomacy. *Foreign Affairs*, 76 (5), 172–182.

Yalcintas, A., and Alizadeh, N. 2020. Digital protectionism and national planning in the age of the internet: The case of Iran. *Journal of Institutional Economics*, 16 (4), 519–536.

Zhang, C. 2020. Who bypasses the Great Firewall in China? *First Monday*, 24 (1). https://firstmonday.org/ojs/index.php/fm/article/download/10256/9409#author.

3

DIGITAL DIVIDES

James B. Pick and Avijit Sarkar

Digital divides constitute a foundational element of media geographies. Social media, radio, TV, streaming video and the spatial web depend on communications and increasingly on the broadband internet. These media services are provided over varied platforms ranging from wearables, to cell phones, tablets, laptops and servers. Messaging is sent by radio signals, fiber optic and other cables, and satellite transmission.

Against this backdrop, media content is generated and digitized. This chapter focuses primarily on digital divides as they relate to digital media. In various manifestations, digital media may include digital images, digital video, digital audio, digital audiovisual media, computer games, video games, digital books, digital text and the like. In 2020, digital media users spanning video games, electronic publishing, video-on-demand and digital music numbered 6.42 billion globally. This number is forecasted to balloon to almost 8 billion by 2025 (Statista 2020a). Global digital media revenue is estimated to be almost USD 200 billion in 2020, and projected to grow to USD 255 billion by 2025. While this projected growth has been somewhat stymied by the COVID-19 pandemic, the pandemic has nonetheless catalyzed the overall growth in digital media content and use in 2020. The crucial metric of success for media is whether or not the media users can read, understand and benefit from the content, and the extent to which they apply the content for entertainment, knowledge acquisition, business operations, management decision-making or leadership purposes. Users vary in their benefits from media, for instance a corporate leader might benefit by weighing a decision based on a news media, while a student might benefit by learning from it, or a retiree might be kept up-to-date about investments and retirement portfolios. Although media have undergone exponential growth, based on rapid technological advances, persistent issues continue to be the inequalities in availability, access, use, and the educational differences that imply varying levels of understanding and application of the content.

The digital divide is defined as the gap between individuals, households, businesses and geographic areas at different socio-economic levels with regard both to their opportunities to access information technology and media and to their use of the internet for a wide variety of activities (modified from OECD 2011). The concept of the digital divide (NTIA 1999) has become widely studied and applied, with thousands of studies of it and hundreds of national and regional governments applying it in practice. It does not imply a dichotomy of the technology rich and poor, but is viewed rather as a continuum of levels of digital access, use

DOI: 10.4324/9781003039068-4

and outcomes (van Dijk 2005; 2020). The digital divide is a complex phenomenon, not only involving technology infrastructure, but also people's motivation, skills, goals and outcomes, as well as the content of information, the cultural setting, and social and economic forces.

Has the digital divide narrowed so it is no longer relevant? This question is often asked; while some indicators of technology access and use are leveling off at high levels in advanced economies, such as internet access and access to broadband, these indicators have a long way to grow in developing countries. As seen in Figure 3.1, for 2005–2019, the number of individuals using the internet per 100 persons is considerably higher in developed countries than for intermediate (developing) or least developed countries (LDCs).

Although there is a huge and growing disparity in fixed broadband subscriptions between developed and developing countries/LDCs, which is ascribed to the expense in developing countries of installing optical fiber landlines (ITU 2019), the developing nations make up for this by adopting and using smartphones and other mobile devices. This is can be seen in Figure 3.2. By far the greatest increase in information technologies during this period was in mobile broadband, which allows developing nations to "leapfrog," skipping the older technology of fixed broadband and speeding up the attainment of full broadband capability. Another trend that is narrowing the digital divide is that 96 percent of the world population

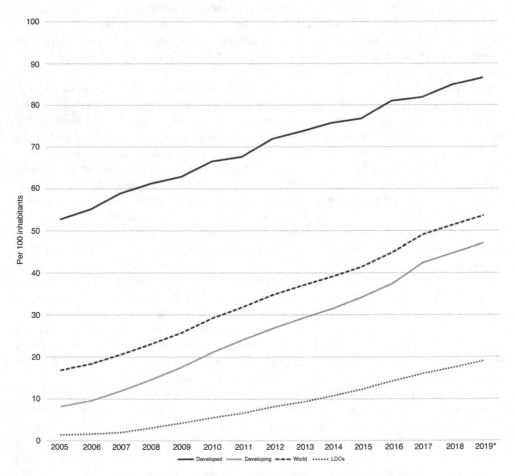

Figure 3.1 Number of individuals using the internet by development status, 2005–2019

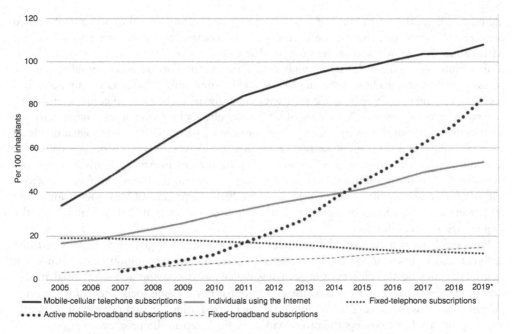

Figure 3.2 Worldwide subscriptions and use of technologies, 2005–2019

in 2018 resided within range of a mobile-cellular network, and 90 percent were within range of a 3G or higher-level network (ITU 2019). Yet only 4.6 billion people worldwide were internet users in 2018, so there is potential for 3 billion more people to access and use the internet (Population Reference Bureau 2018).

Another essential aspect of digital divides is the unit of analysis, which can be measured at different levels between individuals and the planet as a whole (Barzilai-Nahon 2006). Some important sub-national geographic units are states/provinces, counties/districts, neighborhoods and households. At each unit, the digital divide context differs. For instance, the digital divide among households in web streaming is different in meaning than the average web streaming for the states of New York and Missouri. The household differences might be explained by the geography of family structures in a census tract, whereas the state differences would likely be interpreted by effects from their educational systems, urbanization and income levels.

The choice of unit of analysis should be based on the research intent and the results should be interpreted relative to the unit level. In other words, the ecological fallacy applies, which calls for interpretation of associations between variables to reflect the nature of the units. Geographically, when units of analysis change, the modifiable areal unit problem (MAUP) must also be taken into account. MAUP cautions that at different levels of geographic units, for example, for census block groups, tracts and zip codes, the shifting arrangements of unit boundaries may alter average values for the entities within the units, so statistical findings for the same attribute may differ from level to level.

Four stages of digital divide progression over time

Digital divide research early on focused on physical access to technology. The focus later progressed to use, purposeful use and outcomes. This sequence was outlined in an early digital divide model (van Dijk 2006). A subsequent educational digital divide model posited the steps of access, use, specific use by the subjects (youths), and outcomes (Warschauer & Matuchniak 2010). We and others propose that this series of steps is in accord with long-term technological and behavioral trends (van Dijk 2020). For example, consider that the access digital divide made sense in the US in 2000, when only 41% of US households had access to the internet (Newburger 2001), with large disparities between income, educational and ethnic groups. Since in 2018, 85% of US households had internet access (Statista 2020b), with convergence of the groups, there is a reduced need to study the access digital divide in the US. While in 2020 some indicators of access and use were over 90% on average in the US and other advanced economies, most developing nations in Africa, South America and the Middle East today still have limited technology access so, for them, the focus on access and use continues to makes sense. For instance, in 2019, 32 percent of Venezuelan adults and 30 percent of Indian adults do not own a mobile phone (Silver et al. 2019), highlighting the importance of access for them.

Purposeful use is differentiated from access by its specificity of goal. It is beginning to be differentiated in research and can be expected to have greater recognition in the future. An example would be the difference in studying digital divides in the access of social media, measured by subscribers per capita, compared to disparities in using social media for a specific purpose, such as marketing a consumer product, expanding one's professional network, applying for a job, or posting a photo or video. A final step in the progression is outcomes. What impacts or changes have come about as a result of use or purposeful use? Outcomes encompass most of the benefits of the technological activity, but are difficult to study, since an outcome is often delayed and sometimes intangible.

Methods that address the complexity and multiple dimensions of digital divides

As digital divide research has evolved over the past two decades, measurement of the divide has changed from narrow studies of access for a single technology indicator to studies incorporating indices of multiple dimensions of the divide, and to frameworks to explain the digital divide (Barzilai-Nahon 2006). The indices include the Networked Readiness Index (NRI) (WEF 2016), a group of five indicators tied to the UN Sustainable Development Indicators (UN 2015; ITU 2015; 2020), the Digital Divide Index (DDI) (Barzilai-Nahon 2006), several van Dijk models (van Dijk 2005; 2006; 2020) and the Spatially Aware Technology Utilization Model (SATUM) model (Pick & Sarkar 2015; 2016). These indices and models seek to capture the multiple dimensions of the digital divide concept. There is not yet any consensus on a leading or standard index or model. For the investigation of media geographies, these approaches give the researcher flexibility in choosing an approach. These models are all amenable to the inclusion of geography, but only the SATUM model includes geography as an endogenous component. The NRI, van Dijk framework and SATUM model are further explained in this chapter relative to media geographies.

The NRI originated at the World Economic Forum in 2001 and was expanded in 2012. The goal is to encompass a large set of dimensions of digital divide into a single index value, which can be applied for standardized comparisons between nations (WEF 2016). As seen in Figure 3.3, the model is divided into networked-readiness Drivers and Impacts. The index Drivers include infrastructure, affordability and skills, which broadly connote Access as stated earlier, while the usage connotes purposeful use of information and communication technologies by individuals, businesses, and governments – both within and between those

Drivers

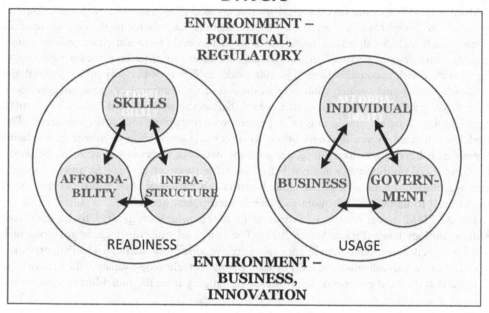

Impact

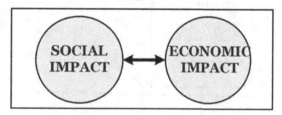

Figure 3.3 Networked readiness index framework

groups. An additional factor not referred to earlier is the NRI's Environment, which refers to the economic, regulatory, political and business macro environment, which influences the readiness and user of technology. Finally, the Impact of the Drivers is divided into economic and social impacts. The NRI has 53 specific indicators, with three to nine indicators comprising each of ten groups, eight of which are shown on the diagram as the grey circles (WEF 2016).

The NRI can be applied at the country level, and networked data were collected from international data sources and from WEF surveys from 2001 to 2016. The prominent NRI and its individual components have constituted a data source for digital divide research. The index has been criticized as not emphasizing and justifying the weighting assumptions for its 53 indicators. Although the WEF did not gather data for smaller units of analysis, the framework might be modifiable for use in studying provinces, cities and other micro units. Ideas from this framework can be useful in assessing the digital divides of media geographies, because many media types are instantiated by some or all of the 53 indicators.

The digital divide model of Jan A.G.M. van Dijk focuses on the individual unit of analysis, emphasizes inequalities and includes multiple constructs, some representing other units of analysis (van Dijk 2005; 2020). The progression of steps for the individual to achieve digital capability follows the steps of motivation, material access, digital skills and usage (see Figure 3.4).

Motivation refers to the psychological motivators as well as inhibitors such as "computer anxiety," "technophobia" and stress. A positive attitude contributes to moving to material access, which includes affordable costs, and next to favorable technical characteristics which combine with resources to lead to use. By another route, material access leads to access to digital skills which encourage usage. The full model in Figure 3.4 also includes personal and positional characteristics which contribute to the resources. Positional categories of education, labor force, geographic unit and social network define the occupational, educational, social capital and geographic positioning of a person, which in turn influences resources. This model was utilized by van Dijk and others to survey and interview individuals about digital disparities. It has the advantages of focusing on the details of a person's motivation, behavior, skills and social positioning, while not being adaptable to larger geographic units of analysis such as counties or nations. It does include location as a positional characteristic so the model would need to be modified to more thoroughly incorporate geography.

The SATUM model consists of independent factors which are posited to be correlated with technology levels (Pick & Sarkar 2015). The independent factors include governmental, economic, education, demographic, societal openness, innovation and social capital attributes, or others known to influence the level of use of a media technology variable. The model can be applied at different geographic units of analysis, ranging from the individual to census tract

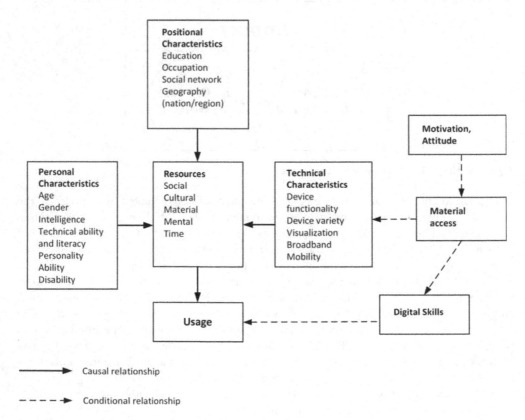

Figure 3.4 van Dijk model of divides of digital media use

all the way to small-scale units such as states/provinces and nations. Unlike the NRI and van Dijk model, the SATUM model implicitly includes mapping and spatial autocorrelation. As seen in Figure 3.5, SATUM performs OLS multiple regression to test models of determinants of levels of media access, use or outcomes, with spatial mapping and Local Indicators of Spatial Autocorrelation (LISA) testing (Anselin 1995) for both the dependent factors and residuals of the multiple regression. Accordingly, spatial bias can be addressed and adjusted for by including independent factors that account for geography. The model can be applied across the continuum of measures of digital divide, including access, purposeful use and outcomes. A weakness of this approach is that it assumes the relationships being studied are linear. Further the model is less suited for strongly behavioral studies at the individual level.

ICT trends worldwide

As shown earlier in Figure 3.2, internet penetration worldwide has steadily increased over the past two decades. This growth has been spurred by gradual improvements in fixed

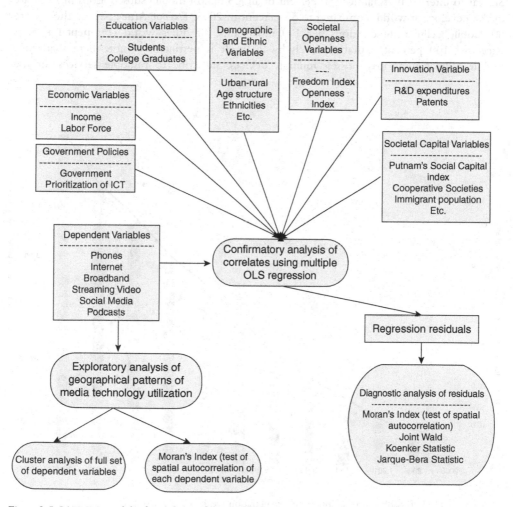

Figure 3.5 SATUM model of correlates of level of technology

infrastructure but more so in the rapid proliferation of mobile-cellular subscriptions, particularly smartphones, coupled with growth in mobile broadband availability. Compared to 2010, when there were 807 million mobile broadband subscribers with global penetration at 11.5 subscriptions per 100 inhabitants worldwide, mobile broadband subscriptions in 2019 reached 6,380 million, with a penetration of 83 (Figure 3.6).

Internet penetration gaps between the developed and developing nations continue to persist. In 2010, 21 individuals per 100 inhabitants used the internet in developing nations compared to 67 in developed nations. In 2019, the corresponding penetration rates were 47 and 87 respectively. Encouragingly however, the base of internet users has almost quadrupled over the past decade in the developing world, from 811 million in 2008 to 3,020 million in 2019, accounting for almost three-quarters of all internet users worldwide.

Mobile-cellular penetration gaps have however narrowed since 2010 between the developed and developing worlds. In 2010, there were 69 mobile-cellular subscribers per 100 inhabitants in the developing world compared to 113 in the developed world. The penetration gap narrowed respectively to 104 and 129 in 2019, yet a substantial gap remains. Similar to internet users, almost 80 percent of all 8.3 billion mobile subscriptions in 2019 are in the developing world, compared to 55 percent in 2005. Except Africa, where there were 80 mobile cellular subscriptions per 100 inhabitants, all other world continental regions exceeded 100 percent saturation levels by 2019. It is pertinent to note here that since mobile-cellular users may possess multiple devices, subscribe to multiple services, or use

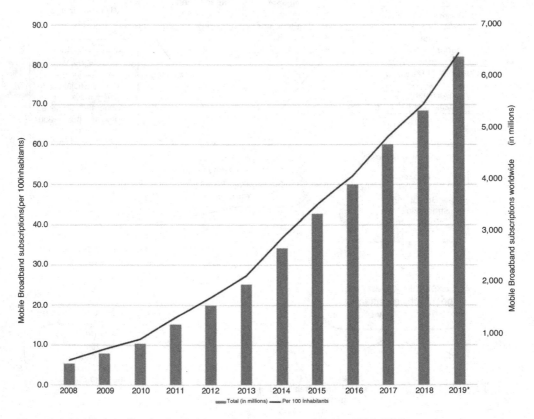

Figure 3.6 Mobile broadband subscriptions and penetration worldwide, 2008–2019

multiple SIM cards for a single device, mobile cellular penetration rates per 100 inhabitants exceeded 100 globally, for both developed and developing nations, in 2019.

The narrowing of the gap in mobile broadband penetration has kept pace with that of mobile cellular penetration. In the developed world, mobile broadband penetration improved approximately three-fold from 45 per 100 inhabitants in 2010 to almost 122 in 2019. The improvement was starker in developing nations where penetration increased from a paltry 4.5 active mobile broadband subscriptions per 100 inhabitants in 2010 to over 75 in 2019. Unlike mobile subscriptions, mobile broadband subscriptions lag in both Africa and the Arab world spanning the Middle East and Sub-Saharan Africa. At 34 mobile broadband subscribers per 100 inhabitants in Africa in 2019, mobile broadband technology holds promise for Africa and the developing world to leapfrog fixed technology and related infrastructural malaise and gradually improve connectivity and bridge digital disparities in this and other developing world regions.

Digital divides of the media

In this section, digital divides as they relate to media geographies are analyzed from several perspectives. The question of free versus paid access for media use relates to digital divides. This is further influenced by international cultural differences in media types and uses. The section finishes with a mini case study in which a radio organization, faced with the COVID-19 pandemic, had to make choices related to the potential to narrow the digital gap.

The prevalence of free versus paid access for some forms of media raises questions of digital inequalities. In key markets in six major nations (US, China, India, Germany, South Korea and UK), a preponderance of news audience (over 80%) and entertainment audience (over 90%) are listeners, viewers or readers of at least 24 hours per week of media (WEF 2020). Moreover, the overall percent of free uses is over 80% for news and over 56% for entertainment. For this massive base of free users, the access and use barriers of the digital divide are largely overcome, so the digital divide focus moves to purposeful use and outcomes. For the base of paid users, there may remain significant barriers to use. For example, a paid digital subscription for *The Economist*, one of the foremost economic and political news outlets, includes, beyond free content, access to entire print editions, online articles, audio editions, daily/weekly newsletters and an archive of over 100,000 articles. Accordingly, the smaller paid audience can be viewed as a source of digital inequality.

A mini-case study illustrates how the balances of equality of usages, geography, demography, cost to the consumer and tradeoffs to the media organization influence a media digital divide. The case involves an anonymous public radio organization, some details of which have been changed to protect their identity. The public radio organization, anonymously referred to as PRNEAST, serves a metropolitan area in the US Northeast and has been sustained mostly by membership contributions at varied levels as well as by underwriting from corporations. For five years, PRNEAST had been transforming from traditional fixed-schedule radio programming of news and music to a mixed format of traditional programming combined with digital offerings such as streaming of some music and initial local news podcasts. The organization had formed a new unit for internet, web and podcast innovation and production. PRNEAST served a traditional older radio audience and diverse younger internet-savvy audience located in fairly distinctive geographic areas within the metropolitan area.

Pre-COVID-19, PRNEAST had the choice to keep this mixed format or to go to a fully online platform, dropping traditional radio programming, which would appeal to its younger listenership but considerably reduce older listeners. It stayed with the mixed format. When the COVID-19 pandemic struck in March of 2020, the new unit's online news services had a tremendous surge of over five-fold in listenership across a wide age range and with extensive geographical reach, providing critical information to homebound individuals and families, hungry for local news, especially in the initial three-month exponential growth of the pandemic. At the same time, PRNEAST had to downsize and deal with reduced membership and underwriting contributions in the pandemic-induced recession, while trying to preserve its digital capabilities. Pre-COVID, if PRNEAST had chosen to go to an entirely online platform, it would have experienced a more severe financial loss during the COVID pandemic, because it would have lost many older members and undermined its base of affluent donors. However, on a generational timeframe, PRNEAST will eventually have to face dropping much of its traditional radio programming—and fortunately it did not go out of business during the pandemic.

This mini-case demonstrates how the balance of digital media usage can be jolted into a narrowing of the digital divide by an emergency, how user geographies may be expanded as a consequence and how a media organization can likewise be transformed to put more emphasis on widespread free services while taking a financial hit.

Social, economic, political determinants of digital divides

As reflected earlier in the SATUM model in Figure 3.5, a variety of demographic, economic, social and political factors have influenced access, adoption and utilization of information and communication technologies (ICTs). In the vast digital divide literature, the roles of these factors in explaining digital divides in varying degrees has been examined in multiple contexts—for varied geographies, populations and extents of human development.

Among demographic determinants, important factors are age structure, gender, educational attainment, race/ethnicity and place of residence. All over the world, technology adoption and use are more prevalent among younger populations, particularly the youth. For example, in the United States, in November 2019 (NTIA 2020), 85 percent of those in the age group 15–24 years used the internet compared to 68% among those aged 65 and over. In 2010, the corresponding proportions were 86% and 45% respectively, indicating that the age-based digital divide has shrunk over time, yet significant disparities still remain. For particular types of internet use, these differences are starker. For example, 53% of those aged 65+ use the internet for social networking compared to 88% of those aged 15–24 years. In contrast, only 27% of those aged 65+ use the internet for streaming or downloading music, radio or podcasts, compared to 78% of those aged 15–24 (Pew 2019b; 2019c).

Gender-based differences in internet use often stem from historically patriarchal societies, such as in the Middle East and parts of Asia and Africa. Race/ethnicity-based differences are observed in both developed and developing nations. For example, African Americans have historically trailed Whites in internet use in the United States (76.5% compared to 83.1% in 2019, and 66.5% compared to 78.8% in 2010) (Pew 2019a). Prior digital divide research has consistently explained such race-ethnic disparities as stemming from the groups' differential social positions, particularly relating to income and educational attainment (Campos-Castillo 2015). Such disparities are often exacerbated among African-American and Hispanic households with lower incomes living in rural areas with inadequate access to computers and unreliable broadband connectivity. Educational attainment also plays a defining role in digital disparities, with college-educated populations often exceeding those with high school diplomas in internet use and in general in the use of digital media. Lastly, the increasing extent of urbanization worldwide often magnifies urban-rural digital disparities. In rural Japanese prefectures, both internet and social media use were found to lag considerably behind those of large megacities such as Tokyo and its surrounding areas (Nishida et al. 2014). In the United States, urban users have a higher intensity of internet use to access online entertainment (watching videos online, streaming digital content online) compared to their rural counterparts, with a per capita gap of 10% in 2019. Infrastructural malaise in rural areas coupled with lower levels of income and educational attainment offer explanations for urban-rural digital disparities. Physical social interconnectedness at the community level, often measured by social capital, bridges digital gaps between those who have access to digital technologies such as the internet and possess related skills, and those who do not. Rural people may enjoy richer non-technology socializing, which might differ by life cycle stage—these are questions calling for future behavioral research.

Among economic factors, the role of income has been explored extensively in explaining global digital divides. Greater personal income is found to make technology more affordable for individuals, while higher national income enables communities, businesses, governments and individuals to invest in ICTs leading to higher per capita use (Pick & Sarkar 2015). Income is part of the well known, inter-correlated income-education-urbanization triad, and the roles of these interrelated factors have been explored by digital divide researchers and policy experts to shape policies to bridge ICT access and usage gaps.

Among notable social factors influencing digital divides are social capital and societal openness. Social capital, often thought of as the social interconnectedness of communities, has been increasingly found to be associated with ICT access and use including digital media. For example, in the United States, bonding social capital (i.e. social capital with strong personal ties) was found to be moderately associated with internet access and intensity of online communications (Chen 2013). Recent studies have found social capital to be associated with technology access in both technologically advanced and emerging nations such as the United States and India (Pick & Sarkar 2015). In the United States, social capital has been found to be positively associated with internet usage and broadband access in the household, and also with the use of Twitter. Oftentimes, social interconnectedness that is part of social capital facilitates access to ICTs including digital social media, and at other times puts those without the technology expertise and know-how in contact with those who are tech-savvy. This bridges the knowledge gap and provides training in an amenable social setting.

Among political factors, internet censorship has been examined especially in those contexts where authoritarian regimes resist democratic norms and often censor internet use (Warf 2011). Internet censorship restricts the free flow of information, curtails electronic communication between people and impedes societal openness.

Country and regional examples

Studies of the digital divide investigate digital patterns and disparities at various units—at the levels of individuals, businesses, households and geographic areas such as nations, national agglomerations, regions, states or provinces. Studies are often based on population and internet censuses conducted by national governments, with facts and figures reported at diverse levels of geographic resolution. One common unit of analysis is the state or provincial level. In this section, we first provide results from recent research into the digital divide in the United States at the state and county levels. Then, as a contrast, we present national comparisons of ICT adoption and utilization from two different parts of the world—Latin America and the Caribbean, and Africa. These major world regions account for almost 1 billion internet users as of May 2020 (Internet World Stats 2020) and have experienced the highest growth in the population of internet users since 2000, outside of the Middle East. Yet, alongside Asia, internet penetration (users as percent of total population) is among the lowest when compared to other world regions, representing major frontiers of the global digital access and use divide.

As an advanced digitally connected nation, the digital divide discourse in the United States has gradually evolved and shifted its focus from ICT access to purposeful usage. This is not to say that access gaps have been completely bridged. Lack of internet availability and inadequate access to computers in the household are among the main reasons for no online connectivity in some parts of the country. Yet, as smartphones have proliferated and mobile broadband penetration has increased, the spectrum of online activities among American internet users has grown and diversified. Americans use the internet to access education and entertainment, undertake financial transactions and purchasing activities such as e-commerce, access services and health-related information, work remotely, effect electronic communication including social networking, and connect with household devices embedded with sensors. For example, Figure 3.7 shows spatial patterns of internet use by Americans to watch videos online (2017).

It is evident from Figure 3.7 that the Rocky Mountain States and the West Coast states have some of the higher per capita use of the internet for e-entertainment purposes, while usage lags significantly in the Southern states and Appalachia—a region that has previously lagged in internet access as well. Surprisingly though, spatial patterns of internet use for social networking (Figure 3.8) show moderate to high intensity in the Southern states, sometimes slightly surpassing the digitally advanced states on the Eastern seaboard. Zooming in further, a recent study has found evidence that the major determinants of social media use in US counties are demographic factors, service occupations, ethnicities and urban location (Pick et al. 2019). These contrary findings—that social media use is moderate to high in rural areas of Southern states while urban location is positively associated with social media use, can be reconciled by observing that high social media use in traditionally rural areas (counties shown in grey) is often spurred by the presence of "tech islands" (counties shown in black) such as universities, medical and research institutions, and military installations, as shown in Figure 3.9. The four categories are comprised of statistically significant hotspots (High-High category—a county with high levels of social media use surrounded by similar high use counties), coldspots (Low-Low category) and outliers—a county with high levels of social media use surrounded by counties with low levels of social media use (High-Low), and vice-versa.

Compared to the United States, Latin America and Africa present significant contrasts. In Latin America and the Caribbean, English is often not the dominant local language and poses a language barrier for internet adoption and diffusion. In fact, a recent study (Pick et al. 2020) found that the factors "English as a primary language," "human development" and "civil liberties," influence ICT adoption and use in this world region, indicating socio-economic, language and societal openness dimensions of the digital divide here. The language barrier in the Latin American ICT landscape is being overcome by the development of web content and apps in Spanish, Portuguese and other native languages. This has spurred continued growth of social media in Latin America where Facebook penetration (63.4% in 2020, Internet World Stats) only lags behind North America (68.5% in 2020, Internet World Stats) among world regions. Figure 3.10 shows spatial patterns and agglomerations of Facebook penetration in Latin America and the Caribbean in the period 2013–2015. Despite the worldwide popularity of Facebook, other social media platforms such as WhatsApp and

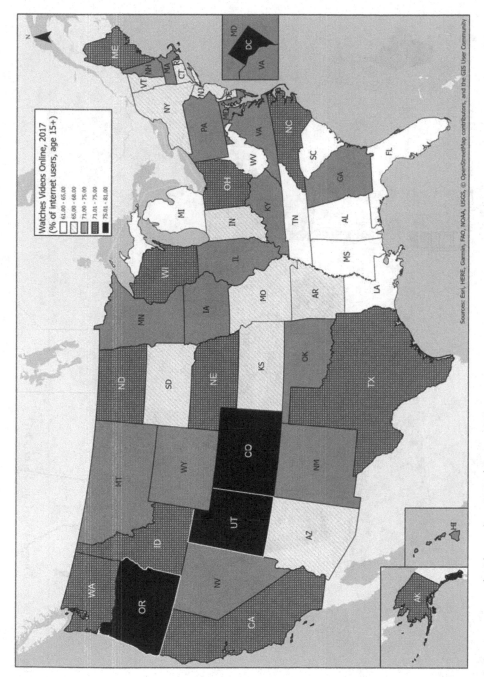

Figure 3.7 Internet use to watch videos online, United States, 2017

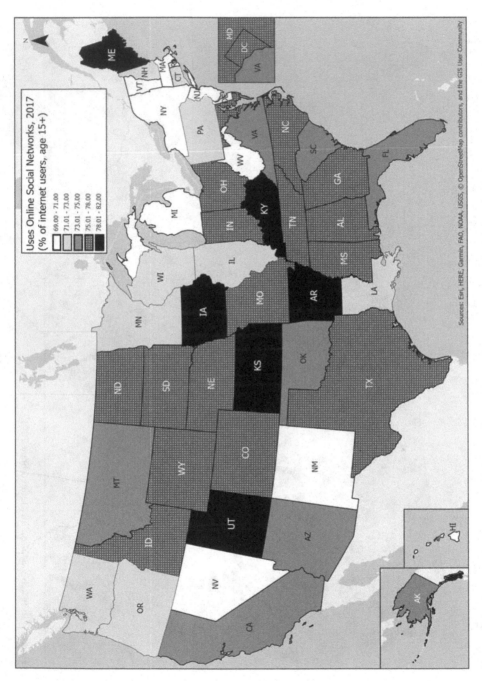

Figure 3.8 Internet use for social networking, United States, 2017

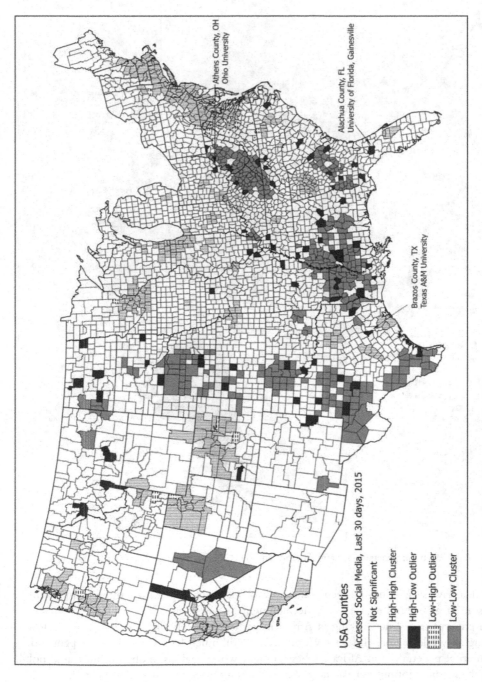

Figure 3.9 Internet use for social networking, United States counties, 2015

Figure 3.10 Facebook penetration, Latin America and the Caribbean, 2013–2015

WeChat have surged in prominence and use and are likely to provide a more comprehensive measure of global and regional ICT development.

In contrast, social networking in Africa, in particular Facebook penetration, is the lowest among all world regions (Internet World Stats 2020), much like overall internet penetration. As internet penetration in Africa catches up other world regions, mobile-cellular and mobile broadband subscriptions per capita in Africa continue to lag significantly behind. This presents impediments for bridging Africa's digital divide compared to the rest of the world. Within Africa, prior studies have found North Africa (Egypt, Libya, Morocco and Tunisia) and some southern African nations (Mauritius, South Africa, Seychelles) to be ICT leaders, while western, central and large parts of sub-Saharan Africa are at the low end of the ICT utilization spectrum.

A study of Africa has also revealed that laws that relate to the use of ICTs are a dominant predictor of all forms of ICT utilization. The effectiveness of a national parliament/congress as a lawmaking institution has been found to be significantly associated with modern (broadband) as well as legacy forms (fixed telephones) of ICT (Pick & Sarkar 2015). This is consistent with prior literature which stresses the need for national governments in Africa to frame ICT sector policies for investment, privatization, deregulation and providing access in underserved areas.

Digital divides and government policies

Digital divides, including for media, can be influenced by government policies at the international, national, state/province and local levels. Governments can set goals and regulations that favor greater or lesser equality of access and use, influence the openness of content, provide resources to support infrastructure, training and education, or implement fair access rulings. Nearly all these issues are controversial, so the goal of this section is to raise the issues and opportunities for governments to influence the digital divide, but not to advocate a uniform solution.

Since the advent of the World Wide Web in 1991, national governments worldwide have struggled with the issues of internet openness, costs and privacy as seen in early comprehensive studies (National Research Council 1991). The internet grew in importance, and it was recognized as a human right by the UN in 2016 with the addition of Article 19 to the Universal Declaration of Human Rights (United Nations 2020), which states: "Everyone has the right to freedom of opinion and expression; this right includes freedom to hold opinions without interference and to seek, receive, and impart information and ideas through any media and regardless of frontiers" (UN 2016).

Although established as a UN worldwide goal, nations have taken their own approaches; for instance since 2013 China has had a strong policy of internet censorship which restricts the viewing of censored information online and actively monitors microblogs, social media and messaging for information it considers threatening to the state (Economy 2018; New York Times 2019). An in-depth study (Zittrain et al. 2017) found that state-sponsored filtering and censoring of internet information were widespread in 26 nations, although it takes different forms, and a censoring government often shifts over time which content sources are being censored. The impact on the digital divide for media is that in many circumstances worldwide, even though access is widely available, some purposeful uses are barred, which restricts societal outcomes. It is important to mention that 80 percent of the world's national governments do not have censorship policies, encouraging the flow of information and offering the opportunity for deprived segments of the population across wide geographies to benefit from media technologies.

In the US, at the advent of the web in the early 1990s, it was an open platform. Jumping forward to the past 12 years, there have been shifts that relate to broadband access and net neutrality policies. In the Obama Administration, starting in 2009, the internet policy was to extend broadband access to technologically disadvantaged segments of the population through infrastructure investments and training (Obama White House Archives 2013; ICT Monitor Worldwide 2016). The Administration also sought to establish a federal policy of net neutrality, which was designed to prevent Internet Service Providers (ISPs) from discriminating on delivery of internet content, equalizing access for all constituents, ranging from individuals to large corporations. Although the internet rollout to the disadvantaged had moderate success, net neutrality was approved and implemented by the Federal Communications Commission in 2015. In contrast, the Trump Administration shifted the thrust of its internet policy to emphasize rural broadband infrastructure (FCC 2020), while the FCC reversed course and overturned net neutrality in 2017, although California and some other states have since restored it. The Biden Administration supports providing broadband infrastructure to communities nationwide that have endured limited access to it.

The fits and starts in government policies and support have contributed to an uneven geography of broadband in the US and to regional inconsistencies about net neutrality. This uneven fabric has resulted in digital divide disparities at the county level. Accordingly, policy-making that influences digital divides in the US is partly done by metropolitan, county and state governments. At the county level, policies suggested by two studies (Sarkar et al. 2018; Pick et al. 2019) were the following: seek to establish free public broadband capabilities; support training programs for technologically deprived citizens; encourage citizens to leverage education and training to the next step which might be hiring or transfering into a job that centers on technologies including media; attract professional, scientific and technical workers; and support ways to broaden social capital and social networks that emphasize IT including media technology and its content production.

Summary

The COVID-19 pandemic has renewed attention on the digital divide. In much of the developed world, disparities in ICT and media access have steadily waned, particularly as smartphones have proliferated and mobile broadband connectivity has become ubiquitous. Yet, as the pandemic shuttered businesses and schools worldwide, millions transitioned to remote working, video conferencing and remote learning. Healthcare practitioners and patients transitioned to telehealth; millions of people shifted to live-streaming and podcasts for entertainment and the news, while millions of others expanded their social media presence. Consumers shopped with mobile apps for household supplies and essentials online. All these online activities imposed unprecedented demands on access to technologies as well as digital infrastructure even in developed countries.

In many metropolitan areas across the United States, school districts scrambled to provide laptops, tablets and portable hotspot devices to students in impoverished communities in their districts. For students in rural areas, similar challenges became an impediment to continuing learning and participating in schoolwork. Apart from the homework gap (Vogels et al. 2020), lower-income Americans also expressed concern about their ability to pay for a smartphone and digital media services, due to large-scale loss of employment reflecting underlying economic disparities that exacerbate digital divides. Unprecedented demand on bandwidth due to remote working, video conferencing, streaming and the sudden surge in demand to consume online entertainment and news has also slowed broadband speeds in many parts of the United States (Broom 2020). In the developing world, already existing access gaps became even more pronounced as households struggled with loss of employment, high data costs (in Africa) and unreliable digital infrastructure. In some nations such as South Africa, national governments remediated cost-related issues by making their COVID-19 website free of charge, with no data or airtime required, and local broadband providers doing the same for educational websites.

As governments, organizations, businesses and economic areas tackle the lack of internet and video conferencing skills, address gender disparities among digital communities, and improve fixed-broadband infrastructure and global internet penetration to meet the United Nations Sustainable Development Goals as part of accelerating human development worldwide, the pandemic is renewing attention on the existing dimensions of the digital divide and on pre-pandemic policies that had been formulated to bridge gaps in access and use of digital media and information.

Additional research is required to examine how such policies fared during the pandemic and what weaknesses and strengths have emerged as the pandemic recedes. As internet service providers all over the world experienced tremendous surges in demand due to teleworking, distance education and voracious consumption of e-entertainment content, it is essential to understand how existing regulations contributed to or relieved network stress. This will help inform additional regulatory actions and policy interventions that might be required in the post-pandemic world to best assist people who might continue to remain on the fringes of a global, networked society, or perhaps have now been pushed to the fringes. These are a handful of the issues that challenge digital divide researchers and policy experts. As the fallout from the COVID-19 pandemic continues to emerge worldwide, the value of information and digital media and the impacts of the digital divide have come to the forefront. It remains to be seen what the pandemic and the hurried public policy responses to it have taught us about information and media policies in the US and elsewhere.

References

Anderson, M. 2019. *Mobile technology and home broadband 2019*. Washington, DC: Pew Research Center.

Anselin, L. 1995. Local indicators of spatial association—LISA. *Geographical Analysis*, 27 (2), 93–115. https://doi.org/10.1111/j.1538-4632.1995.tb00338.x.

Barzilai-Nahon, K. 2006. Gaps and bits: Conceptualizing measurements for digital divide/s. *The Information Society*, 22 (5), 269–278. https://doi.org/10.1080/01972240600903953.

Broom, D. 2020. *Coronavirus has exposed the digital divide like never before*. www.weforum.org/agenda/2020/04/coronavirus-covid-19-pandemic-digital-divide-internet-data-broadband-mobbile/.

Campos-Castillo, C. 2015. Revisiting the first-level digital divide in the United States: Gender and race/ethnicity patterns, 2007–2012. *Social Science Computer Review*, 33 (4), 423–439. https://doi.org/10.1177/0894439314547617.

Chen, W. 2013. The implications of social capital for the digital divides in America. *The Information Society*, 29 (1), 13–25. https://doi.org/10.1080/01972243.2012.739265.

FCC. 2020. *FCC launches $20 billion rural digital opportunity fund to expand rural broadband deployment: Represents FCC's largest investment event to close digital divide*. Washington, DC: Federal Communications Commission.

ICT Monitor Worldwide. 2016. Obama Administration announces $400 million funding push for 5G wireless tech. *ICT Monitor Worldwide* (July 16). https://ezproxy.redlands.edu/login?url=https://www-proquest-com.ezproxy.redlands.edu/wire-feeds/obama-administration-announces-400-million/docview/1804599380/se-2?accountid=14729.

Internet World Stats. 2016. *2016 internet statistics for Latin America*. www.internetworldstats.com/stats.htm.

Internet World Stats. 2020a. *World Internet usage and population statistics, 2020 year-Q1 estimates*. www.internetworldstats.com/stats.htm.

Internet World Stats. 2020b. *Facebook users in the world*. www.internetworldstats.com/facebook.htm.

ITU. 2018. *Measuring the Information Society Report*, Vol. 1. Geneva: International Telecommunications Union.

National Research Council. 1991. *The internet's coming of age*. Washington, DC: National Academies Press.

Newburger, E. C. 2001. *Home computers and internet use in the United States, August 2000*. Washington, DC: US Bureau of the Census.

Nishida, T., Pick, J. B., and Sarkar, A. 2014. Japan's prefectural digital divide: Multivariate and spatial analysis. *Telecommunications Policy*, 38 (11), 992–1010. https://doi.org/10.1016/j.telpol.2014.05.004.

NTIA. 1999. *Falling through the net, defining the digital divide.* Washington, DC: National Telecommunications and Information Administration.

NTIA. 2020. *Digital nation data explorer.* www.ntia.doc.gov/data/digital-nation-data-explorer.

Obama White House Archives. 2013. *ConnectED Initiative.* https://obamawhitehouse.archives.gov/issues/education/k-12/connected.

OECD. 2011. *Understanding the digital divide.* Paris: OECD Publishing.

Pew Research Center. 2019a. *Internet/broadband fact sheet.* Washington, DC: Pew Research Center.

Pew Research Center. 2019b. *Mobile fact sheet.* Washington, DC: Pew Research Center.

Pew Research Center. 2019c. *Social media fact sheet.* Washington, DC: Pew Research Center.

Pick, J. B., and Sarkar, A. 2015. *The global digital divides: Explaining change.* Heidelberg: Springer-Verlag.

Pick, J. B., Sarkar, A., and Rosales, J. 2019. Social media use in American counties: Geography and determinants. *ISPRS International Journal of Geo-Information*, 8 (9), 424. https://doi.org/10.3390/ijgi8090424.

Pick, J. B., Sarkar, A., and Parrish, E. 2020. The Latin American and Caribbean digital divide: A geospatial and multivariate analysis. *Information Technology for Development*, 27 (2), 235–262..

Population Reference Bureau. 2018. *World population data sheet 2018.* Washington, DC: Population Reference Bureau.

Sarkar, A., Pick, J. B., and Rosales, J. 2016. Multivariate and geospatial analysis of technology utilization in US counties. In *Proceedings of 2016 Americas Conference on Information Systems.* Atlanta, GA: Association for Information Systems.

Sarkar, A., Pick, J. B., and Moss, G. 2017. Geographic patterns and socio-economic influences on mobile internet access and use in US counties. In *Proceedings of the 50th Hawaiian International Conference on System Sciences.* Washington, DC: IEEE.

Sarkar, A., Pick, J. B., and Rosales, J. 2018. ICT-enabled e-entertainment services in United States counties: Socio-economic determinants and geographic patterns. In J. Choudrie, S. Kurnia and P. Tsatsou (eds.) *Social inclusion and usability of ICT-enabled services*, pp. 115–146. London: Routledge.

Silver, L., Vogels, E. A., Mordecai M., Cha, J., Rasmussen, R., and Lee, R. 2019. *Mobile divides in emerging economies.* Washington, DC: Pew Research Center.

Statista. 2020a. *Digital media.* www.statista.com/outlook/200/100/digital-media/worldwide#market-users.

Statista. 2020b. *Percentage of households with internet use in the United States.* www.statista.com/statistics/189349/us-households-home-internet-connection-subscription/.

UN. 2020. *The universal declaration of human rights.* www.un.org/en/universal-declaration-human-rights/#:~:text=Article%2019.,media%20and%20regardless%20of%20frontiers.

van Dijk, J. A. G. M. 2005. *The deepening divide: Inequality in the information society*: London: Sage Publications.

van Dijk, J. A. G. M. 2006. Digital divide research, achievements, and shortcomings. *Poetics*, 34, 221–235.

van Dijk, J. A. G. M. 2020. *The digital divide.* Cambridge: Polity Press.

Vogels, E., Perrin, A. Rainie, L., and Anderson, M. 2020. *53% of Americans say the internet has been essential during the COVID-19 outbreak.* www.pewresearch.org/internet/2020/04/30/53-of-americans-say-the-internet-has-been-essential-during-the-covid-19-outbreak/.

Warf, B. 2011. Geographies of global internet censorship. *GeoJournal*, 76, 1–23.

Warschauer, M., and Matuchniak, T. 2010. New technology and digital worlds: Analyzing evidence of equity in access, use, and outcomes. *Review of Research in Education*, 34 (1), 179–225. https://doi.org/10.3102/0091732X09349791.

WEF. 2016. *The global information technology report 2016.* Geneva: World Economic Forum and INSEAD.

WEF. 2020. *Understanding value in media: Perspectives from consumers and industry.* Geneva: World Economic Forum.

Zittrain, J. L., Faris, R., Noman, H., Clark, J., Tilton, C., and Morrison-Westphal, R. 2017. *The shifting landscape of global internet censorship.* Cambridge, MA: Berkman Klein Center for Internet and Society, Harvard.

4

HACKING IN DIGITAL ENVIRONMENTS

Mareile Kaufmann

Hacking in digital environments tends to be portrayed as a criminal or an activist practice. Though less simplistic analyses of hacking exist (e.g. Coleman 2015; Jordan 2017), online image searches still present us with dark hoodies and masks, movies tend to typecast hackers as illegal intruders (e.g. in the movies *Skyfall; Hackers; Untraceable*) or heroic helpers (e.g. *The Imitation Game; The Matrix Reloaded*) who play evil games, seek revenge or occupy moral high ground. News stories largely cover calamitous hack-attacks, whereas hackathons are organized as events to develop and compare valuable technical and analytic skills. As we shall see below, governmental and scholarly analyses, too, lean towards understanding hacking in terms of binaries. Within today's digital media landscapes, however, hacking occupies many places. Calls for "more complete pictures of what hacking is like" (Flick 2018) are on the rise (see also Adana 2017; WIRED Technique Critique 2018). This chapter discusses hacking as a material practice that defies simple categorizations. It invites researchers to investigate the specificities and multiplicities of different hacking cultures and groups. This is a novel and relevant contribution to media geography, not least because the media practice of hacking entertains a unique relationship to geography and territory. On the one hand, hacking activities transcend traditional geographic territories, because they relate to digital spaces and material connectivities. On the other hand, geography may still influence material connectivity, hacking network topologies as well as hacking behavior (Rechavi et al. 2015). Keeping these diverse territorializations of hacking in mind, the contribution of the chapter is to point out that hacking even varies *within* specific spaces, cultures and media. In order to illustrate these multiplicities in hacking, the chapter turns to a case study geographically located in Germany, Austria and Switzerland. Empirical insights are gained via a qualitative interview study with hackers who were recruited via the Chaos Computer Club, a German hacker association that has a tradition of engaging with questions of privacy and security. The sample reflects the aim of the study, namely to analyze how those who navigate the socio-technical details of dataveillance choose to engage with it, how they hack and redefine it. As will be shown, different techniques can be used to hack dataveillance. These techniques are vividly discussed among hackers—not least because each technique can be associated with specific socio-political standpoints. This is why the study also illustrates that hacking is always a practice of meaning-making that varies in this case from mundane and pragmatic engagements to performing identity work or articulating critique. In providing concrete examples

DOI: 10.4324/9781003039068-5

from the interview material, the chapter finally argues that all these practices also tie in with affective and aesthetic aspects. Dataveillance hacks are also instances of pleasure and play. Further, dataveillance causes concrete emotions and affects that serve as the onset for hacking. The case study is thus a call to step away from simplistic portrayals and to let these different dimensions of hacking speak to each other. When we allow this to happen, we can establish analytics of hacking that have more explanatory power, understanding it as anything between pleasureful pragmatism and crafty critique.

The chapter will first summarize scholarly and political discourses about hacking, where hacking tends to be presented in terms of binaries and "hats" that lead to rather truncated portrayals. It will then argue for analyzing hacking as an embodied, material practice instead. The central part of the chapter then turns to the case study of *hacking dataveillance*. Here, the context of dataveillance is explained, methodological challenges and opportunities of the specific case study are reflected on and hacking is analyzed as a matter of technique, meaning-making and embodied experience. In conclusion, the chapter reiterates the invitation to step away from hats and binaries, and to multiply analytic concepts and embrace the plurality of what hacking is and can be.

A diverse terrain captured by binaries and hats

Scholarly and governmental discourses about hacking are dominated by specific topoi. Though the academic literature about hacking has become more varied over the past decades[1] the portrayal of hacking as a political practice is easily dominated by a rhetoric of binaries. Hackers corrupt or protect (Steinmetz & Gerber 2015), they are a threat (Furnell & Warren 1999; Rost & Glass 2010) or produce social value (Nissenbaum 2004). As an engagement with digital governance hacking tends to be either lauded or denounced, as Gabriella Coleman and Alex Golub (2008) observe. Especially in legal discourses and national strategies for cybersecurity, hacking is associated with criminal or illegal activity and is assessed in terms of informational and monetary loss (for a study of legal discourses see Kaufmann 2018). The notion of hacktivism (e.g. Wray 1998) introduced a more diverse understanding of hackers' political influence, framing hacktivism in terms of netwars (Denning 2001) or cyberwarfare, but also as a social movement and political protest (Jordan & Taylor 2004; Hampson 2012; Maxigas 2017), or as data activism and advocacy (Schrock 2016). Similarly, "ethical hacking" (Palmer 2001) and the penetration-testing of infrastructures to enhance security (Engebretson 2013) broadened the conceptual landscape. Discussions about the diverse ethical positions hackers inhabit, subcultural studies about their material practices, or dynamics and characteristics of specific hacker collectives also gave rise to the model of "hats." Hats are an attempt to make sense of this cultural diversity: a distinction of ethically minded and desired *white hat hackers* (e.g. Caldwell 2011) from malicious *black hats* or crackers also led to the more complex *gray hats*, the descriptions of which combine the language of "social engineering" and "prevention" with "exploits" and "insider attacks" (cf. Harper et al. 2011). *Blue hats* (e.g. Gold 2014) hack as a part of quality testing in commercial contexts, *red hats* are vigilantes who take revenge on black hats, *green hats* want to learn more about hacking, but do not yet have recognition in the hacker community they seek to be part of.

The model of hats clarifies that binary categorizations do not capture the diversity of the practice: hacking ties in with different ethical positions and subcultures, hacks can lead to security and insecurity, there is vanity and cults, but also benevolence and a culture of passing on knowledge. Hackers can change their hats, too. However, it is at this point that the theoretical value and explanatory power of *hats* comes to an end. In fact, the steadily growing amount of hats rather emphasizes that such attempts to categorize do not do justice to the diversity of the practice after all. Hats, too, encourage conceptual reductions and neat distinctions. As that, they capture and make sense of the "surface" of hacking. Hats can indicate directions, but they can also mislead analyses and miss out on depth and diversity.

Understanding hacking as a material practice

Structuring hacking in terms of dichotomies obscures rather than helps grasping what hacking can be. Hats may help us navigate the vast landscape of hacking discourses, but they still only focus on the hacker, that is the person. Hats foreground the hackers' societal role, their status and intentions. Thinking in hats does not encourage the study of hacking as a material practice. Hats do not invite us to see the technologies and techniques, the equipment and skills that co-shape hacking.

If we think of hacking as redefining the borders of digital environments, then taking account of the materiality and spatiality of these environments is key. Indeed, we cannot resort to theories of "embodied subjects situated and interacting in environments curiously lacking specific material constraints" (Blanchette 2011, 1055). Digital environments as well as the data that move in and through them require material expression (Drucker 2009; Gitelman & Jackson 2013). They can be divorced neither from the infrastructures nor the social practices that shape and re-shape them. Digital environments and data are coined by their many material, social, bodily and ethical points of contact, each of which leaves its residue. Acts of hacking, then, necessarily engage with all of these points of contact. Thus, hacking is dependent on these materialities and the hackers' knowledge about them. This is why Tim Jordan defines hacking as "different, sometimes incompatible, material practices" (Jordan 2017, 528). What is more, hackers not only interact with and redefine materialities as well as the social and ethical contact points that constitute them, but the socio-materiality also shapes how an act of hacking comes about or may even render a hack impossible. That is to say: hackers are not the only ones with agency in the process of hacking, but hackers and the material environments they interact with co-construct each other. Hacking, then, also has an unpredictable dimension as material environments "act back" and sometimes create consequences that were not intended by hackers.

Paying attention to the material dimension of hacking also involves studying its situatedness.[2] As argued above, hacking is always in contact with material, social, bodily and ethical dimensions. It is embedded in a set of relations. Precisely because hackers and digital materialities are shaped by social and ethical forces, one would have to understand these latter, too, if one wanted to grasp what a hack, specific technique or hacker collective is about.

If we think of hacking as embodied, as embedded in a set of relations and as a material practice, we take distance from overly simplified categories or hats that strip hacking of its context. Instead, we can trace networks of social, bodily, ethical and material relationships. This also pushes us to acknowledge the diverse aesthetic and cultural dimensions of hacking. In hacking, we find self-expression and romantic individualism, which invoke Nietzschean notions of power (Coleman & Golub 2008) and masculinity (Hunsiger & Schrock 2016; Jordan 2017), but also feminist values of inclusion (Goode 2015; SSL Nagbot 2016) and collective forms of action (Söderberg 2017). There is not just one rationale for conducting a specific hack. A hack, as we will see, can combine multiple potentially contradictory positions and purposes, aesthetic aspects and affects. For example, hacking can include moments and affects of authority and sharing (Powell 2016) at the same time.

With this contribution, I want to encourage researchers to (re)discover the specificities of a complex practice that can have multiple dimensions at the same time. As mentioned above, hacking is not (yet) a standard media practice. A case study can grant us insight into a field that is relatively new to media geography, which is why I will now turn to examples from my own research work. I want to show that there is variety in the media practice of hacking—even when all the hackers I have interviewed have the common vantage point of criticizing digital surveillance. Thus, in order to develop this argument about multiplicity, there is a need for empirical material to substantiate and illustrate these points. Furthermore, my empirical insights were shaped by methodical choices—not least because the media I used for interviews already co-determined the type of questions I would be able to ask hackers. This is why the following part will also dedicate some space to methodical considerations.

Hacking dataveillance: A case study with empirical insights from Germany, Austria and Switzerland

All digital environments are integrated with dataveillance. Any digital data comes from surveillance—whether it is generated by users or sensors, fed into digital systems by hand, or synthetically produced by machines that were trained on data that originally comes from veillance. Not only does digital data come from surveillance, but it also opens up new types of surveillance as digital data is by design traceable, storable, searchable, transferable and networkable (Kaufmann & Jeandesboz 2017). This character of digital data and the growing possibilities for analysis have given rise to social and political shifts. Practices of dataveillance and data analysis nurture ideas about social and biological engineering (Rouvroy 2012). They have given rise to data economies (Zuboff 2019) where the value of data can outreach the exchange value of money (O'Dwyer 2016). Digital data change social interaction and our relationships to the body and biology in many ways (e.g. Oravec 2020). They change our understanding of privacy and autonomy. Ultimately, dataveillance and analysis open up a range of new societal developments at the same time as they involve substantial inequalities in access to information and ultimately the governance of personal, social and political life. Thus, it does not come as a surprise that hacking in digital environments is always an engagement with dataveillance. This engagement can emerge out of a critical or an affirmative context; it can address practices of dataveillance that qualify as sur-veillance (dataveillance from above), sous-veillance (dataveillance from below), or equi- or peer-to-peer-veillance (the ways in which we watch and collect check data about each other). As practices of dataveillance vary according to media geographies and territories, we shall see that ideas and practices of hacking dataveillance do, too.

The same characteristics that allow for dataveillance, namely that data is material, traceable, storable, searchable, transferable and networkable, also enable hackers to question and redefine data flows, values and forms of governance. These characteristics enable hackers to redesign with data. I have studied how and why hackers engage with dataveillance. Thus, the focus of my study was not to understand the hacks of specific collectives or the idea of what it means to penetrate systems; the study was concerned neither with regulatory nor ethical questions. Rather, the specific focus of my study was to interview those who navigate the complex material, technical, social, bodily and political aspects of dataveillance and to find out how they choose to engage with these. A large proportion of my interviewees initially understood their hacking practices as a critique of dataveillance. What is crucial, however, is that the same interviewees also identify with the digital environments they hack. They partly built these environments themselves at the same time as they critically reflect on the extent to which this environment is integrated with and dependent on surveillance. Even though their critique on different politics and practices of dataveillance can be very outspoken, hackers do not contemplate leaving digital services (Burgess 2018), "un-facebooking" (Evans 2014) or "non-participation" (Casemajor et al. 2015). Rather than denying digital environments, they would remain in contact with, hack and shape them. They would "de- and re-construct" technologies (*Bl4ckb0x*[3]) in order to "discover" (*Crypsis*), "define the undefined" (*Panoptipwned*), "test" (*heisenbugwatch*), "reinvent" (*LOLveillance*), "create" (*3x3cute*) or "divert" (*GCSgateway*).

Methodical considerations

Studying hacking as a situated practice can be done in many ways. Observation, for example, can provide direct insight into interactions between hackers and technologies. At the same time, observation impacts the anonymity of the participants. Since privacy and the engagement with one's digital identity were central themes of the study, I chose anonymity over the merits of observation. My research design was based on qualitative interviews instead, none of which were conducted face-to-face. Here, dataveillance was not only a theme, but also became a methodical challenge of the study as the interviewees and I communicated via different apps and programs, each of which had their respective levels of integrated dataveillance. Thus, the mediation of the interview situation became quite important: some interviewees suggested using commercial services that had a low level of anonymity or data security. Others preferred open source software for political reasons and one third would only agree to be interviewed in writing via end-to-end encrypted connections. Especially the relative anonymity of the written interview situation allowed hackers to share thoughts and experiences that they otherwise would not have. More than that, their choice of interview medium already expressed something about their views and knowledge, their practices of engaging with dataveillance. Methodical aspects, then, became already a source of insight into the topic of dataveillance. Vis-à-vis observation, interviews also gave more room for asking questions about the why and how of hacking, about the relationships between technologies of dataveillance, politics, the social and the body. The 22 interviews varied from 45 minutes to 2.5 hours and included participants who were loosely associated with hacker clubs to more technically experienced hackers. Almost all interviews were conducted one-to-one, while some preferred a group setting.

Another methodical choice shaped the type of insights I would get from my interview study—that is the interviewees' geographies. Their geographic situation again illustrates the embedded aspect of studying hacking. My entry point for recruiting interviewees was the Chaos Computer Club (CCC). Since the 1980s the CCC has become an established civil society organization "dealing with the security and privacy aspects of technology in the German-speaking world" (Chaos Computer Club Website n.d.). The club is organized in 25 regional hackerspaces and smaller groups, where hackers work in a decentralized fashion. From the CCC onwards I snowballed my sample by means of recommendations and invitations. Thus, a relatively distinct geography characterized my group of interviewees. Most of my interviewees spoke German and I deduced that I must have been in contact with hackers in Germany, Austria and Switzerland. The geographic spread of the sample is relevant, because these countries have distinct histories of state surveillance, not least during and in the aftermath of World War II. These histories still influence the countries' debates on privacy and surveillance today. The CCC especially is known for their commitment to online privacy and many members support the club's policy of sharing knowledge and giving advice in workshops and local meetings. While many different types of hacking and ethical positions co-exist in these countries, using the CCC as a vantage point for a snowball sample was a fruitful choice, because I wanted to speak to interviewees who would engage with dataveillance.

Gaining meaningful access, however, was not always easy. Not only does the functioning of interview technologies co-determine the interview situation and the flow of the conversation, but before I even got to establish an interview situation I had to pass "social captcha tests" (Kaufmann & Tzanetakis 2020). By answering questions about data politics and fulfilling technical tasks, I had to gain credibility before I was allowed to conduct my interview. This test is, according to interviewee *DataD14709*, a typical characteristic of hacker communities: "In the same way that you'd complete a captcha test online, you also have to pass a captcha test as a human being, a hacker, a member of a social network. Every community has their own test and you have to find out how to do it." These tests were in their own way descriptive of the interviewee's stances on surveillance, privacy and trust. Thus, already methodical and practical interview encounters are part of giving a situated account of hacking dataveillance.

Matters of technique, meaning-making and embodied experience

When analyzing the practice of *hacking dataveillance* it becomes clear that hackers neither embrace nor resist dataveillance. They do not undo or agree with the surveillance inherent in online technologies. Rather, hacking dataveillance is an ongoing engagement with the materialities and politics of surveillance that varies across geographic and subcultural contexts. When I now turn to the range of techniques that are used to hack dataveillance, to the social and political meanings that such hacks assume, as well as their relevance as an embodied experience, we bear witness to a multiplicity that can hardly be captured by binaries and hats.

Computer hacking is often framed as a technical skill with advanced knowledge of coding and technological expertise (e.g. Lee 2008). In the context of dataveillance, however, there are many ways in which hackers take control over the data associated with themselves. Most of my interviewees mentioned that it does not require much technical skill to articulate the value of data that concerns their online persona. To them, hacking starts out with an awareness of dataveillance (which is much in line with the CCC's self-conception) and with a practice that I term data-minimalism (Kaufmann 2020). *Re-ID*, for example, argues that "once shared, information exists out there and cannot be retracted anymore." Thus, an everyday practice of hackers is to keep the personal information that one shares online to a minimum. Data-minimalism can be subject to constant reflection, but it can also become what some interviewees described as a body automatism. Limiting ones' own presence on social media, avoiding mediated conversations that one can have face-to-face, not using cloud or commercial services, so my interviewees argue, limits the quality and quantity of data available for surveillance.

When they do share online, my interviewees resort to encrypted peer-to-peer networks, where information is encrypted before sending and decrypted upon arrival. Only those with a matching set of en-/decryption keys have access to the meaningful content of the message. Another way to enhance the effort needed to surveil data are technical solutions that transmit data through alternative infrastructures, such as overlay networks or tunnel protocols. While the use of overlay networks and tunnel protocols is not illegal, they tend to be associated with the darknet or criminal behavior. This makes some hackers stay away from such solutions. *Crypsis*, for example, comes "to the conclusion that effort to monitor those presumably not surveilled spaces is spent exactly there," that is, spaces associated with criminal activity. Others mention that choosing encryption keys needs to be done with care as keys can get stolen, lost or be commercial. What is hacked when one uses encryption, overlay networks or tunnel protocols is mainly the standard use of communication technologies. Instead of using widely accepted communication technologies, hackers turn to alternatives. However, these "alternative" means of communication have by now become quite widespread, too. That is to say, in order to hack the standard use of communication technology one merely needs practical user-knowledge. Using alternative communication means does not necessarily require programming or other technical skills, because different encryption technologies and overlay networks already exist, ready to be used.

In addition to hacking the standard use of communication technologies, we find techniques of hacking social and sociotechnical communication standards. These hacks are characterized by opacity (Glissant 1997) and obfuscation (Brunton & Nissenbaum 2011), because the main aim here is to render one's online persona or "data double" (Haggerty & Ericson 2000) unidentifiable, to dissociate data from a person or to conceal the fact that a message is being sent. This can be done by hiding messages in a large amount of unimportant data, which then means time and resources are required from the surveilling parties to identify the meaningful part of a message. Others try to ambiguate their entire data double by feeding browsers conflicting information about themselves. Yet a different hack of social communication standards is to make information available for the intended parties only, for example via riddles and the use of symbols. *DataD14709* explains how hackers use steganography—that is, communication in social codes: "There are many ways: visual ways, languages, different logical connections to transmit a message that cannot be digitized... a chat via two different chat providers, where you only send a part via each program." Hackers discuss types, effects and ethics of hacking social and sociotechnical communication standards amongst each other. Taken together, these techniques illustrate that engaging with dataveillance does not necessarily require technical skill, but analytic creativity.

Some interviewees move, however, from hacking sociotechnical communication standards towards diversifying or building alternative hardware, which requires technical equipment and potentially more skill. *GCSgateway*, for example, starts the computer's operating system from a USB stick so that the system cannot be accessed by surveilling parties. Others, for example, build local networks for data-sharing that run disconnected from the internet. *Re-ID* explains: "The raspberry pi creates a WLAN that is completely secluded from the internet—that means nobody goes online with this and no data are shared via the internet. That means anything you communicate via a pirate box, that you upload—pictures, photos, videos, any data—are local and that is an interesting concept."

Yet a different type of hacking dataveillance is characterized by practices that are meant to be noticed. These are hacks that render critique visible and have a tendency to be more aggressive than hacks that dissociate data traffic from oneself. Barricading online traffic (often called distributed denial of service attacks), changing the appearance of a website or leaking information are part of this category. Some of these techniques interact quite directly with surveillers. Amongst hackers, these techniques are controversial. Many interviewees pointed to their destructive effects and the ways in which one can also make a "crusade" with small-scale and less aggressive practices (*AceOfPlays*). A less visible yet significant example of hacking dataveillance is to turn surveillance against itself. The hacker *=Overview* does so by reverse-engineering surveillance algorithms. Once she understands how they work she uses this knowledge for her own protection, but also to exploit the algorithms' weaknesses and strengths.

The above examples illustrate that *hacking dataveillance* is done in many different ways. Each of them is an engagement with one's own data-double or an attempt to redefine dataveillance practices. Thus, they are all instances of meaning-making. Any choice of technique is personal as it communicates a hacker's practical, ethical and political standpoints. The meanings that hacks assume are as diverse as the hacking techniques. For example, many interviewees hack dataveillance technologies on an everyday basis. The techniques of data-minimalism and encryption, for example, demonstrate that hacking does not need to be based in the extraordinary. Many hackers react to dataveillance in mundane ways: they improvise, they make small, spontaneous choices that become routines over time by which they slowly change dataveillance habits. In these cases, hacking surveillance is a pragmatic engagement, a strategy or tactic that stays within the realm of the given characteristics of the internet, but yet strives to effect change (Kaufmann et al. 2020; cf. de Certau 1984).

Other techniques, especially those that involve the hacking of social communication standards, can be understood as a way of giving hacking a personal note. By coming up with riddles and languages hackers communicate their creativity and make the hack their own. Hacking as a form of communication can become very concrete and detailed: as we have seen in the examples above, hacks are not only a hacker's "handwriting," but they can become an actual language with vocabulary and syntax. This language can be spoken loudly or gently. When spoken deliberatively, hacking dataveillance becomes a reflexive practice, an instance of critique and analytical creativity that is deeply political. Here, hacking is a practice of re-appropriation that does not take the technically inscribed rules of dataveillance as a given. By engaging with these rules, hacks become a form of critical reflection, a dispute expressed in craft. Thus, many hacks take a dialogical form: they are a material interplay between surveillers' practices and those that redefine them. This dialogue necessarily shapes future forms of dataveillance and data politics, but it is neither a coordinated activity, nor does it follow a unified ethical code. Rather, this part has argued that the diverse techniques tie in with instances of meaning-making that can range from banal and pragmatic strategies to the extraordinary, from the passive to the aggressive; they can be instances of identity work or outspoken criticism.

Finally, hacking dataveillance is more than a matter of technique and of meaning-making. As an embodied practice, hacking dataveillance also ties in with affective and aesthetic aspects. For example, the interviewees mention the pleasures of tinkering with software, of solving riddles or of facing a challenge. The system of rules that characterizes dataveillance is approached in a playful way: rules are identified, cracked, circumvented and redefined. This playfulness and analytic creativity is typical for hacker cultures (Richterich & Wenz 2017). Not only is hacking a pleasurable practice, but play, so Manuel Sicart (2014) argues, is an important form of appropriation. To hack is then not just to "play dataveillance" by having understood and redefined its underlying rules. To hack is also to "play *with* dataveillance." It is to create riddles and hideaways, to lure surveillance algorithms into different directions and to trick them by becoming invisible while remaining online. This aesthetic of play also creates physical experiences of excitement and thrill. This type of embodied experience is not unique to hacking dataveillance. It is remarkable, however, that the interviewees did speak about dataveillance in terms of specific affects and described how these affects literally became the onset for action (see affect theory, Massumi 1995). Hackers described the affective character of dataveillance in terms of "powerlessness" (*AceOfPlays*), "fear" and "creeping pain levels" (*Crypsis*), while the act of decoding and redefining these "pressures" (*Bl4ckb0x*) is experienced as "activism" (*Kate90r13*), as "being awake" (*Numbercruncha*), "willpower" (*Filterer*) or as "venting anger" (*heisenbugwatch*).

Conclusion

Hacking dataveillance is not just a matter of technique, but a practice of meaning-making with specific aesthetic and affective dimensions. It is a playful re-appropriation of the dataveillance rules that structure online politics. Some of these practices can be silent, harmless and affirmative, others are intrusive or deliberately located at the border of legality. What is more, the techniques of hacking dataveillance, the instances of meaning-making and embodied experience influence each other. Thus, in hacking dataveillance we find political playfulness, pleasureful pragmatism, embodied opacity, crafty critique or aggressive identity work. When we acknowledge the multiplicity and specificity of hacking as a material practice we arrive at new analytic concepts and with more explanatory power than binaries and hats can offer.

This chapter is an invitation to analyze the specificities of different geographic and cultural contexts of hacking. This approach was illustrated with a case study on *hacking dataveillance* that deliberately used the geography of the CCC as an entry point for sampling. In analyzing this example case, the chapter seeks to foster an attitude of identifying the many practices and places that hacking can and will inhabit in the global media landscape.

Notes

1 Some scholars, for example, frame hacking as constructive pleasure (Hannemyr 1999) or a mindset and needed creativity (Snook 2014).
2 For calls for situatedness and specificity see Ruppert et al. 2013; Kaufmann & Jeandesboz 2017.
3 For all references to interviewed hackers I use imagined handles. Any similarities with existing handles are unintentional.

References

Adana, K. 2017. *Why can't films and TV accurately portray hackers?*www.bbc.com/future/article/20170802-why-cant-films-and-tv-accurately-portray-hackers.

Blanchette, J. F. 2011. A material history of bits. *Journal of the American Society for Information Science and Technology*, 62 (2), 1042–1057.

Brunton, F., and Nissenbaum, H. 2011. Vernacular resistance to data collection and analysis: A political theory of obfuscation. *First Monday*, 16 (5). https://doi.org/10.5210/fm.v16i5.3493.

Caldwell, T. 2011. Ethical hackers: Putting on the white hat. *Network Security*, 7, 10–13.

Chaos Computer Club. n.d. *Chaos Computer Club.* www.ccc.de/en/club.

Coleman, G. 2015. *Hacker, hoaxer, whistleblower, spy: The many faces of Anonymous.* London: Verso.

Coleman, G., and Golub, A. 2008. Hacker practice: Moral genres and the cultural articulation of liberalism. *Anthropological Theory*, 8 (3), 255–277.

De Certeau, M. 1984. *The practice of everyday life.* Berkeley, CA and Los Angeles, CA: University of California Press.

Denning, D. E. 2001. Activism, hacktivism, and cyberterrorism: The internet as a tool for influencing foreign policy. In J. Arquilla and D. F. Ronfeldt (eds.) *Networks and netwars: The future of terror, crime, and militancy*, pp. 239–288. Santa Monica, CA: RAND Corporation.

Drucker, J. 2009. *SpecLab: Digital aesthetics and projects in speculative computing.* Chicago, IL: University of Chicago Press.

Engebretson, P. 2013. *The basics of hacking and penetration testing: Ethical hacking and penetration testing made easy.* Waltham: Elsevier.

Flick, C. 2018. *What Hollywood gets right and wrong about hacking.* https://theconversation.com/what-hollywood-gets-right-and-wrong-about-hacking-100126.

Furnell, S., and Warren, M. 1999. Computer hacking and cyber terrorism: The real threats in the new millennium? *Computers & Security*, 18 (1), 28–34.

Gitelman, L., and Jackson, V. 2013. Introduction. In L. Gitelman (ed.) *"Raw Data" Is an Oxymoron*, pp. 1–14. Cambridge, MA: MIT Press.

Glissant, E. 1997. *Poetics of relation*. Ann Arbor, MI: The University of Michigan Press.

Gold, S. 2014. Get your head around hacker psychology. *Engineering & Technology*, 9 (1), 76–80.

Goode, L. 2015. Anonymous and the political ethos of hacktivism. *Popular Communication*, 13 (1), 74–86.

Haggerty, K. D., and Ericson, R. V. 2000. The surveillant assemblage. *British Journal of Sociology*, 51 (4), 605–622.

Hampson, N. C. N. 2012. Hacktivism: A new breed of protest in a networked world. *Boston College International & Comparative Law Review*, 35, 511–542.

Hannemyr, G. 1999. Technology and pleasure: Considering hacking constructive. *First Monday*, 4 (2). http s://doi.org/10.5210/fm.v4i2.647.

Harper, A., Harris, S., Ness, J., Eagle, C., Lenkey, G., and T. Williams. 2011. *Gray hat hacking: The ethical hacker's handbook*. New York: McGraw-Hill Osborne Media.

Hunsiger, J., and Schrock, A. 2016. The democratization of hacking and making. *New Media & Society*, 18 (4), 535–538.

Jordan, T. 2017. A genealogy of hacking. *Convergence*, 23 (5), 528–544.

Jordan, T., and Taylor, P. 2004. *Hacktivism and cyberwars: Rebels with a cause*. New York: Routledge.

Kaufmann, M. 2018. "Now you see me—now you don't!" Practices and purposes of hacking online surveillance. *Mediatization Studies*, 1, 85–101.

Kaufmann, M. 2020. Hacking surveillance. *First Monday*, 25 (5). https://doi.org/10.5210/fm.v25i5.10006.

Kaufmann, M., and Jeandesboz, J. 2017. Politics and "the digital": From singularity to specificity. *European Journal of Social Theory*, 20 (3), 309–328.

Kaufmann, M., and Tzanetakis, M. 2020. Doing internet research with hard-to-reach communities: Methodological reflections about gaining meaningful access. *Qualitative Research*. https://doi.org/10. 1177/1468794120904898.

Kaufmann, M., Leander, A., and Thylstrup, N. B. 2020. Beyond cyberutopia and digital disenchantment: Pragmatic engagements with and from within the internet. *First Monday* 25 (5). https://doi.org/10. 5210/fm.v25i5.10617.

Lee, J. C. 2008. Hacking the Nintendo Wii Remote. *IEEE Pervasive Computing*, 7 (3), 39–45.

Massumi, B. 1995. The autonomy of affect. *Cultural Critique*, 31 (2), 83–109.

Maxigas. 2017. Hackers against technology: Critique and recuperation in technological cycles. *Social Studies of Science*, 47 (6), 841–860.

Nissenbaum, H. 2004. Hackers and the ontology of cyberspace. *New Media & Society*, 6 (2), 195–217.

O'Dwyer, R. 2016. *Where's the money?*www.kingsreview.co.uk/new-page-26.

Oravec, J. A. 2020. Digital iatrogenesis and workplace marginalization: Some ethical issues involving self-tracking medical technologies. *Information, Communication & Society*. doi:10.1080/ 1369118X.2020.1718178.

Palmer, C. C. 2001. Ethical hacking. *IBM Systems Journal*, 40 (3), 769–780.

Powell, A. 2016. Hacking in the public interest: Authority, legitimacy, means, and ends. *New Media & Society*, 18 (4), 600–616.

Rechavi, A., Berenblum T., Maimon, D., and Sevilla, I. S. 2015. Hackers topology matter geography: Mapping the dynamics of repeated system trespassing events networks. In *ASONAM '15: Proceedings of the 2015 IEEE/ACM International Conference on Advances in Social Networks Analysis and Mining 2015*. New York: ACM. https://doi.org/10.1145/2808797.2808873.

Richterich, A., and Wenz, K. 2017. Introduction: Making and hacking. *Digital Culture and Society*, 3 (1), 5–21.

Rost, J., and Glass, R. 2010. Hacking. In J. Rost and R. Glass (eds.) *The dark side of software engineering*, pp. 113–156. Hoboken, NJ: Wiley.

Rouvroy, A. 2012. The end(s) of critique: Data-behaviourism vs. due-process. In M. Hildebrandt and K. de Vries (eds.) *Privacy, due process and the computational turn: Philosophers of law meet philosophers of technology*, pp. 143–167. New York and London: Routledge.

Ruppert, E., Law, J., and Savage, M. 2013. Reassembling social science methods: The challenge of digital devices. *Theory, Culture & Society*, 30 (4), 22–46.

Schrock, A. 2016. Civic hacking as data activism and advocacy: A history from publicity to open government data. *New Media & Society*, 18 (4), 581–599.

Sicart, M. 2014. *Play matters*. Cambridge, MA: MIT Press.

Snook, T. 2014. *Hacking is a mindset, not a skillset: Why civic hacking is key for contemporary creativity*. https:// blogs.lse.ac.uk/impactofsocialsciences/2014/01/16/hacking-is-a-mindset-not-a-skillset/.

Söderberg, J. 2017. Inquiring hacking as politics: A new departure in hacker studies? *Science, Technology & Human Values*, 42 (5), 969–980.

SSL Nagbot. 2016. Feminist hacking/making: Exploring new gender horizons of possibility. *Journal of Peer Production*, 8. http://peerproduction.net/issues/issue-8-feminism-and-unhacking-2/feminist-hackingma king-exploring-new-gender-horizons-of-possibility/.

Steinmetz, K., and Gerber, J. 2015. "It doesn't have to be this way": Hacker perspectives on privacy. *Social Justice*, 41 (3), 29–51.

WIRED Technique Critique. 2018. *Hacker breaks down 26 hacking scenes from movies & TV*. www.youtube. com/watch?v=SZQz9tkEHIg.

Wray, S. 1998. Electronic civil disobedience and the World Wide Web of hacktivism: A mapping of extraparliamentarian direct action net politics. *Switch: New Media Journal*, 4 (2). http://switch.sjsu.edu/ web/v4n2/stefan/index.html.

Zuboff, S. 2019. *The age of surveillance capitalism*. London: Profile Books.

5

THE INTERNET MEDIA IN CHINA

Xiang Zhang

In the past few decades, China's advancement in economic development has attracted researchers from all over the world and different fields to study the phenomenal rise of a new power in global politics (Démurger 2001). Along with the economic achievement, media in China have also enjoyed substantial growth during this time. The technological progress generates new forms and instruments from media industry. Channels to receive news and information have been significantly changed, from newspapers, to radio, to television, and now to news websites and social media on the internet. Media in China experienced a thorough structural transformation during the Chinese economic reform period (Zhang 2011). Some state-run media received aspects of marketization and commercialization in these neoliberal reforms. Securitization has become a major method for media outlets and companies to raise money. Private capital is allowed to hold part of the share in state-run media and official outlets. The government is no longer the only provider of mass media in the country but a powerful supervisor and controller over the information dissemination process (Zhao 2008). Individuals and small businesses are allowed to join the media market and open personal media outlets on different social media platforms. In addition, the equipment used to receive news and information from media outlets has been mobilized, evolving from stereo radio receivers, to televisions, to desktop computers, and now to mobile electronic devices such as smartphones and tablets.

One major change in the growth of the Chinese media market is the increasing popularity of and participation in the internet and internet-based media. The internet in China has continued to boom in recent years, as indicated by the constant growth in internet users and penetration of mobile internet (Figures 5.1 and 5.2). In the past decade, the number of internet users in China has substantially grown to over 900 million, more than all the G7 countries' populations combined. The internet penetration rate increased from 40% to 64%. Among these internet users, most of them now access the internet via mobile devices including smartphones, tablets, laptops, etc. The percentage of mobile internet users grew from 80% in 2013 to 99.3% in the most recent survey in March 2020. Simultaneously, the increasing penetration of mobile internet access facilitates the use of mobile devices as the major source for news and information. According the Chinese Academy of Social Sciences (2020) report that, by mid-2019, about 620 million internet users in China have installed at least one news app on their smartphones; 55% of these app users have subscribed to at least one media outlet or blogger to receive news and information regularly; and 60% of users

DOI: 10.4324/9781003039068-6

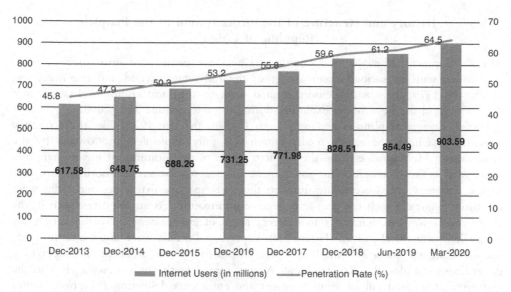

Figure 5.1 Number of internet users in China, December 2013 – March 2020

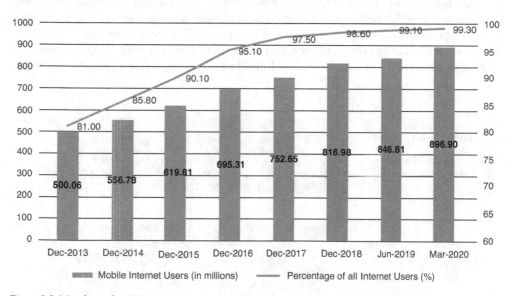

Figure 5.2 Number of mobile internet users in China, December 2013 – March 2020

normally use short breaks in working and daily life to search and review news and information via these apps.

This internet penetration and ubiquity of the mobile phone not only mark an outstanding feature of media in China in this new era, but also challenge the traditional governance of media, as media is overly controlled by the government in China. The next section provides background information about the media system in China and its development over the past few decades.

History and structure of the media system in the People's Republic of China

The development of media in Communist China can be divided into three stages in accordance with its socioeconomic progress: the fully state-owned and controlled era before the 1979 Open and Reform; limited marketization and commercialization after the 1980s; and the new media explosion during the internet age after 2000.

During the era of planned economy from 1949 to the late 1970s, like the economic activities in China, the entire media system was fully controlled and owned by the government. Media outlets were the de facto agents of the Communist government and functioned as the mouthpiece for government policies, regulations and notifications. Due to the underdeveloped economic situation then, newspapers, magazines and radio were the three major channels for people to receive information. As an important unit in the party, media was incorporated into different levels of government following a centrally controlled hierarchical pattern (Figure 5.3). Paper-based media were distributed in a rigorously designed quota system. Radio was scheduled by the order of the supervising government. Besides, the Xinhua News Agency served as the only news agency in the country and scripted official announcement and comments delivering the government's view on both domestic and international affairs.

From the late 1970s, restrictions on media marketization and commercialization were gradually loosened in China, led by the introduction of commercial advertising on paper-based media. Official media was permitted to transform from government agencies into either for-profit organizations or state-owned enterprises. This process sped up in the 1990s along with liberalization process of the Chinese economy (Zhao 2008b). Direct government funding to state-owned media declined and qualified private capital was permitted to invest

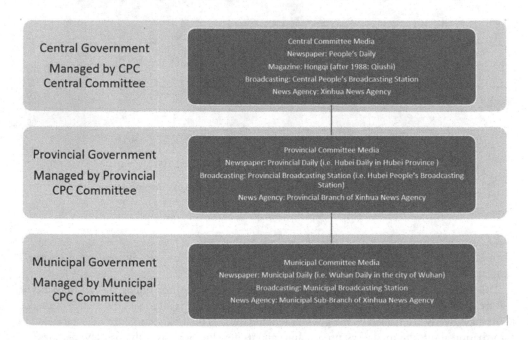

Figure 5.3 The hierarchical structure of state media in China

in the media sector. Newspaper and magazine agencies began to develop new titles and volumes for the market in a new commercialized distribution system. Retailing and subscription replaced the quota system. The general public also had greater access to news and information due to the diversification of the press market as media outlets were allowed to produce reports and comments themselves. At the same time, television became popular among Chinese families, and emerged as the most popular way to access news and information in China. News reports sourced from foreign news agencies were allowed under some permitted circumstances, and the Xinhua News Agency was no longer the only source of foreign events for Chinese people (Shirk 2010). However, the government still has strict regulations on investment in media and no foreign entities are allowed to hold any media in China.

Since 2000, the internet has begun to step into daily Chinese life. Cyberspace became the new battlefield of the competitive market of Chinese media. Internet companies such as Sina, Netease, Tencent and Sohu became new media giants as each developed a popular news web portal in China and enjoyed a noticeable share in the entire media market. As the internet has the advantage of timeliness and geographical coverage (Almeida & Lichbach 2003), these news web portals soon began to challenge the role of traditional media. State-owned media began to build up their own websites and tried to compete with these web portals.

Chinese media has undergone enormous changes as social media has emerged as an essential part of internet users' lives in China in recent years (Chiu et al. 2012). New information dissemination channels such as microblogging (Weibo developed by Sina) and social networking (WeChat Moment developed by Tencent) via social media apps soon attracted millions of users, as well as the attention of various media outlets in China. These new channels embedded within social media apps allow targeted pushing of news and information which enables users to see what they are most interested in (Jacobson et al. 2020). In addition, the capability to include multimedia attachments soon helped social media replace the role of traditional paper-based and telecommunications-based media and became the most popular way to receive news and information among Chinese internet users (Chinese Academy of Social Sciences 2020).

However, as an authoritarian regime, the Communist government holds hegemonic power monitoring all media activities in China (Lei 2019). A key department in the operation of the Communist Party is the Central Propaganda Committee, now named the Publicity Department of the Communist Party of China (CPC), which was established in 1921. As part of the core ideological control over Chinese society, the CPC Publicity Department has the ultimate power to oversee all media activity in China and supervise information and content across different media formats, as it has designated units to monitor the various media operations in the country. In the 2018 national government reform, the duties and responsibilities of the Publicity Department were revised and it became the top censoring and supervising administration of all the media affairs in China. Currently, the Publicity Department has four units, as shown in Figure 5.4, controlling the press, film and television in China, and coorporates closely with the National Radio and Television Administration surveilling the traditional media.

Simultaneously as part of the 2018 government reform, a new agency, the Cyberspace Administration of China, has been established to monitor the increasingly expanded internet-based new media, which uses a set of tools including contextual filtering, content and activity censorship, and network firewalls to monitor internet activity in China. As all telecommunication companies and networks are state-owned in China, the government holds

Figure 5.4 Governance structure of the CPC Publicity Department

the ultimate power to control the internet. This feature of internet telecommunications infrastructure in China facilitates the government's capability to control accessibility and monitor content for censorship purposes as much as possible, as the government has the ultimate power to inspect all the online traffic (Qiang 2019). As the internet emerges to be a more popular and influential channel for Chinese people to receive information, censorship and surveillance of online activities have intensified in recent years and blocklists of forbidden words and websites continue to grow (Hobbs & Roberts 2018). From an international perspective, the freedom of the press, media and internet have been deteriorating in recent decades and there is still no sign of any loosening of the restriction or regulation of the internet and media (Reporters Without Borders 2019; Freedom House 2020).

New internet media and power in authoritarian society

Knowledge, media and power

Foucault's framework on the production of power and discourse (1990) outlines the relationship between knowledge and power in a cyclical process: power reproduces knowledge and power is exercised through knowledge. In this model, the dissemination of knowledge acts as the expansion of power and those empowered are embedded with the capability to produce discourses in society. Therefore, the role of media is critical in practicing power and producing new knowledge as media disseminate information and knowledge to the public (Couldry 2010). In other words, media are the producer and conveyor of discourse in society. For the empowered group or social class in power, media would serve an important role in expressing their values and opinions to the public as the discourse developed by these empowered groups would be more likely to become enforced during this process.

Following the Foucaudian framework, media act as the key distributor of knowledge and hence are a crucial component in building the power structure in society. When media functions as an interpreter of information and knowledge to the public, media indeed are engaged with the power to influence the public's perception of events and subjects, which in return would affect public opinions and awareness on the reported events and subjects and ultimately help to shape the ideology of the society (Brookfield 1996; Cotterrell 1999). Media act as curators of the contemporary plural society by providing different interpretations to the public in a country with less media regulation and restriction. In contrast, in a country with more media regulation and restriction, the spread of pluralistic values and thoughts would be difficult as media enjoy less freedom to express a value that is different from the mainstream, and face suppression from the empowered group upholding mainstream ideas.

New power structure of new internet media: A theoretical perspective

In a spatial perspective, traditional mass media were deeply embedded with geographical scales (Figure 5.5). The physical distribution and accessibility of newspapers and magazines in the traditional business model is highly correlated with geographical location and the coverage is limited (Zimmermann 2007). The transmission of radio and television signals is also subject to topological terrains. Therefore, for traditional media, such a coverage restraint has limited the accessibility of information for the public, and moreover, the outreach of power from the empowered social class. In the case of China, such limitations have affected the control and governance between central government and local residents in remote areas.

Compared to traditional mass media such as newspapers, magazines and television, the internet media enjoy several advantages. First, internet media have incomparable timeliness in transmitting information as they makes instant breaking news a real-time experience in various ways, from text-only to live broadcasting (Risley 2000). Second, internet media, assisted by the penetration of mobile devices, have significantly enhanced our mobility in obtaining information (Andersson & Mantsinen 1980; Gohdes 2020). We can access news and information via mobile devices anywhere, unless there is no signal coverage. Third, the internet creates database of unprecedented size for users to search and obtain useful information and sources, empowering the grassroot class with abundant materials for discourse-making. Thus, internet media shall challenge the tradition power relation and knowledge production as the internet disseminates the originally concentrated media power to individuals. The timeliness and coverage have reached unprecedent levels of convenience for the empowered to disseminate the discourse and for the masses to receive the information. With improved infrastructure, particularly the optical fiber network and telecommunication-based internet access, the increasing penetration of the internet marks an increasing influence of internet media, as well as an elevated capability to dominate knowledge and power structures in society.

Figure 5.5 A comparison across tradition media

As social media allow individuals to report news and information on the internet, power structures in society would be modified as a new relationship between the government and the general public arises. Individuals can now establish social media channels to express personal opinions. Therefore, the new internet media would promote functions of providing transparent information, facilitate effective communication and encourage public participation in political affairs, all of which could benefit China's development towards a civil society.

First, new participants enter the media sector in China via the internet and information can be produced not only from state-run or state-monitored professional media outlets, but also independent grassroots social media accounts. One new rising source of power in this process is celebrity accounts, or cewebrities (social media celebrities, originated from WeMedia + Celebrity, or Web celebrity, "wang hong" in Chinese) (Stokes & Price 2017). These are the news and information media accounts edited by independent reporters and commentators that have a noticeable number of subscribers or followers, hence have the capability to influence certain people's view and ideology. In the meantime, the state-owned media and other government outlets have also joined the social media platforms and broadcast official voices in this new channel as they try to retain their influence with the public. Theoretically, the extended participation in media would provide a new direct channel to receive information from the government and an alternative method to collect public opinions on social events from the comments and messages. The original single source power in controlling information and knowledge is challenged by the introduction of these new participants.

Second, the new internet media creates a new power structure that challenges the government to alter its original policy on media issues. From a progressive perspective, the new media enables users to query the government directly, which endows individuals in the public with the potential to produce discourses and influence each other's opinion. In return, this process would undermine government power in society and put government behaviors under the supervision of these independent media subscribers and viewers as the public tends to have more negotiating power in contesting with the government (Saeed 2009). Over time, the unbalanced pattern of power in an authoritarian society would be adjusted and the general public would have the opportunity to be empowered and enabled for the improvement and enhancement of civil rights.

However, the introduction of new media and new discourse-makers in cyberspace might not be a realm of pro-democratic reform or movement. Instead, it could also create another enclosed social space due to the control and censorship exercised by the authoritarian government, and erode the freedom enjoyed by the public (Stoycheff et al. 2020). The government could also use new media and censorship tricks to eliminate different opinions and maintain its authority with existing administrative power. Similar to the social disorder during the Cultural Revolution under Mao's regime, new internet media, with its unprecedent penetration power among the public, could be used as a weapon to attack unfavorable opinions and discourses in the name of patriotism, nationalism and populism. If such a scenario happened, Chinese society would be in a dangerous situation with highly polarized groups and intensified class conflicts.

Another possibility is that the pluralization of thoughts and ideologies sparkled by new internet media and mass participation could generate new incentives for an authoritarian regime to revise its censorship and surveillance policy, which would lead to further drawing back of media control, and other social freedoms (Dick 2012). Hypothetically, an authoritarian government could make the use of the internet and its control over media and technology companies to create a tailored cyberspace for citizens: unfavorable information and news could easily be removed; people with different opinions could be tracked, monitored and suppressed; cewebrities could be ordered to post designated materials and content to influence fans and followers which would uniformize the voice, opinion and comment online.

Through the build-up of a power-relation framework of media in society, it is true that the new technology could be a double-edged sword in modifying existing social structures. In a progressive environment, new internet media and increasing numbers of participants in discourse and information production could benefit the society by providing multi-dimensional approaches to interpreting events and exchanging thoughts and ideas. However, in a suppressive society with less democratic rights, new media and mass participation can be manipulated by the government and lead to an Orwellian dystopia, which every civil society should avoid (Richard 2012).

New internet media in China

Similar to the market structure of e-commerce in China, the Chinese internet media market is also highly polarized and controlled by a few giant platforms (Miao 2020), including WeChat, the most popular social media platform and SMS application owned by Tencent; Weibo, the Chinese counterpart of Twitter owned by the gateway website Sina; Toutiao, an open platform of news and information content owned by ByteDance; Xigua Video, Byte-Dance's video-sharing platform; and Douyin, the Chinese version of Tiktok also developed by ByteDance. According to recent market research, the three companies—Tencent, Sina Weibo, and ByteDance—have occupied over 90% of the share of the entire Chinese internet media market (Forward Business Info 2019). Among these, WeChat, as the most popular social media and SMS app in China, has occupied the largest share of the media market, and generated two-thirds of all profits in Chinese internet media. And Weibo, the microblogging platform introduced by Sina in 2009, is the second most popular platform for internet media. By the end of 2018, WeChat and Weibo together occupied 80% of the total number of view counts of all Chinese internet media.

WeChat-based social media platform Moments adopted a restricted access feature for users. No access will be granted to the shared content until a user is verified and added as a friend. Such security and privacy features soon made it a popular platform to share information among a more intimate group of people, which helped to strengthen small but dynamic circle cultures among groups sharing some similarities (Lucero 2017). Theoretically, this increased level of intimacy helps to create a comfortable zone for users to share and comment on online materials and serves as a unique feature of the WeChat platform, that Moments is a relatively safe space to share more sensitive and private views and opinions.

Started in 2013, Moments soon experienced an explosive growth in popularity. Differing from other public platforms, the semi-private Moments allows the new media participant to write much more detailed opinions with more personal comments (Kietzmann et al. 2012). The stickiness does not only imply a consistent number of subscribers and viewers of a new media, but also a consistent political view or positionality on news and information. In another words, as each Moments user establishes a personal intimate network of reliable friends and acquaintances, all users in this group tend to have more commonalities in social and political views. When one user subscribes to a new media on Moments and share the article to their friends, members in this circle tend to share the article and new media further when they agree with the article. In return, writers and commentators tend to retain their fans and followers by providing analysis and comments in the same way. The subscriber and fan group of each new media account indeed have established a semi-closed knowledge and fandom sphere (Highfield et al. 2013). Therefore, when looking into statistics on WeChat new media accounts, it shows that subscribers have a higher level of stickiness to the subscription and the subscribers. Table 5.1 shows the most recent statistics on top WeChat new media accounts.

In contrast, considered as the Chinese counterpart of Twitter, Weibo is an open-access platform for sharing bite size microblog information. Hence users have access to an enormous range of content providers and are able to subscribe to as many materials as they like. Thus,

Table 5.1 Top news and information accounts on the WeChat platform (June 2020)

Media name	No. of active subscribers	Average view count	Description
	(in millions)	(per article)	
Zhanhao (占豪 zhanhao668)	17.04	Over 100,000	Patriotic nationalist political commentator
Webmaster Feng's Home (冯站长之家 Fgzadmin)	14.99	Over 100,000	Briefer of news from official sources
Yuanfang Qingmu (远方青木 YFqingmu)	6.17	Over 100,000	Patriotic nationalist political commentator
Ping's Talk Today (今日平说 zg5201949)	11.98	98,761	Patriotic nationalist political commentator
Youyou Luming (呦呦鹿鸣 youyouluming99)	5.41	98,553	Business and economic commentator
Shoulouchu (兽楼处 ishoulc)	4.95	Over 100,000	Real estate market analyst
Lukewen's Studio (卢克文工作室 lukewen1982)	7.39	Over 100,000	International politics commentator
China Railway 12306 (铁路 12306 CRTT12306)	4.32	Over 100,000	Online rail ticket service
Xueyin (血饮 caojianming1989)	7.04	Over 100,000	Patriotic nationalist political commentator
China Super Dad (超级学爸 chinasuperdad)	3.65	88,601	Patriotic nationalist political commentator
Zhengshitang (政事堂 zhengshitang2019)	2.74	Over 100,000	Political commentator
Ning Nanshan (宁南山 ningnanshan2017)	3.22	Over 100,000	Business and economic analyst
Hu Xijin's View (胡锡进观察 huxijinguancha)	3.91	98,746	Semi-official nationalist political commentator
Housha (后沙 HSYGLGJ)	3.26	78,785	Patriotic nationalist political commentator
North American College Daily (北美留学生日报 collegedaily)	2.14	98,595	Patriotic news info for overseas Chinese students
Overall average	**6.55**	**N/A**	

Source: Xigua Data (Retrieved http://data.xiguaji.com)

Weibo is a one-directional information acquisition platform, rather than a place to express personal and sensational opinions (Gao et al. 2012). Table 5.2 shows the most recent statistics on numbers of active subscribers for top Weibo information and news accounts, which ranges from politics to entertainment and lifestyle fashions. Most of them record a much larger subscriber number than media on WeChat.

The statistics on the two platforms illustrate two critical trends in current Chinese society. First, internet media is closely connected to the political life in China, and patriotism seems

Table 5.2 Top news and information accounts on the Weibo platform (June 2020)

Media name	No. of active subscribers	Average no. of likes	Description
	(in millions)	(per article in thousands)	
People's Daily 人民日报	118.65	26.2	Official CPC Newspaper
CCTV News 央视新闻	110.98	24.6	Official Chinese Television
Hu Xijin 胡锡进	23.02	17.0	Chief Editor of nationalist *Global Times*
I am Jerry Kowal 我是郭杰瑞	2.69	34.8	American blogger living in China
Panda Guardian 熊猫守护者	17.99	5.02	Wildlife conservation NGO
Weibo Mission 微博任务	73.34	1.35	News and promotions on the Weibo platform
Breaking News 头条新闻	79.20	11.1	News and information managed by Weibo's parent company Sina
Call Me "Mr. Hot" 请叫我热门君	18.14	3.09	Independent news and information blogger
Korea Me2Day 韩国me2day	26.51	14.8	Independent K-pop and fandom blogger
JD.com 京东	4.48	6.90	Official account of the e-commerce giant JD.com
Headline Qiwen 头条奇闻网	18.57	2.77	Official account of the news website Qi520W
Global Headline 头条全球	18.84	2.76	Entertainment head-line news
Fashion Life Headline 时尚生活头条榜	13.29	3.16	High-end life and fashion information
Call Me "Mr. News" 请叫我新闻君	14.85	3.12	Independent news and information blogger
Shanghai Fashion Headline 上海时尚生活头条	11.76	2.96	High-end life and fashion information
Overall average	**36.82**	**10.64**	

Source: Xigua Data (Retrieved http://data.xiguaji.com)

to be the dominant ideology among internet media users: seven out of the top 15 Wechat media, totaling more than 50 subscriptions, claim themselves patriotic reporters on political issues. Among Weibo media, all top three are either official accounts of state-run media or individuals with an official background. Political information remains the most popular content on the internet. And the state still plays a significant role in internet media. Second, when tracking the content published by these accounts on Moments, it shows that most of these "patriotic" news commentators or writers are also igniting hatred towards Western countries and democratic ideologies, which signals an increasing nationalism and xenophobia in China at this stage, given the contemporary conflict between China and Western countries over several issues.

Media in China have long been criticized for the lack of basic freedom of the press and speech. The government has adopted complex censoring and filtering systems for both traditional and internet media, claiming the necessity of these measures in order to prevent "the dissemination of misinformation and news" and preserve a "positive power" for the society to maintain the social stability. However, the excessive control of information and news on all media has triggered several social problems across the country and created information asymmetry for media users in China, as negative information from oversea media outlets would have been automatically removed. International information and news on the Chinese media have been vastly censored and sorted by the government in order to depict a favorable image of the Communist Party among the public, and this deliberately constructed information asymmetry creates a widening gap of information and perception between Chinese ideas and Western ideology. Western liberal democratic systems are labelled as the puppets of capitalism and a disaster in managing social development (Boix & Svolik 2013). Only approved information can be disseminated freely by internet media. For example, during the global COVID-19 crisis, media in China had extensively spread the "Western failure" to contain the virus and used the pandemic in Europe and the US as a foil to illustrate the success of the Chinese way of controlling the disease, as well as the advantage of the authoritarian decision-making process by the CPC government. At the same time. Information and news in China about COVID-19 have been strictly controlled and only government approved information can be published across different platforms. According to CitizenLab (2020), a research cohort on the Chinese internet and social media from the University of Toronto, the Chinese government promptly expanded its blocklist of forbidden words in February 2020 and negative information about the government's COVID-19 response tended to be deleted.

The Wuhan-based[1] Chinese writer Fang Fang and her Wuhan Diary is another good example showing government intervention in new internet media. During the Wuhan lockdown between January 25 and March 25, as an established writer with more than 4 million subscribers to her Weibo account, Fang Fang posted her diary about the life in Wuhan during the lockdown period to Weibo. Her diary was based on both witnessed fact and learned information from others about the situation in Wuhan. Some diaries with unconfirmed information of the disease were censored and deleted from Weibo at the discretion of the internet censorship agency. Her Weibo account was once withheld and then reinstated early in February after she posted the first few chapters online. Her diary was appraised by both Chinese and foreign readers due to its informative and spiritual value. On March 30, Hu Xijin published a comment on his *Global Times* account positively confirming the value of Fang Fang's diary, as it reflected the solidarity and love among Wuhan residents. On April 8, the US press HarperCollins translated and published all of her 60 diaries during the lockdown period, which soon ignited a fierce wave of criticism of Fang Fang in China, led by several nationalist cewebrities. For example, Hu Xijin soon made a U-turn from his previous view by claiming Fang Fang's work deteriorates the image of China and the Chinese people as it records the dark side of Wuhan during the lockdown and contains unconfirmed stories and information. Later, in mid-April, Wuhan local news reported that Fang Fang had experienced cyber bullying from nationalist internet users and received death threats in her daily life.

The case of Fang Fang does echo several characteristics of new internet media in China. First, it is a new channel for individuals to express personal ideas and thoughts. Therefore, people would be able to access alternative information and viewpoints differing from official statements and notifications by the state-run media. However, such civil rights are merely accessible to the public in theory as the censorship system would delete unfavorable information and messages without anyone's consent. Second, the cyber ideology could be easily manipulated when needed as both Chinese government and the pro-government media can clean out different opinions quickly by various cyber bullying tricks such as verbally insulting and abusing. Third, state-run internet media have attempted to gain full control over online speeches and posts. Instead of providing a less restricted platform for the public to exchange thoughts and ideas, the internet media platforms are becoming bustling fields for censorship and information filtering.

Concluding notes

The internet and social media have significantly changed communication and information for most Chinese people. In cyberspace, new internet media do benefit Chinese society in many aspects such as coverage of information, convenience of acquisition, channel of communication and interaction and networking. Several pieces of research show that the new internet media provides new channels to solve various social problems such as property rights conflicts (Huang & Sun 2014), complaints about some government agencies (Nip & Fu 2016), environmental concerns about air pollution (Jiang et al. 2015), etc. However, the internet in China is still heavily controlled and censored by the government. Signals of tightening media control can be spotted in party policies and government regulations. Posts by new internet media writers are still in danger of being removed, withheld or deleted if posted information is considered unfavorable by government standards. At the same time, a wave of increasing pro-Communist and pro-government media is blowing into this new space of knowledge and information, marking a leftward turn towards nationalism and patriotism in China, which could lead the country on a unique ideological pathway that is totally different from the Western liberal ideal and might shed the prospect of democratic reform and human rights improvement in the country.

Note

1 The first COVID-19 outbreak was recorded in Wuhan, China.

References

Almeida, P., and Lichbach, M. 2003. To the internet, from the internet: Comparative media coverage of transnational protests. *Mobilization: An International Quarterly*, 8 (3), 249–272.

Andersson, Å. E., and Mantsinen, J. 1980. Mobility of resources, accessibility of knowledge, and economic growth. *Behavioral Science*, 25 (5), 353–366.

Boix, C., and Svolik, M. W. 2013. The foundations of limited authoritarian government: Institutions, commitment, and power-sharing in dictatorships. *The Journal of Politics*, 75 (2), 300–316.

Brookfield, S. 1986. Media power and the development of media literacy: An adult educational interpretation. *Harvard Educational Review*, 56 (2), 151–171.

Chinese Academy of Social Sciences. 2020. *Annual report on the development of new media in China No. 11 (2020)* [in Chinese]. Beijing: Social Sciences Academic Press.

Chiu, C., Ip, C., and Silverman, A. 2012. Understanding social media in China. *McKinsey Quarterly*, 2, 78–81.

CNNIC. 2020. *The 45th statistical report on internet development in China*. Beijing: China Internet Network Information Center (CNNIC).

Constantiou, I. D., and Kallinikos, J. 2015. New games, new rules: Big data and the changing context of strategy. *Journal of Information Technology*, 30 (1), 44–57.

Cotterrell, R. 1999. Transparency, mass media, ideology and community. *Journal for Cultural Research*, 3 (4), 414–426.

Couldry, N. 2010. Theorising media as practice. In B. Bräuchler and J. Postill (eds.) *Theorising media and practice*, Vol. 4. pp. 35–54. New York: Berghahn Books.

Démurger, S. 2001. Infrastructure development and economic growth: An explanation for regional disparities in China? *Journal of Comparative Economics*, 29 (1), 95–117.

Dick, A. L. 2012. Established democracies, internet censorship and the social media test. *Information Development*, 28 (4), 259–260.

Forward Business Information. 2019. *Report of the development path and investment strategy planning on China media convergence (2020–2025)* [in Chinese]. Shenzhen: Forward Business Information.

Gao, Q., Abel, F., Houben, G. J., and Yu, Y. 2012. A comparative study of users' microblogging behavior on Sina Weibo and Twitter. In *International Conference on User Modeling, Adaptation, and Personalization*. Berlin and Heidelberg: Springer.

Gohdes, A. R. 2020. Repression technology: Internet accessibility and state violence. *American Journal of Political Science*. https://doi.org/10.1111/ajps.12509.

Highfield, T., Harrington, S., and Bruns, A. 2013. Twitter as a technology for audiencing and fandom: The Eurovision phenomenon. *Information, Communication & Society*, 16 (3), 315–339.

Hobbs, W. R., and Roberts, M. E. 2018. How sudden censorship can increase access to information. *American Political Science Review*, 112 (3), 621–636.

Huang, R., and Sun, X. 2014. Weibo network, information diffusion and implications for collective action in China. *Information, Communication & Society*, 17 (1), 86–104.

Jacobson, J., Gruzd, A., and Hernández-García, Á. 2020. Social media marketing: Who is watching the watchers? *Journal of Retailing and Consumer Services*. https://doi.org/10.1016/j.jretconser.2019.03.001.

Jiang, W., Wang, Y., Tsou, M. H., and Fu, X. 2015. Using social media to detect outdoor air pollution and monitor air quality index (AQI): A geo-targeted spatiotemporal analysis framework with Sina Weibo (Chinese Twitter). *PloS One*, 10 (10), e0141185.

Kietzmann, J. H., Silvestre, B. S., McCarthy, I. P., and Pitt, L. F. 2012. Unpacking the social media phenomenon: Towards a research agenda. *Journal of Public Affairs*, 12 (2), 109–119.

Lei, Y. W. 2019. *The contentious public sphere: Law, media, and authoritarian rule in China*, Vol. 2. Princeton, NJ: Princeton University Press.

Lucero, L. 2017. Safe spaces in online places: Social media and LGBTQ youth. *Multicultural Education Review*, 9 (2), 117–128.

Miao, Y. 2020. Can China be populist? Grassroot populist narratives in the Chinese cyberspace. *Contemporary Politics*, 26 (3), 1–20.

Nip, J. Y., and Fu, K. W. 2016. Challenging official propaganda: Public opinion leaders on Sina Weibo. *The China Quarterly*, 225, 122–144.

Nocentini, A., Calmaestra, J., Schultze-Krumbholz, A., Scheithauer, H., Ortega, R., and Menesini, E. 2010. Cyberbullying: Labels, behaviours and definition in three European countries. *Australian Journal of Guidance and Counselling*, 20 (2), 129.

Qiang, X. 2019. The road to digital unfreedom: President Xi's surveillance state. *Journal of Democracy*, 30 (1), 53–67.

Richards, N. M. 2012. The dangers of surveillance. *Harvard Law Review*, 126, 1934.

Risley, F. 2000. Newspapers and timeliness: The impact of the telegraph and the internet. *American Journalism*, 17 (4), 97–103.

Saeed, S. 2009. Negotiating power: Community media, democracy, and the public sphere. *Development in Practice*, 19 (4–5), 466–478.

Shin, J., Jian, L., Driscoll, K., and Bar, F. 2018. The diffusion of misinformation on social media: Temporal pattern, message, and source. *Computers in Human Behavior*, 83, 278–287.

Shirk, S. L. (ed.) 2011. *Changing media, changing China*. Oxford: Oxford University Press.

Stokes, J., and Price, B. 2017. Social media, visual culture and contemporary identity. In *11th international multi-conference on society, cybernetics and informatics*. www.iiis.org/CDs2017/CD2017Summer/papers/EA876TF.pdf.

Stoycheff, E., Burgess, G. S., and Martucci, M. C. 2020. Online censorship and digital surveillance: The relationship between suppression technologies and democratization across countries. *Information, Communication & Society*, 23 (4), 474–490.

Zhang, X. 2011. *The transformation of political communication in China: From propaganda to hegemony*. Singapore: World Scientific.

Zhao, Y. 2008a. Neoliberal strategies, socialist legacies: Communication and state transformation in China. In P. Chakravartty and Y. Zhao (eds.) *Global communications: Toward a transcultural political economy*, pp. 23–50. London: Rowman & Littlefield.

Zhao, Y. 2008b. *Communication in China: Political economy, power, and conflict*. Lanham, MD: Rowman & Littlefield.

Zimmermann, S. 2007. Media geographies: Always part of the game. *Aether: The Journal of Media Geography*, 1, 59–62.

6

DIGITAL MEDIA AND PERSONS WITH VISUAL IMPAIRMENT OR BLINDNESS

Susanne Zimmerman-Janschitz

The dynamic evolvement of information and communication technologies (ICT) offers both opportunities and challenges for persons with visual impairment or blindness (VIB). Digital information appears to be available 365 days a year, 24 hours a day; but this is only true for parts of (Western) society. Some people are still excluded. There are limitations concerning availability of the internet and technological resources in terms of cost, digital divide, social exclusion and/or accessibility. This is especially true for persons with VIB. What is intended to support unlimited access to information and therefore to enhance independency and self-determination often contains hurdles or barriers. From a geographical perspective the focus is on space-related information, because spatial data or maps are important to enhance personal mobility—in digital as well as physical environments.

It might be assumed that persons with VIB are generally disadvantaged or even excluded in terms of using spatial information. Thinking about mobility in this context, the widely imagined scene is of a person with a long white cane walking along a street. Yet, the reality is far more complex and so are the strategies to cope with limitations in access of spatial information. However, the individual and therefore unique needs of persons with VIB regarding information can seem contradictory. Some persons with remaining sight use buildings reflecting the light (e.g. with glass fronts) to orient themselves, others avoid high reflecting surfaces because of glare. Persons with color vision deficiency might have problems distinguishing colored symbols on a map, while for persons with vision loss, colors are helpful to differentiate symbols due to a reduced ability to identify details. These examples illustrate that persons with VIB are a very heterogeneous group (Huebner 2000, 55; Zimmermann-Janschitz et al. 2017, 68), requiring different approaches in scientific theory as well as in practice.

From a practical viewpoint, solutions can be constructed upon, for example, (1) the existing disability or limitation (e.g. based on the International Classification of Functioning, Disability and Health (WHO 2001)), (2) the sense used to complement or replace vision (auditory, kinesthetic, tactile, olfactory or combination) or (3) the technical approach for the assistive tools or technologies used (GPS, GIS, etc.). The theoretical perspective is based on two main research foci (Worth 2013, 575). Golledge (1993) as well as Kitchin and Jacobson (1997), amongst others, discuss the cognitive representation of spatial settings seen from the personal limitation, and hence follow the medical model of disability. Scholars around Butler

DOI: 10.4324/9781003039068-7

(Butler & Bowlby 1997, 421) shift the attention towards the social experience in public space, following the social model of disability which identifies the society as the obstacle to accessibility. The advancement of the social model shifts with Worth (2013) and Macpherson (2010) towards a relational perspective, where people "engage and perform their embodiment and in so doing re/produce and transform both themselves and their surroundings" (Hall & Wilton 2017, 728).

This chapter will give an overview of the variety of scientific approaches to supporting accessibility of (spatial) information, starting off with a look at the legislation(s) enhancing accessibility in ICT. Two core aspects, the accessibility of information and the information about accessibility of the built environment to support orientation and wayfinding, will be discussed. Current research trends next to future research issues are intended to reflect the findings.

Legislation as framework

Visual impairment and blindness in numbers

The World Health Organization states that 2.2 billion people around the globe are visually impaired or blind (WHO 2019, 26), including also mild forms of visual impairment. For the year 2020 Bourne et al. (2017, e894) present an estimated number of 237 million people with moderate or severe visual impairments around the globe, of whom 38.5 million persons are blind. This number corresponds with a percentage of 3.06% of persons with visual impairment and 0.50% of persons with blindness worldwide. In the United States, Varma et al. (2016, 806) compute the number of persons with visual impairments (aged 40+) for the year 2020 at 3.67 million persons, and 1.12 million people with (legal) blindness. Remarkably, there is a lack of up-to-date and comparable statistical data due to differing models of surveying and projection as well as contrasting definitions of impairment. It is noteworthy, though, that there are no comparable figures available for the European Union (EU).

Legislative frameworks

The increasing number of individuals with VIB and the shift in the paradigm of disabilities has lead to numerous legislations, standards and recommendations at different institutional levels. At the international level a milestone for the implementation of equality of persons with disabilities (PWD) was achieved with the "Convention on Rights of Persons with Disabilities (CRPD)," adopted by the United Nations General Assembly on December 13, 2006 (United Nations (UN) 2006). Article 4 promotes ICT with the integration of Universal Design (para. f), and "mobility aids, devices and assistive technologies suitable for PWD, giving priority to technologies at an affordable cost (para. g)" (UN 2006, 6). Article 9 (UN 2006, 9) focuses on accessibility of ICT and the built environment. Despite being binding, the CRDP needs (supra-)national laws and standards for guideline realization.

The EU provides support to get the CRPD into practice at the supra-national level. While the "European Disability Strategy 2010–2020" focuses mainly on the reduction of barriers (European Commission 2010), the "European Accessibility Act" promotes employment and occupation. Extra emphasis is given to the accessibility of websites and mobile apps of public sector bodies (European Commission 2015). European Standards define specifications and technical details, e.g. the accessibility of ICT (EN 301 549), of websites and mobile applications (EN 301 549), of the built environment (EN 17210), or "Design for all" standards (EN

17161) (European Commission 2020). The subjects covered by national laws and their enforcement vary significantly between member states, although basic strategies are adopted in all EU countries (Waddington & Lawson 2009, 54).

Pioneering in 1990, the United States established the Americans with Disability Act (ADA). This protects the rights of PWD and ensures their access to employment, goods and services, places of accommodation and telecommunications. Title II specifies that "individuals who are blind, deaf-blind, or visually impaired may not be denied full and equal enjoyment of the goods, services, facilities, privileges, advantages, or accommodations provided by a state or local government or place of public accommodation" (American Foundation of the Blind 2020). Explicitly, ADA covers the elimination of communication barriers by providing auxiliary aids and services to persons with VIB and removal of structural communication barriers through the Americans with Disabilities Accessibility Guidelines (ADAAG) (Joffee 1999, 8–10). The ADAAG fosters the accessibility of facilities, public buildings and public rights-of-way like accessibility of street furnishings, curb ramps, crosswalks or pedestrian signals (United States Access Board 2020). Section 508 of the Rehabilitation Act of 1973 particularly addresses accessibility of ICT provided by Federal governments (GSA 2019). To give some examples, braille visualizing signs, large print signage, audio displays, new ICT devices, detectible warnings at curb ramps, etc. can be indicated. Similar to the EU, state laws complement and/or strengthen the ADA.

Why laws are not enough

Even if standards and laws are well established, there is a gap between the legislative framework and reality in daily life. For example, laws on the protection of historical buildings in European cities with their historical heritage overrule accessibility of newly constructed or refurbished buildings. Regardless of a detailed design of public space, curbs, sidewalk width, the amount of space to guarantee barrier-free movement, etc. often fails to provide an accessible urban setting. Access to the built environment is also limited due to the attitudes of people and the "barriers" in their minds. Moveable obstacles like parked bicycles or pavement signs in front of shops often block the way. Next to construction sites they define essential barriers for persons with VIB, but this cannot be governed by law (Zimmermann-Janschitz et al. 2017). Legislation can also be contradictory for different disabilities: while persons using wheelchairs need low or no curbs in terms of access, persons with VIB often use curbs to orient themselves, e.g. when using a long white cane.

Information in digital media as key to independence

Availability and accessibility of information are key elements to independence for persons with disability, as they support self-determined participation in daily life. For example, the menu in a restaurant or information at the destination board of a bus line are inaccessible to persons with VIB if not available digitally or in braille. Common orientation hints like streets signs, visual landmarks or maps are barely perceptible, underpinning the importance of access to information. Consequently, wayfinding challenges in particular can be considered as problems of information quality and access (May & Casey 2018, 84).

A matter of format and content

Accessibility of information in general is a two-sided matter in which each aspect is dependent on the other; it can be defined by the (1) *format of its presentation* as well as by the (2) *content* itself (European Agency for Special Needs and Inclusive Education 2020). While the format must address at least two senses, haptics and/or sound in the context of VIB, the content needs to allow easy orientation.

Alternative *formats* of presentation in ICT are well established and several scholars give overviews and critical reflections on techniques and tools to complement or replace the visual sense (see e.g. Jansson 2008; Power & Jürgenssen 2010; Fuglerud 2011; Bhowmick & Hazarika 2017). Although digital forms of (inter-)action from e-business to social networks increasingly shape daily life, "traditional" and analogue formats of information stay crucial (Janschitz 2012, 73). Braille is an analogue way of displaying text, e.g. on automated teller machines, medicine boxes or in elevators. In regards to geography, tactile and talking maps (Miele et al. 2006; Stampach & Mulickova 2016), 3D models and tactile graphics (Völkel et al. 2008) have to be mentioned. However, braille displays are serving as a bridge to assistive technologies and digital forms of information access. Moreover, ICT finds counterparts to large print for persons with low vision with screen magnifiers.

The other sense used by persons with VIB is sound. Although olfactory hints may be useful especially in terms of orientation, they can hardly be digitized and are therefore omitted in this chapter. Parallel to haptics, a shift towards digital tools can be indicated. Traditional methods like audiobooks, auxiliary audio channels (second audio program) on TV or voice instructions in the elevator, etc. are extended by software solutions like screen readers (e.g. JAWS, VoiceOver, NVDA, etc.), the sonification of maps (Kaklanis et al. 2013) and hardware approaches with assistive devices like electronic white canes or mobile phones (see discussion below in the section "Types of assistive devices").

Aside from addressing alternative senses, the *structure of the content* is of equal importance. Screen reader software converts the content of a screen into synthesized speech. Since the navigation is based on using the keyboard instead of the mouse, navigation elements need to be defined in documents (e.g. headers, font, etc.). Feedback of websites is achieved by ear-cons, e.g. sound illustrating the opening of a document (Power & Jürgenssen 2010, 98). Next to the navigation, the structured content and the reduction of redundancy adds to the accessibility of content. Silva et al. (2017, 155) along with Giraud et al. (2018, 26) therefore developed key items and guidelines to increase accessibility.

More information—less accessibility

With technological development, media involves more visual-based information like pictures, videos and animations (Kern 2008, 144). Even maps gain new importance. Although ICT context, tools and techniques have improved, access to the digital information environment for persons with VIB has not advanced to the same degree. When it comes to a critical evaluation of progress, various gaps can be identified.

After investigating the content accessibility of 50 popular websites, Sullivan and Matson (2000, 144) state that despite availability and knowledge of guidelines, accessibility is hardly implemented: 82% of sites analyzed are inaccessible regarding at least one accessibility criteria. WebAIM analyzed one million webpages, 98% of them failed the Web Content Accessibility Guidelines 2.0. These guidelines define legislations and standards to guarantee accessibility of (public) webpages (WAI 2020). More than half of the pages show errors due to low contrast text (86%), missing alt text for images (66%) or empty links (60%) (WebAIM 2020).

Croll (2009) mentioned that graphical as well as video elements are inaccessible due to missing alternative formats. This is ever more true for social media. Hence, increasing the accessibility of photos and videos is gaining importance in recent research (Morris et al.

2016). Next to technical issues (keyboard accessible elements on the webpage, mosaic maps), Calle-Jimenez and Lujan-Mora (2016, 81) address the differentiation of colors and text in image format as the most important hindrances. Ducasse et al. (2018) offer an overview of hybrid solutions, combining sound and touch. Richardson et al. (2014, 664) reflect the importance of colors and contrasts when capturing websites.

Another accessibility dimension opens up with the touch interface of smartphones. Rodriguez-Sanchez et al. (2014, 7212) especially identify problems with wayfinding applications: "the information is not dynamic, the design is not universal, the interface is not adapted to different users and preferences, blind users need assistance to open the application, the typical screen reader is not applicable, they need to install a special screen reader or the auditory feedback is not enough." Damaceno et al. (2018, 429–432) categorize problems in seven dimensions (buttons, data input, gesture-based interaction, screen reader, screen size, user feedback and voice command), and additionally offer solution strategies based on an extensive literature review.

Finally, and next to the technical barriers, the digital divide refers to a barrier in accessing the digital information environment. Financial, technical and economic barriers (Kern 2008, 144–145), age and (dis-)abilities, but also aspects of literacy detain people from retrieving information (Goggin 2017, 76–77). For insights refer to Chapter 2 in this book.

Digital media and spatial accessibility: Dimensions of research

In terms of information and accessibility, the (geo)spatial component plays an important role (Anselin 1989; Goodchild 2001, 1). However, spatial cognition and spatial perception vary between persons. Guidice (2018, 269) identifies more limits for persons with VIB regarding their spatial competences than the disability itself. Spatial accessibility or access to the built environment and the accessibility of information on the built environment have to be core components to increase mobility and independency. Addressing the access of the built environment is predominantly associated with challenges in wayfinding, and comes along with issues of localization, orientation and navigation. Analogue and digital maps, (geographic) information systems, navigation systems and 3D models support these processes.

From the perspective of applications, individual mobility issues need further research regarding for example urban/public transport, health/emergency management and spatial planning. Access to transportation hubs (airport, train and bus terminals) (Guerreio et al. 2019), availability of means of transport and the development of assistive (electronic) devices supporting mobility (Strumillo 2010) are overlapping with challenges of public transportation (Ghalleger et al. 2011). The perspective in emergency, disaster and health management shifts from the individual with VIB to institutional (urban/regional) decision and planning support systems (Arai et al. 2013). Spatial information and information about the built environment serve for management decisions, for simulation purposes and for designing more inclusive settings by reducing barriers.

Moving forward from the areas of implementation towards structuring scopes and research topics, accessibility of the environment and mobility issues can be seen from diverse dimensions. Granting that these dimensions are linked and interconnected to each other (Figure 6.1), emphasis can be given to the:

- spatial context, looking at outdoor or indoor settings or a combination of both
- orientation elements implemented in the wayfinding process and obstacle detection
- senses used to complement or supplement vision
- methods and/or technologies applied
- types of assistive device customized

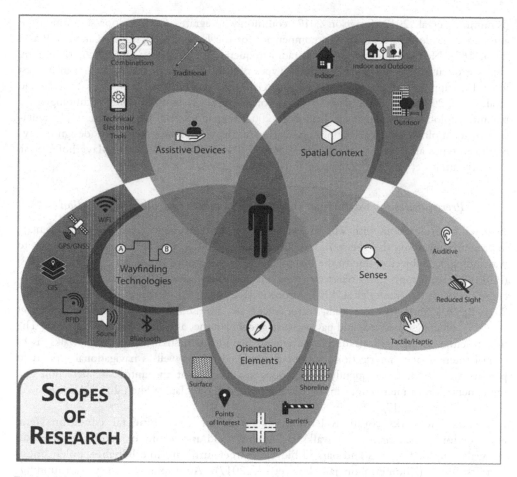

Figure 6.1 Essential research topics in the context of orientation and navigation for persons with VIB

Spatial context: Outdoor and indoor environments

Golledge (1993) was a forerunner in investigating orientation and navigation in urban settings with a focus on persons with visual impairments and their spatial cognition. His work is a milestone in, if not the beginning of research on, wayfinding of persons with VIB in geographical context. It was the outdoor environment where scholars started research. Outdoor orientation and navigation differ from indoor environments due to the utilization of orientation elements, technologies and/or assistive devices. Hints for outdoor orientation used by sighted persons combine directions, street names, landmarks or points of interest (POI) like unique buildings, while in indoor settings, hallways, floor levels, exits, stairs and rooms dominate descriptions for wayfinding. These references are integrated in standard navigation tools, but are not sufficient for persons with VIB, who need additional and more detailed descriptions for wayfinding. This includes alternative features like walls, fences, acoustic pedestrian signals, tactile pavement, etc.

The evolution of Global Positioning System (GPS), or more precisely global navigation satellite system (GNSS), navigation systems and web-maps (Google, etc.) intensified the discussion on wayfinding for persons with VIB (Real & Araujo 2019, 1). Outdoor wayfinding

(Kammoun et al. 2012; Emerson 2017) commonly integrates GPS; indoor solutions need alternative modes of positioning to compensate for the lack of GPS signal (Lakde & Prasad 2015, 167). New technologies such as radio-frequency identification (RFID-9), Bluetooth, etc. (see discussion below in the section "Types of assistive devices") led to a rise in interest for indoor applications after the millennium (Gallagher et al. 2011; Miao et al. 2011; Serrão et al. 2012; 2015; Murata et al. 2019). Additionally, some systems combine technologies for in- and outdoor wayfinding (e.g. Ran et al. 2004; Fernandes et al. 2014). The implementation of technologies for localization, orientation and navigation has led to the design of several prototypes and assistive devices (Ran et al. 2004; Wilson et al. 2007; Mayerhofer et al. 2008; Strumillo et al. 2018; Zimmermann-Janschitz 2019).

Prerequisites for wayfinding: Orientation elements and obstacle detection

As stated before, persons with VIB have demand for alternative, additional and supplementary information regarding their personal mobility (Dias et al. 2015, 147). Moreover, approaches to wayfinding are subjective due to the heterogeneity of the group. Landmarks, points of interest (POI)—like bus stops— and olfactory cues—e.g. trash bins—are indicated as important orientation elements (Strothotte et al. 1995; Afrooz et al. 2012, 1086; Serrao et al., 2014; Park et al. 2015, 24470). Shorelines like walls, curbs, fences, etc. are used as guiding features (Koester et al. 2017). Changes in surface and pavement (e.g. cobblestone, asphalt) support orientation. The infrastructure at intersections (tactile pavement, acoustic pedestrian signals, crosswalks) is of special interest, since intersections pose both amplified risk as well as navigational support to persons with VIB. Consequently, applications are looking at the automatic detection and implementation of intersections in the routing process (Coughlan & Shen 2013; Ahmetovic et al. 2017; Bentzen 2017).

Missing landmarks count as barriers, as do urban environmental conditions (e.g. construction sites, narrow sidewalks or squares) and moveable or fixed obstacles on sidewalks (unloading cars and parked bicycles, street furniture like benches, poles, boom barriers, etc.) (Zimmermann-Janschitz et al. 2017). Additionally, noisy surroundings interfere with the orientation of persons with VIB. Bergner et al. (2011) add an emotional component to navigation, measuring the stress of persons with VIB when approaching an obstacle.

Next to the subjective adequate number of orientation cues, orientation needs to be based on sidewalks instead of streets. Hence, the spatial accuracy of street-centerlines used in commercial navigation tools designed for motorized vehicles is insufficient (Neis & Zeilstra 2014, 70). Pedestrian networks allow the integration of impassable segments in networks (e.g. pedestrian zones, shortcuts, etc.) as well as the differentiation between left- and right-hand sides of the street, since persons with VIB orient themselves towards traffic. The automated generation of pedestrian networks as the basis for navigation is therefore important (Ballester et al. 2011; Karimi & Kasemsuppakorn 2013, 958). Factors contributing to the accessibility of sidewalks and influencing the route choices of persons with VIB are evaluated by Tajgardoon and Karimi (2015, 85).

Landmarks and POI serve as upgrades for step-by-step wayfinding directions, while reference to obstacles increases safety issues (Rice et al. 2013; Park et al. 2015). Both static and permanent barriers (e.g. post boxes, boom barriers) as well as movable obstacles (e.g. bikes) pose a risk to persons with VIB of getting injured when moving independently. Obstacle detection includes not only the recognition of obstacles, but the development of collision avoidance systems (Bhowmick & Hazarika 2017, 164).

Apparently, the amount of data involved in the wayfinding process, the availability of the information, georeferenced data on sidewalks, possibilities to gather information as well as the method of presentation are crucial components in the design of a tool to support wayfinding processes.

Senses used to complement or supplement vision

Similar to the individual needs regarding the content of step-by-step directions, the presentation of spatial information for orientation and navigation purposes has to be distinct from conventional, commercial navigation aids. At this point the focus is set to the spatial component with a crossover to technologies and assistive devices.

For persons who are visually impaired or blind, navigation tools and route planners need to transfer textual step-by-step directions into speech and maps need to be made accessible via touch and/or sound. Persons with remaining sight want an appropriate presentation of the map regarding extent, labeling, contrast and/or colors (Jenny & Kelso 2007, 64; Calle-Jimenez & Luján-Mora 2016, 81; Henning et al. 2017, 17). Tactile maps, used if remaining vision is not sufficient to fully discover maps, have a long tradition in cartography. Zeng and Weber (2011, 5) present a classification based on the mode of exploration of tactile maps (braille tactile map, virtual tactile map, virtual acoustic map, printable tactile map, augmented paper-based tactile map), indicating the complexity of approaches as well as the move towards a combination of touch and sound with the generation of web-maps.

The presentation of tactile maps and attempts at their automated production (Miele et al. 2006) as well as the potential of technological developments made in recent decades is not yet fully exploited (Lobben 2015). Rather, with the increasing number of online maps, the problem of inaccessibility rises. However, there are numerous approaches to overcome this problem and progress is being made in touch interfaces, integrating haptic or tactile information. Haptic information refers to an (inter)active capturing of information on objects and typically involves feedback to the object, while tactile information is perceived passively (Hersh & Johnson 2008, 137). A detailed description of accessible interactive maps is given in Ducasse et al. (2017). For wayfinding purposes, Wang, Zheng and Fan (2017, 721) offer different patterns of vibration feedback in terms of intensity and frequency of feedback to help staying on the route. Wang, Li and Li (2012, 98) for example designed a prototype to transfer online maps for navigation to scalable vector graphics (SVG) which can be printed as tactile maps. Other scholars use devices (e.g. belts, pointers) as spatial tactile displays to give vibration feedback regarding direction of movement (Marston et al. 2007, 205; Heuten et al. 2008, 175).

An important support for the communication of spatial information is the use of sound. This approach covers voice, audio and music in- and output tools with a strong focus on the output. Speech recognition is used sparingly (see Roentgen et al. 2008, 714–718). Interactive maps present in addition to touch features, acoustic feedback to users with VIB through audio and/or voice annotations (Siekierska 2003, 487). Navigating the map, sound represents orientation features, while annotations on objects present additional information with voice. Feedback via sound is also used to represent obstacles (Moreno et al. 2012, 80). A completely different approach to read maps is made via sound and music. Bearman and Fisher (2012, 159–160) turn information in maps into music by assigning for example changes in elevation to a sound scale. The sonification of statistical data in maps is clarified by Zhao et al. (2005, 2).

When it comes to navigation, text-to-speech can be seen as the most common form of presenting step-by-step directions to users, regardless of using screen reader software or recorded audio (Brock et al. 2015, 161). The combination of voice recognition, sound feedback as earcons and speech synthesis can be found in Parente and Bishop (2003) or O'Sullivan et al. (2015). Several scholars combine touch and sound elements to a multimodal experience in interactive, talking maps (Parente & Bishop 2003; Kaklanis et al. 2013) as well as in navigation tools and devices (Rodriguez-Sanchez et al. 2014; O'Sullivan et al. 2015; Wang et al. 2017). Some projects extend sound and touch to the virtual space using virtual or augmented reality. A three-dimensional virtual reality serves as a basis for mental maps and supports the acquisition of information on objects. This is consequently transferred to real space by the user and serves as a source for navigation (Lahav & Mioduser 2008; Katz et al. 2012).

Presenting the information: Methods and technologies applied

Alternative modes to make information available and accessible for persons with VIB require not only other types of information (as discussed above), but also additional technology to gather and display this information. The localization and tracking of a person with VIB along the route have to be very precise in terms of safety. For instance, if road works occur along a route, there should be the possibility to avoid the construction site. While GPS in general is sufficient in outdoor surroundings, differential GPS (DGPS) offers higher precision. Still, GPS signals can be shielded by trees or reflected in street canyons, followed by signal delay or coarse positioning (Ladke & Prasad 2015). The algorithm of Ivanov (2012, 1560–1562) simplifies GPS tracking by including only GPS points which indicate changes in the direction. Accessible GPS comprises solutions integrating haptic and/or sound interfaces to retrieve position and tracking information en route (May & Casey 2018).

In indoor environments, beacons, sensors and markers based on radio-frequency, infrared-, ultrasound signals, or using existing wireless networks (e.g. WiFi), are used to define positions. Yanez et al. (2016, 349) equip persons with several ultrasound sensors to detect obstacles in the surroundings, distributing the relevant information via WiFi. Regardless of whether Bluetooth is used (Castillo-Cara et al. 2016, 692) or RFID (Fernandes et al. 2014), beacons provide the spatial information in buildings and have to be placed on strategic locations. RFID determines the position utilizing tags on the floor, which are read with a corresponding device, e.g. an electronic white cane. Fernandes et al. (2014, 6–7) implement RFID in combination with a Geographic Information System (GIS) to store and analyze additional environmental information. Finally, cameras—either those attached to external equipment or smartphone cameras—are integrated to capture obstacles, landmarks or other environmental settings (Moreno et al. 2012; Serrao et al. 2015; Ahmetovich et al. 2017).

Shifting the focus from localization and tracking technologies towards analyzing spatial information for navigation purposes, GIS comes into play. Even though GIS is mainly used to store, manipulate and visualize spatial data in maps, its core part and intrinsic potential, analyzing spatial data, is not fully exploited. In the 1990s Jacobson and Kitchin (1997) gave a reflection of the potential of GIS for navigation and orientation of persons with VIB. Golledge et al. (1998, 728) utilize the full potential of GIS, using it to calculate routes and store for example landmarks and present them to the user while keep the user on track. Other scholars focus on the database to manage the amount of data characterizing the environment (Fernandes et al. 2012; Serrao et al. 2015). The analytical component is applied to analyze and examine networks, calculate routes or integrate parameters as the basis for decisions. Zeng and Weber (2012) developed a user interface based on braille and sound to overcome accessibility barriers. Bearman and Fisher (2012) turn information into sound. Zimmermann-

Janschitz (2019) implemented a web-GIS on a medium-sized city scale to support wayfinding and closest facility search (see case study "ways2see").

Case study: ways2see (text by Simon Landauer)

ways2see is a web-based wayfinding application assisting people with visual impairment or blindness when preparing and planning routes and shorter trips of everyday life in an urban outdoor environment. Focusing on the concept of availability and accessibility of information, the detail and selection of content are designed in accordance with the format of presentation.

Concerning wayfinding technology, ways2see uses GIS to manage and process necessary data for routing of pathways. The wayfinding itself is based on a specifically designed pedestrian routing network, incorporating type of way-use (sidewalk, footway, cycle way, etc.), auditory and tactile features of crossing, obstacles and barriers (boom barrier, pole etc.) as well as shorelines (house wall, fence, etc.) and surfaces (pavement, cobblestone, etc.). Results are consequently returned as turn-by-turn directions, where each segment holds details on the aforementioned information (see Figure 6.2). Aside from addresses, POI (medical facilities, supermarkets and drug stores, education infrastructure, entertainment, etc.) can be chosen as destinations according to their distance from the starting point. The tool further addresses the issue of user heterogeneity by introducing three user profiles and one open profile, determining the routing according to a pre-defined ruleset. Yet, the application does not feature on-site routing via GPS/GNSS due to the focus on pre-trip navigation as well as issues of GPS signal accuracy in the context of VIB. Consequently, the application is considered an assistive planning tool, supporting the use of traditional assistive devices along with new technology.

In terms of format, the accessibility of content is provided through the screen reader compatibility of the entire application and navigation through keyboard-only input. Use of high contrast coloring for visual aspects and maps allows for additional access of output for users with reduced vision. Furthermore, keeping maps features limited to essential elements enhances usability and performance. The same holds true for the cartographic as well as overall symbology used. Predominantly designed for desktop use, implementation on mobile devices is considered challenging.

For further information: https://barrierefrei.uni-graz.at/ways2see/

Types of assistive devices

Orientation, navigation and wayfinding research has produced a wide range of assistive devices over the last few decades. Several studies, evaluating the topic from diverse perspectives, document scientific progress and drawbacks. Roentgen et al. (2009; 2011) investigate the performance of obstacle detection and navigation tools. The focus of Hakobyan et al. (2013) is on mobile devices covering orientation and navigation next to smart homes and robotics. Lakde and Prasad (2015) investigate indoor and outdoor navigation systems. Tapu et al. (2018) and Real and Araujo (2019) give a holistic view on tools and devices supporting navigation from the application side, as Bhowmick and Hazarika (2017) approach assistive technology with the goal of identifying research networks and clusters.

Additionally, assistive devices not directly associated with orientation or navigation have to be mentioned, since they often are an integrative part of assistive tools to support mobility. Amongst these are screen reader software, refreshable braille displays, raised pin pads and sensitive tablets.

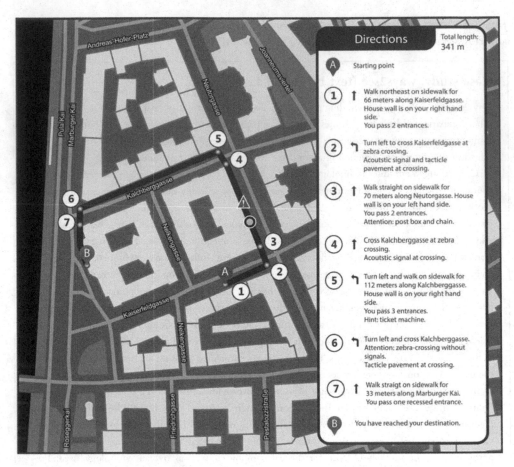

Figure 6.2 Detailed turn-by-turn directions in ways2see for people with VIB

"Traditional" types of assistive devices to support individual mobility of persons with VIB are the long white cane, the guide dog as well as sighted guides (orientation and mobility trainers, friends or family members). In terms of independency, the long white cane is indicated as most important (Ladke & Prasad 2015, 166). For this reason, electronic adaptions and extensions of the long white cane are significant. An overview of haptic and/or sound features for long white canes is given in Khan et al. (2018). Next to the electronic cane numerous new devices have been designed, combining sensors, cameras, canes and computers or phones in complex assistive devices. Cameras or GPS attached to a hat, computers in backpacks, sensors integrated in belts, bracelets or all over the body, even tablets used for communicating with the system can be found as described above (examples in: Roentgen et al. 2009, 2011; Tapu et al. 2018; Real & Araujo 2019). Nowadays, applications for smartphones have gained importance, offering multisensory equipment in one device (Rodriguez-Sanchez 2014; Murata et al. 2019).

"Nothing about us without us": Challenges and future research

The chapter illustrates the variety of digital media as well as their prerequisites for persons with VIB. The development of assistive devices and tools supporting their daily life, particularly enhancing their personal mobility, is an emerging research topic. Numerous articles, approaches and applications go together with the progress made in ICT. During the last two decades, the increasing number of evaluation papers shows an accelerating trend.

It is remarkable, and has been stated many times, that the improvement for persons with VIB—generally and related in terms of mobility—has been limited (e.g. Chandler & Worsfold 2013, 920; Rodrigues-Sanchez et al. 2014, 7211). Bhowmick and Hazarika (2017, 164) express: "Although the advancement of technology is evident, only a limited number of assistive technology solutions have emerged to make a social or economic impact and improve quality of life." Since digital media and ICT offer tremendous potential regarding what can be achieved for persons with VIB, it is not sufficient at this point to end with an overview of this topic and a snapshot of the development. There is need for a final critical reflection of ongoing research and a glimpse of future research strands.

Data as critical input

From a technical perspective, specific data is the most crucial part in the design of mobility support for persons with VIB. Therefore, many applications have a narrow spatial extent and are limited to university campuses, city blocks or single buildings or remain prototypes (Roentgen et al. 2009, 745; Murata et al. 2019, 14). These "lab conditions" use a limited data set. Many factors play a role regarding data: the data volume, for example data included in wayfinding (cues, barriers etc.) or network data as the basis for the navigation process. The elements and objects relevant for the navigation process of people with VIB are assessed by interviews or questionnaires (Chandler & Worsfold 2013), but the necessity for individuality (Cuturi et al. 2016) is barely achieved. Ballester et al. (2011) certify the need for pedestrian networks offering an automatization of network generation.

There are different attempts to overcome data limitations. Open source and open government data, volunteered geographic information and crowdsourced data are intended to reduce costs. Free data make applications available for wider spatial extent and raise the chance to integrate appropriate data for individual solutions (Kaklanis et al. 2013; Rice et al. 2013; Zeng et al. 2017). However, open data generate additional challenges like spatial accuracy, reliability or coverage. Emerging new technologies, e.g. big data, the internet of things (IoT) or laser scanning, promise to provide data previously not available.

Technological development

Fast-evolving information technology is an integral part of life, especially for the youth. ICT, including assistive devices for PWD, are not commonly accepted the same way. Söderström and Ytterhus (2010, 307) found that young individuals with visual impairment try to avoid the use of assistive devices, since they perceive them as stigmatizing and excluding. People who are blind rather embrace assistive devices as a chance to be integrated into (ICT) society. Eventually, ICT can be perceived as both opportunity and threat. Obvious obstacles are affordability, complexity and usability. Moreover, map literacy and technical literacy hinder persons with VIB from using assistive devices. Additionally, a certain amount of training is needed (Tapu et al. 2018). Cuturi et al. (2016, 241, 248) criticize the amount of information involved in complex assistive devices and tools. The development towards smartphone-based solutions is promising, since this device is easy to use, affordable and has the potential to be adapted to individual needs (screen reader etc.) with low effort.

Participation and empowerment to reduce "barriers in mind"

Finally, there is a need to look at the humans concerned: persons with visual impairment or blindness. There is no doubt that mobility is a human right and increasing individual mobility provides a step towards a self-determined and independent daily life. Technology and media are an important part of achieving this goal, but nevertheless they are only one part. Inclusive and integrative research design (Butler 1994, 368) and the empowerment of persons with VIB leads to wider acceptance and implementation of research results. Guidelines, laws and standards are available, but if not implemented, they are inoperable. The advancement of ICT has the potential to support society towards inclusion, raise awareness and generate sensibility in society for PWD. But there is still a long way ahead to reduce remaining "barriers in our minds."

References

Afrooz, A., Toktam, H., and Parolin, B. 2012. Wayfinding performance of visually impaired pedestrians in an urban area. In M. Schrenk, V. V. Popovich, P. Zeile, and P. Elisei (eds.) *Re-Mixing the City, Real Corp Proceedings*, pp. 1081–1091. Vienna: Real Corp.

Ahmetovic, D., Manduchi, R., Coughlan, J., and Mascetti, S. 2017. Mind your crossings: Mining GIS imagery for crosswalk localization. *ACM Transactions on Accessible Computing (TACCESS)*, 9 (4), 1–25. https://doi.org/10.1145/3046790.

American Foundation for the Blind (AFB). 2020. *Disability rights resources for people with vision loss.* www.afb. org/blindness-and-low-vision/disability-rights/advocacy-resources/disability-rights-resources#laws.

Anselin, L. 1989. *What is special about spatial data? Alternative perspectives on spatial data analysis.* https:// escholarship.org/uc/item/3ph5k0d4.

Arai, K., and Sang, T. X. 2013. Decision making and emergency communication system in rescue simulation for people with disabilities. *International Journal of Advanced Research in Artificial Intelligence*, 2 (3), 77–85.

Ballester, M. G., Pérez, M. R., and Stuiver, J. 2011. Automatic pedestrian network generation. The 14th AGILE International Conference on Geographic Information Science, 18–21 April, Utrecht. https://a gile-online.org/conference_paper/cds/agile_2011/contents/pdf/shortpapers/sp_116.pdf.

Bearman, N., and Fisher, P. F. 2012. Using sound to represent spatial data in ArcGIS. *Computers and Geosciences*, 46, 157–163. https://doi.org/10.1016/j.cageo.2011.12.001.

Bentzen, B. L., Barlow, J. M., Scott, A. C., Guth, D., Long, R., and Graham, J. 2017. Wayfinding problems for blind pedestrians at noncorner crosswalks: Novel solution. *Transportation research record*, 2661 (1), 120–125. https://doi.org/10.3141/2661-14.

Bergner, B. S., Zeile, P., Papastefanou, G., Rech, W., and Streich, B. 2011. Emotional Barrier-GIS: A new approach to integrate barrier-free planning in urban planning processes. In M. Schrenk, V. V. Popovich and P. Zeile (eds.) *Change for stability, Real Corp proceedings*, pp. 247–258. Vienna: Real Corp.

Bernhard, J., and Kelso, N. V. 2007. Color design for the color vision impaired. *Cartographic Perspectives*, 58, 61–67. https://doi.org/10.14714/CP58.270.

Bhowmick, A., and Hazarika, S. M. 2017. An insight into assistive technology for the visually impaired and blind people: State-of-the-art and future trends. *Journal on Multimodal User Interfaces*, 11 (2), 149–172. https://doi.org/10.1007/s12193-016-0235-6.

Bourne, R. R., Flaxman, S. R., Braithwaite, T., Cicinelli, M. V., Das, A., Jonas, J. B., Keeffe, J., Kempen, J. H., Leasher, J., Limburg, H., and Naidoo, K. 2017. Magnitude, temporal trends, and projections of the global prevalence of blindness and distance and near vision impairment: A systematic review and meta-analysis. *Lancet Global Health*, 5 (9), e888–897. https://doi.org/10.1016/S2214-109X(17)30293-0.

Brock, A. M., Truillet, P., Oriola, B., Picard, D., and Jouffrais, C. 2015. Interactivity improves usability of geographic maps for visually impaired people. *Human–Computer Interaction*, 30 (2), 156–194. https://doi. org/10.1080/07370024.2014.924412.

Butler, R. 1994. Geography and vision-impaired and blind populations. *Transactions of the Institute of British Geographers*, 19 (3), 366–368. https://doi.org/10.2307/622329.

Butler, R., and Bowlby, S. 1997. Bodies and spaces: An exploration of disabled people's experiences of public space. *Environment and Planning D: Society and Space*, 15 (4), 411–433.

Calle-Jimenz, T., and Lujan-Mora, S. 2016. Web accessibility barriers in geographic maps. *International Journal of Computer Theory and Engineering*, 8 (1), 80–87. https://doi.org/10.7763/IJCTE.2016.V8.1024.

Castillo-Cara, M., Huaranga-Junco, E., Mondragón-Ruiz, G., Salazar, A, Orozco-Barbosa, L., and Antúnez, E. A. 2016. Ray: Smart indoor/outdoor routes for the blind using Bluetooth 4.0 BLE. *ANT/SEIT, Procedia Computer Science*, 83, 690–694.

CEN-CENELEC (European Committee for Standardization and European Committee for Electrotechnical Standardization). 2011. *Accessibility in built environment: Draft joint report—CEN/BT WG 207 (PT A and PT B)*. Brussels: CEN and CENELEC. ftp://ftp.cen.eu/CEN/Sectors/Accessibility/Rep ortAccessibilityBuiltEnvironment%20Final.pdf.

Chandler, E., and Worsfold, J. 2013. Understanding the requirements of geographical data for blind and partially sighted people to make journeys more independently. *Applied Ergonomics*, 44 (6), 919–928. http s://doi.org/10.1016/j.apergo.2013.03.030.

Coughlan, J. M., and Shen, H. 2013. Crosswatch: A system for providing guidance to visually impaired travelers at traffic intersections. *Journal of Assistive Technologies*, 7 (2). https://doi.org/10.1108/17549451311328808.

Croll, J. 2009. Internet-Digitale Integration mit allen Sinnen [in German]. In W. Christ (ed.) *Access for All: Zugänge zur gebauten Umwelt*, pp. 158–170. Basel, Boston, Berlin: Birkhäuser Verlag AG.

Cuturi, L. F., Aggius-Vella, E., Campus, C., Parmiggiani, A., and Gori, M. 2016. From science to technology: Orientation and mobility in blind children and adults. *Neuroscience and Biobehavioral Reviews*, 71, 240–251. http://dx.doi.org/10.1016/j.neubiorev.2016.08.019.

Damaceno, R. J. P., Braga, J. C., and Mena-Chalco, J. P. 2018. Mobile device accessibility for the visually impaired: Problems mapping and recommendations. *Universal Access in the Information Society*, 17, 421–435. https://doi.org/10.1007/s10209-017-0540-1.

Dias, M. B., Teves, E. A., Zimmerman, G. J., Gedawy, H. K., Belousov, S. M., and Dias, M. B. 2015. Indoor navigation challenges for visually impaired people. In H. A. Karimi (ed.) *Indoor wayfinding and navigation*, pp. 141–164. New York: CRC Press.

Ducasse, J., Brock, A. M., and Jouffrais, C. 2018. Accessible interactive maps for visually impaired users. In E. Pissaloux and R. Velazquez (eds.) *Mobility of visually impaired people*, pp. 537–584. Cham: Springer. https://doi.org/10.1007/978-3-319-54446-5.

Emerson, R. W. 2017. Outdoor wayfinding and navigation for people who are blind: Accessing the built environment. In *International conference on universal access in human-computer interaction*. Cham: Springer.

European Agency for Special Needs and Inclusive Education. 2020. *Guidelines for Accessible Information*. www.european-agency.org/resources/publications/guidelines-accessible-information.

European Commission. 2010. *European disability strategy 2010–2020: A renewed commitment to a barrier-free Europe*. https://eur-lex.europa.eu/LexUriServ/LexUriServ.do?uri=COM%3A2010%3A0636%3AFIN%3Aen%3APDF.

European Commission. 2015. *Directive of the European Parliament and of the Council on the approximation of the laws, regulations and administrative provisions of the Member States as regards the accessibility requirements for products and services*. https://eur-lex.europa.eu/legal-content/EN/TXT/HTML/?uri=CELEX:52015PC0615&from=EN.

European Commission. 2020. *European innovation partnership on active and healthy ageing. standards*. https://ec.europa.eu/eip/ageing/standards_en.

Fernandes, H., Filipe, V., Costa, P., and Barroso, J. 2014. Location based services for the blind supported by RFID technology. *Procedia Computer Science*, 27, 2–8. https://doi.org/10.1016/j.procs.2014.02.002.

Fuglerud, K. S. 2011. The barriers to and benefits of use of ICT for people with visual impairment. In C. Stephanidis (ed.) *Universal access in human-computer interaction: Design for all and eInclusion*. Berlin-Heidelberg: Springer. https://doi.org/10.1007/978-3-642-21672-5_49.

Gallagher, B. A. M., Hart, P. M., O'Brien, C., Stevenson, M. R., and Jackson, A. J. 2011. Mobility and access to transport issues as experienced by people with vision impairment living in urban and rural Ireland. *Disability and Rehabilitation*, 33 (12), 979–988. https://doi.org/10.3109/09638288.2010.516786.

Giraud, S., Thérouanne, P., and Steiner, D. D. 2018. Web accessibility: Filtering redundant and irrelevant information improves website usability for blind users. *International Journal of Human-Computer Studies*, 111, 23–35. https://doi.org/10.1016/j.ijhcs.2017.10.011.

Giudice, N. A. 2018. Navigating without vision: Principles of blind spatial cognition. In D. R. Montello (ed.) *Handbook of behavioral and cognitive geography*. Cheltenham: Edward Elgar Publishing. https://doi.org/10.4337/9781784717544.00024.

Goggin, G. 2017. Disability and digital inequalities: Rethinking digital divides with disability theory. In M. Ragnedda and G. W. Muschert (ed.) *Theorizing digital divides*, pp. 69–80. New York: Routledge.

Golledge, R. G. 1993. Geography and the disabled: A survey with special reference to vision impaired and blind populations. *Transactions of the Institute of British Geographers*, 18 (1), 63–85. https://doi.org/10.2307/623069.

Golledge, R. G., Klatzky, R. L., Loomis, J. M., Speigle, J., and Tietz, J. 1998. A geographical information system for a GPS based personal guidance system. *International Journal of Geographical Information Science*, 12 (7), 727–749. https://doi.org/10.1080/136588198241635.

Goodchild, M. F. 2001. A geographer looks at spatial information theory. In *International conference on spatial information theory*. Berlin-Heidelberg: Springer. https://doi.org/10.1007/3-540-45424-1_1.

Guerreio, J., Ahmetovic, D., Sato, D., Kitani, K. M., and Asakawa, C. 2019. Airport accessibility and navigation assistance for people with visual impairments. In *Proceedings of the 2019 CHI conference on human factors in computing systems*. New York: Association for Computing Machinery. https://doi.org/10.1145/3290605.3300246.

Hakobyan, L., Lumsden, J., O'Sullivan, D., and Bartlett, H. 2013. Mobile assistive technologies for the visually impaired. *Survey of Ophthalmology*, 58 (6), 513–528. https://doi.org/10.1016/j.survophthal.2012.10.004.

Hall, E., and Wilton, R. 2017. Towards a relational geography of disability. *Progress in Human Geography*, 41 (6), 727–744. https://doi.org/10.1177/0309132516659705.

Hennig, S., Zobl, F., and Wasserburger, W. W. 2017. Accessible web maps for visually impaired users: Recommendations and example solutions. *Cartographic Perspectives*, 88, 6–27. https://doi.org/10.14714/CP88.1391.

Hersh, M. A., and Johnson, M. A. (eds.) 2008. *Assistive technology for visually impaired and blind people*. London: Springer.

Heuten, W., Henze, N., Boll, S., and Pielot, M. 2008. Tactile wayfinder: A non-visual support system for wayfinding. In *Proceedings of the 5th Nordic conference on human-computer interaction: Building bridges (NordiCHI '08)*. New York: Association for Computing Machinery. https://doi.org/10.1145/1463160.1463179.

Huebner, K. M. 2000. Visual impairment. In M. C. Holbrook and A. J. Koenig (eds.) *Foundations of education, history and theory of teaching children and youths with visual impairments*, pp. 55–76. New York: AFB Press.

Ivanov, R. 2012. Real-time GPS track simplification algorithm for outdoor navigation of visually impaired. *Journal of Network and Computer Applications*, 35 (5), 1559–1567. https://doi.org/10.1016/j.jnca.2012.02.002.

Jacobson, R. D., and Kitchin, R. 1997. GIS and people with visual impairments or blindness: Exploring the potential for education, orientation and navigation, *Transactions in GIS*, 2 (4), 315–332. https://doi.org/10.1111/j.1467-9671.1997.tb00060.x.

Janschitz, S. 2012. *Von Barrieren in unseren Köpfen und "Karten ohne Grenzen" Geographische Informationssysteme im Diskurs der Barrierefreiheit—ein Widerspruch in sich oder unerkanntes Potenzial* [in German]. Vienna: LIT.

Jansson, G. 2008. Haptics as a substitute for vision. In M. A. Hersh and M. A. Johnson (eds.) *Assistive technology for visually impaired and blind people*, pp. 135–166. London: Springer. https://doi.org/10.1007/978-1-84628-867-8_4.

Joffee, E. 1999. *A practical guide to the ADA and visual impairment*. New York: American Foundation for the Blind Press.

Kaklanis, N., Votis, K., and Tzovaras, D. 2013. Open touch/sound maps: A system to convey street data through haptic and auditory feedback. *Computers and Geosciences*, 57, 59–67. https://doi.org/10.1016/j.cageo.2013.03.005.

Kammoun, S., Parseihian, G., Gutierrez, O., Brilhault, A., Serpa, A., Raynal, M., Oriola, B., Macé, M. M., Auvray, M., Denis, M., Thorpe, S. J., Truillet, P., Katz, B. F. G., and Jouffrais, C. 2012. Navigation and space perception assistance for the visually impaired: The NAVIG project. *IRBM*, 33 (2), 182–189. https://doi.org/10.1016/j.irbm.2012.01.009.

Karimi, H. A., and Kasemsuppakorn, P. 2013. Pedestrian network map generation approaches and recommendation. *International Journal of Geographical Information Science*, 27 (5), 947–962. https://doi.org/10.1080/13658816.2012.730148.

Katz, B. F., Kammoun, S., Parseihian, G., Gutierrez, O., Brilhault, A., Auvray, M., Truillet, P., Denis, M., Thorpe, S., and Jouffrais, C. 2012. NAVIG: Augmented reality guidance system for the visually impaired. *Virtual Reality*, 16 (4), 253–269. https://doi.org/10.1007/s10055-012-0213-6.

Kern, W. 2008. Web 2.0: End of accessibility? Analysis of most common problems with Web 2.0 based applications regarding Web accessibility. *International Journal of Public Information Systems*, 4 (2), 131–154.

Kitchin, R., and Jacobson, R. D. 1997. Techniques to collect and analyze the cognitive map knowledge of persons with visual impairment or blindness: Issues of validity. *Journal of Visual Impairment and Blindness*, 91 (4), 360–376.

Koester, D., Awiszus, M., and Stiefelhagen, R. 2017. Mind the gap: Virtual shorelines for blind and partially sighted people. In *Proceedings of the IEEE international conference on computer vision workshops*. Washington, DC, Brussels, Tokyo: IEEE. https://doi.org/10.1109/ICCVW.2017.171.

Lahav, O., and Mioduser, D. 2008. Construction of cognitive maps of unknown spaces using a multisensory virtual environment for people who are blind. *Computers in Human Behavior*, 24 (3), 1139–1155. https://doi.org/doi:10.1016/j.chb.2007.04.003.

Lakde, C. K., and Prasad, P. S. 2015. Review paper on navigation system for visually impaired people. *International Journal of Advanced Research in Computer and Communication Engineering*, 4 (1), 166–168. https://doi.org/10.17148/IJARCCE.2015.4134.

Lobben, A. 2015. Tactile maps and mapping. *Journal of Blindness Innovation and Research*, 15 (1). www.nfb.org/images/nfb/publications/jbir/jbir15/jbir050102.html.

Macpherson, H. 2010. Non-representational approaches to body–landscape relations. *Geography Compass*, 4 (1), 1–13. https://doi.org/10.1111/j.1749-8198.2009.00276.x.

Marston, J. R., Loomis, J. M., Klatzky, R. L., and Golledge, R. G. 2007. Nonvisual route following with guidance from a simple haptic or auditory display. *Journal of Visual Impairment and Blindness*, 101 (4), 203–211. https://doi.org/10.1177/0145482X0710100403.

May, M., and Casey, K. 2018. Accessible global positioning systems. In R. Manduchi and S. Kurniawan (eds.) *Assistive technology for blindness and low vision*, pp. 99–122. Boca Raton: CRC Press.

Mayerhofer, B., Pressl, B., and Wieser, M. 2008. ODILIA: A mobility concept for the visually impaired. In K. Miesenberger, J. Klaus, W. Zagler and A. Karshmer (eds.) *Lecture notes in computer science (LNCS) 5105*, pp. 1109–1116. Berlin and Heidelberg: Springer. https://doi.org/10.1007/978-3-540-70540-6_166.

Miao, M., Spindler, M., and Weber, G. 2011. Requirements of indoor navigation system from blind users. In *Symposium of the Austrian HCI and usability group*. Berlin and Heidelberg: Springer. https://doi.org/10.1007/978-3-642-25364-5_48.

Miele, J. A., Landau, S., and Gilden, D. 2006. Talking TMAP: Automated generation of audio-tactile maps using Smith-Kettlewell's TMAP software. *British Journal of Visual Impairment*, 24 (2), 93–100. https://doi.org/10.1177/0264619606064436.

Moreno, M., Shahrabadi, S., José, J., du Buf, J. H., and Rodrigues, J. M. 2012. Realtime local navigation for the blind: Detection of lateral doors and sound interface. *Procedia Computer Science*, 14, 74–82. https://doi.org/10.1016/j.procs.2012.10.009.

Morris, M. R., Zolyomi, A., Yao, C., Bahram, S., Bigham, J. P., and Kane, S. K. 2016. "With most of it being pictures now, I rarely use it": Understanding Twitter's evolving accessibility to blind users. *Proceedings of the 2016 CHI conference on human factors in computing systems*. New York: Association for Computing Machinery. https://doi.org/10.1145/2858036.2858116.

Murata, M., Ahmetovic, D., Sato, D., Takagi, H., Kitani, K. M., and Asakawa, C. 2019. Smartphone-based localization for blind navigation in building-scale indoor environments. *Pervasive and Mobile Computing*, 57, 14–32. https://doi.org/10.1016/j.pmcj.2019.04.003.

Neis, P., and Zielstra, D. 2014. Generation of a tailored routing network for disabled people based on collaboratively collected geodata. *Applied Geography*, 47, 70–77. https://doi.org/10.1016/j.apgeog.2013.12.004.

O'Sullivan, L., Picinali, L., Gerino, A., and Cawthorne, D. 2015. A prototype audio-tactile map system with an advanced auditory display. In *International Journal of Mobile Human Computer Interaction*, 7 (4), 53–75. https://doi.org/10.4018/IJMHCI.2015100104.

Parente, P., and Bishop, G. 2003. BATS: The blind audio-tactile mapping system. In *Proceedings of the ACM southeast regional conference*. New York: Association for Computing Machinery.

Park, S., Bang, Y., and Yu, K. 2015. Techniques for updating pedestrian network data including facilities and obstructions information for transportation of vulnerable people, *Sensors*, 15 (9), 24466–24486. https://doi.org/10.3390/s150924466.

Power, C., and Jürgensen, H. 2010. Accessible presentation of information for people with visual disabilities. *Universal Access Information Society*, 9, 97–119. https://doi.org/10.1007/s10209-009-0164-1.

Ran, L., Helal, S., and Moore, S. 2004. Drishti: An integrated indoor/outdoor blind navigation system and service. In *Proceedings of the second IEEE annual conference on pervasive computing and communications (PERCOMW'04)*. Washington, DC, Brussels, Tokyo: IEEE. https://doi.org/10.1109/PERCOM.2004.1276842.

Real, S., and Araujo, A. 2019. Navigation systems for the blind and visually impaired: Past work, challenges, and open problems, *Sensors*, 19 (15), 3404. https://doi.org/10.3390/s19153404.

Rice, M. T., Jacobson, R. D., Caldwell, D. R., McDermott, S. D., Paez, F. I., Aburizaiza, A. O., Curtin, K. M., Stefanidis, A., and Qin, H. 2013. Crowdsourcing techniques for augmenting traditional accessibility maps with transitory obstacle information. *Cartography and Geographic Information Science*, 40 (3), 210–219. https://doi.org/10.1080/15230406.2013.799737.

Richardson, R. T., Drexler, T. L., and Delparte, D. M. 2014. Color and contrast in e-learning design: A review of the literature and recommendations for instructional designers and web developers. *Journal of Online Learning and Teaching*, 10 (4), 657–670.

Rodriguez-Sanchez, M. C., Moreno-Alvarez, M. A., Martin, E., Borromeo, S., and Hernandez-Tamames, J. A. 2014. Accessible smartphones for blind users: A case study for a wayfinding system, *Expert Systems with Applications*, 41 (16), 7210–7222. https://doi.org/10.1016/j.eswa.2014.05.031.

Roentgen, U. R., Gelderblom, G. J., Soede, M., and De Witte, L. P. 2008. Inventory of electronic mobility aids for persons with visual impairments: A literature review. *Journal of Visual Impairment and Blindness*, 102 (11), 702–724. https://doi.org/10.1177/0145482X0810201105.

Roentgen, U. R., Gelderblom, G. J., Soede, M., and De Witte, L. P. 2009. The impact of electronic mobility devices for persons who are visually impaired: A systematic review of effects and effectiveness. *Journal of Visual Impairment and Blindness*, 103 (11), 743–753. https://doi.org/10.1177/0145482X0910301104.

Roentgen, U. R., Gelderblom, G. J., and de Witte, L. P. 2011. Users' evaluation of four electronic travel aids aimed at navigation for persons who are visually impaired. *Journal of Visual Impairment and Blindness*, 105 (10), 612–623. https://doi.org/10.1177/0145482X1110501008.

Serrão, M., Rodrigues, J. M., Rodrigues, J. I., and du Buf, J. H. 2012. Indoor localization and navigation for blind persons using visual landmarks and a GIS. *Procedia Computer Science*, 14, 65–73. https://doi.org/10.1016/j.procs.2012.10.008.

Serrão, M., Shahrabadi, S., Moreno, M., José, J. T., Rodrigues, J. I., Rodrigues, J. M., and du Buf, J. H. 2015. Computer vision and GIS for the navigation of blind persons in buildings. *Universal Access in the Information Society*, 14 (1), 67–80. https://doi.org/10.1007/s10209-013-0338-8.

Siekierska, E., Labelle, R., Brunet, L., Mccurdy, B., Pulsifer, P., Rieger, M. K., and O'Neil, L. 2003. Enhancing spatial learning and mobility training of visually impaired people: A technical paper on the internet-based tactile and audio-tactile mapping. *Canadian Geographer/Le Géographe canadien*, 47 (4), 480–493. https://doi.org/10.1111/j.0008-3658.2003.00037.x.

Silva, C. G., Coelho, V., and Silva, M. A. R. 2017. Guideline for designing accessible systems to users with visual impairment: Experience with users and accessibility evaluation tools. In *Proceedings of the 19th international conference on enterprise information systems (ICEIS 2017)*. https://doi.org/10.5220/0006351301510157.

Söderström, S., and Ytterhus, B. 2010. The use and non-use of assistive technologies from the world of information and communication technology by visually impaired young people: A walk on the tightrope of peer inclusion. *Disability and Society*, 25 (3), 303–315.

Stampach, R., and Mulickova, E. 2016. Automated generation of tactile maps. *Journal of Maps*, 12 (supp. 1), 532–540. https://doi.org/10.1080/17445647.2016.1196622.

Strothotte, T., Petrie, H., Johnson, V., and Reichert, L. 1995. MoBIC: User needs and preliminary design for a mobility aid for blind and elderly. In *The European context for assistive technology: Proceedings of the 2nd TIDE congress*. Paris: IOS Press.

Strumillo, P. 2010. Electronic interfaces aiding the visually impaired in environmental access, mobility and navigation. In *3rd international conference on human system interaction, Rzeszow*. https://doi.org/10.1109/HSI.2010.5514595.

Strumillo, P., Bujacz, M., Baranski, P., Skulimowski, P., Korbel, P., Owczarek, M., Tomalczyk, K., Moldoveanu, A., and Unnthorsson, R. 2018. Different approaches to aiding blind persons in mobility and navigation in the "Naviton" and "Sound of Vision" projects. In E. Pissaloux and R. Velazquez (eds.) *Mobility of visually impaired people*, pp. 435–468. Cham: Springer.

Sullivan, T., and Matson, R. 2000. Barriers to use: Usability and content accessibility on the Web's most popular sites. In *Proceedings on the 2000 conference on universal usability*. New York: Association for Computing Machinery. https://doi.org/10.1145/355460.355549.

Tajgardoon, M., and Karimi, H. A. 2015. Simulating and visualizing sidewalk accessibility for wayfinding of people with disabilities. *International Journal of Cartography*, 1 (1), 79–93. https://doi.org/10.1080/23729333.2015.1055646.

Tapu, R., Mocanu, B., and Zaharia, T. 2018. Wearable assistive devices for visually impaired: A state of the art survey. *Pattern Recognition Letters*. https://doi.org/10.1016/j.patrec.2018.10.031.

United Nations (UN). 2006. *Convention on the rights of persons with disabilities.* www.un.org/disabilities/defa ult.asp?id=259.

United States Access Board. 2020. *ADA standards.* www.access-board.gov/guidelines-and-standards/buil dings-and-sites/about-the-ada-standards/ada-standards.

US General Services Administration (GSA), Section 508.gov. 2019. *IT accessibility laws and policies.* www. section508.gov/manage/laws-and-policies.

Varma, R., Vajaranant, T. S., Burkemper, B., Wu, S., Torres, M., Hsu, C., Choudhury, F., and McKean-Cowdin, R. 2016. Visual impairment and blindness in adults in the United States: Demographic and geographic variations from 2015 to 2050. *JAMA Ophthalmology*, 143 (7), 802–809. https://doi.org/10. 1001/jamaophthalmol.2016.1284.

Völkel, T., Weber, G., and Baumann, U. 2008. Tactile graphics revised: The novel brailledis 9000 pin-matrix device with multitouch input. In *Computers helping people with special needs: ICCHP 2008.* Berlin and Heidelberg: Springer. https://doi.org/10.1007/978-3-540-70540-6_124.

Waddington, L., and Lawson, A. 2009. *Disability and non-discrimination law in the European Union: An analysis of disability discrimination law within and beyond the employment field.* https://kmd.al/wp-content/uploa ds/2019/06/Disablity-and-non-discrimination-law-in-EU.pdf.

Wang, G., Zheng, J., and Fan, H. 2017. Route guidance for visually impaired based on haptic technology and their spatial cognition. In *Data science: ICPCSEE 2017.* Singapore: Springer. https://doi.org/10. 1007/978-981-10-6385-5_60.

Wang, Z., Li, N., and Li, B. 2012. Fast and independent access to map directions for people who are blind. *Interacting with Computers*, 24 (2), 91–106.

Web Accessibility Initiative (WAI). 2020. *Web content accessibility guidelines (WCAG) overview.* www.w3. org/WAI/standards-guidelines/wcag/.

WebAIM. 2020. *The WebAIM Million: An annual accessibility analysis of the top 1,000,000 home pages.* Old Main Hill, UT: Center for Persons with Disabilities, Utah State University.

Wilson, J., Walker, B. N., Lindsay, J., Cambias, C., and Dellaert, F. 2007. Swan: System for wearable audio navigation. In *2007 11th IEEE international symposium on wearable computers.* Washington, DC, Brussels, Tokyo: IEEE. https://doi.org/10.1109/ISWC.2007.4373786.

World Health Organization (WHO). 2001. *International classification of functioning, disability and health: ICF.* Geneva: WHO.

World Health Organization (WHO). 2019. *World report on vision.* www.who.int/publications-detail/worl d-report-on-vision.

Worth, N. 2013. Visual impairment in the city: Young people's social strategies for independent mobility. *Urban Studies*, 50 (3), 574–586. https://doi.org/10.1177/0042098012468898.

Yánez, D. V., Marcillo, D., Fernandes, H., Barroso, J., and Pereira, A. 2016. Blind guide: anytime, any-where. In *Proceedings of the 7th international conference on software development and technologies for enhancing accessibility and fighting info-exclusion.* New York: Association for Computing Machinery. https://doi.org/ 10.1145/3019943.3019993.

Zeng, L., and Weber, G. 2010. Audio-haptic browser for a geographical information system. In *International Conference on Computers for Handicapped Persons.* Berlin and Heidelberg: Springer. https://doi.org/ 10.1007/978-3-642-14100-3_70.

Zeng, L., and Weber, G. 2011. Accessible maps for the visually impaired. In *Proceedings of IFIP INTER-ACT 2011 Workshop on ADDW, CEUR.* http://ceur-ws.org/Vol-792/.

Zeng, L., Kühn, R., and Weber, G. 2017. Improvement in environmental accessibility via volunteered geographic information: A case study. *Universal Access in the Information Society*, 16 (4), 939–949. https:// doi.org/10.1007/s10209-016-0505-9.

Zhao, H., Plaisant, C., and Shneiderman, B. 2005. "I hear the pattern" Interactive sonification of geo-graphical data patterns. In *CHI'05 extended abstracts on human factors in computing systems.* New York: Association for Computing Machinery. https://doi.org/10.1145/1056808.1057052.

Zimmermann-Janschitz, S. 2018. Geographic information systems in the context of disabilities. *Journal of Accessibility and Design for All*, 8 (2), 161–192. https://doi.org/10.17411/jacces.v8i2.171.

Zimmermann-Janschitz, S. 2019. The application of geographic information systems to support wayfinding for people with visual impairments or blindness. In *Visual Impairment and Blindness.* London: IntechO-pen. https://doi.org/10.5772/intechopen.89308.

Zimmermann-Janschitz, S., Mandl, B., and Dückelmann, A. 2017. Clustering the mobility needs of per-sons with visual impairment or legal blindness. *Transportation Research Record*, 2650 (1), 66–73. https:// doi.org/10.3141/2650-08.

PART II
Mass media

7

NEWSPAPERS

Geographic research approaches and future prospects

Paul C. Adams

[T]he real environment is altogether too big, too complex, and too fleeting for direct acquaintance. We are not equipped to deal with so much subtlety, so much variety, so many permutations and combinations. And although we have to act in that environment, we have to reconstruct it on a simpler model before we can manage it.

(Lippmann 1965 [1922], 11)

Thus did renowned journalist and political commentator, Walter Lippmann, justify the need for newspapers. In *Public Opinion*, he provided a defense of journalism as a tool bringing information to democratic citizens and thereby making democracy possible. But the quote above raises questions. What kind of assumptions lie behind his preferred "simpler model" of reality, who are the "we" who simplify the world, and who are (the presumably different) "we" who cannot deal with too much subtlety and variety?

Notwithstanding the questions a century of social theory impels us to ask, Lippmann nonetheless touched on something that needs to be kept in mind. Despite all its flaws, the newspaper serves vital social roles. Ideally, newspapers function as what Thomas Carlyle called the "fourth estate" (2019 [1841]), raising public awareness and bringing public opinion to bear on elected leaders. It has even been argued that "the public cannot come to know itself or defend its interests without journalists" (Iggers 1999, 141). However, neither of these observations ensures that newspapers will defend or define public interest in ways that are just and inclusive. Lippmann did, in fact, recognize that news functions as propaganda, coining the term "manufacture of consent," a fortuitous term that would later be taken up by Edward Herman and Noam Chomsky in their classic *Manufacturing Consent* (1988). Herman and Chomsky's propaganda model showed how corporate ownership, need for advertisers, and dependency on information sources push newspapers to protect the interests of the rich and powerful, though in theory they may be free of government control. Indeed, "the public" addressed by journalists can be envisioned in a way that is narrow and bigoted (e.g. Carlyle 1849).

Journalism serves not only to disseminate information but also to support public opinion, to define the public, and to shape relations between the public and authorities (de Haan et al.

DOI: 10.4324/9781003039068-9

2014, 212). Thus, newspapers have a special social role that distinguishes them from other media: they not only assume the role of translating complex phenomena into simplified stories for consumption by the masses, but they also help to define those masses in ways that are more or less broad-minded, more or less inclined to "manufacture consent," and more or less supportive of genuinely inclusive democratic processes.

These issues have long been recognized by geographers, albeit in somewhat piecemeal fashion, in work that alternately targets, depends on, critiques and examines newspapers. First, newspapers have been treated as a venue to convey geographical research to the public. Indicative of this appeal are the occasional op-ed pieces and letters to the editor from professional geographers. Second, newspapers are used as sources of data for empirical research, furnishing archival data from distant places and times as well as from the societies in which geographers live and work. The third focus of geographic interest is critical and theory-based: newspaper articles, editorials, photos and cartoons can all be critiqued as examples of representational strategies, or *frames*, revealing how power relations shape public discourses. Fourth, newspapers can be treated as social actors that actively intervene in social processes and relations.

Newspapers therefore offer much to the discipline of geography: a venue for findings, a source of data, a source of representations to critique, an actor within the social arena. The chapter therefore is structured around these four strands of research, considering each in turn. After that we look at the transformation of the newspaper toward a post-paper medium—a process rendering "newspaper" an archaic term, as paper tabloids and broadsheets give way to news services inhabiting a digital environment and tied to the imperatives of digital surveillance.

Newspapers for academic outreach

One rarely hears calls for geographers to promote their findings on Facebook or Twitter, to make feature-length movies or musical recordings about their research, or translate their findings into modern dance. Most popular media are seen as too trivial, too complex, or too specialized to assist in disseminating geographical insights. A bestselling book would of course be a fantastic way to broaden the geographic audience for almost any geographic topic (Murphy 2006, 1), but it can take years to write a book, and few geographers master the trick of writing bestselling literature. Newspapers are therefore rather distinct among media as appropriate and accessible means for academics to conduct outreach, with op-ed pieces and letters to the editor helping to deliver geography so it is "understandable, relatable, and useful to a larger world" (Alderman 2017). This is not to say that geographers have an easy time conquering their insecurities about stepping into the public eye, or that it is easy to ignore institutional disincentives to such outreach, make time for it, or cleanse a reputation that has been muddied in public debates (Alderman 2004; Moseley 2010). In any case, newspapers are a favored venue for geographers seeking to face these challenges and reach a broader public.

Compared to other media, newspapers present particular opportunities for outreach. Unlike the social media, newspapers impose strict gatekeeping (Shoemaker et al. 2009). While this means accepting the imposition of editorial authority, a top-down power that must be viewed critically, it does have the benefit of filtering out ad-hominem responses and "flaming." Likewise a decent newspaper can filter out the more virulent forms of nationalism and racism currently affecting public culture while offering more than television's typical sound bite (Alderman & Inwood 2019). Newspapers therefore offer opportunities to develop, shape, and control the channeling of geographical information to the public, and whereas writing an op-ed piece or letter to the editor is vastly different than writing an academic article, at least the adaptation does not involve learning to use a mass of new techniques and technologies.

Newspapers as data sources

Newspapers are often treated by researchers as windows into particular places and times. They profess to tell what happened, who was involved, and what they said and did. For example, geographers seeking a glimpse of Southeast England in the 18th century, Melbourne, Australia in the 19th century, Yorkshire in the early 20th century, Denmark in the late 20th century or Arab countries in the early 21st century have found valuable secondary data about those times and places in newspapers (Wren 2001; Falah et al. 2006; Griffin 2006; Rycroft & Jenness 2012; Coleman 2014). Even if we acknowledge that newspapers are not transparent windows on the world but tools for manufacturing consent, we can still look carefully through these distorted journalistic lenses and draw some valid conclusions regarding what is, or was, actually "out there" in the world at a particular place and time. Newspapers provide a sense of day-to-day life: conflicts that were occurring, details from the lives of particular people, and evidence about how various actors situated themselves in their respective societies, landscapes, geopolitical arenas and environments.

Among the materials that are regularly archived, newspapers hold a special place as geographical data archives, in part because they provide readily available data, whether preserved on microfilm, or microfiche, or in digital format. In addition, and perhaps more importantly, under "print capitalism" (Anderson 1991) the newspaper serves as a record of history. A newspaper folds the worldviews and events of a particular place and time into the world of its readers. In the words of Benedict Anderson: "the very conception of the newspaper implies the refraction" of all events, "even 'world events' into a specific imagined world of vernacular readers" (1991, 63). The audience recognizes itself as a public through the newspaper, and the paper's perspective links a public or publics to a shared national story. Conversely, we can understand publics as social formations constituted through shared practices of reading, viewing or listening, so newspapers help define these publics. "Publics differ from nations, races, professions, or any other groups that, though not requiring co-presence, saturate identity... Merely paying attention can be enough to make you a member" (Warner 2002, 53).

Geographical research has acknowledged the tricky relationship between representation and reality in newspaper accounts. Griffin (2006, 39) argues, for example, that "extant provincial newspapers are by far the most important source in researching histories of popular protest," while explicitly noting how they are biased in terms of their geographical coverage and angle. Falah et al. explain that portrayals of the United States by newspapers of Arab states are "carefully calibrated" to balance "a country's own national political interest and its geopolitical and geoeconomic ties to Washington on the one hand, and the pressure emanating from Arab political and civil society, both national and transnational, on the other" (2006, 152). Coleman (2014) documents how a newspaper editor publicly branded a social activist as an "irrepressible busybody," marginalizing her with gendered and class-based stereotypes. "Played out in the columns of the local daily newspapers... debates were an important site in which ideological constructs of gender and class were contested and reinscribed" (Coleman 2014). Newspapers often take part in such efforts to delegitimize and demonize women, as well as minorities and other marginalized groups, through stigmatizing labels and techniques of Othering (Wren 2001, 156). Journalistic versions of historical reality are repeatedly drawn into these forms of positionality. However, not all journalistic opinions are regressive or reactionary; geographers have also documented newspapers undermining colonial narratives (Rycroft & Jenness 2012, 961). We would do well to recall J.K. Wright's injunction that "we are by no means deluded by all of our illusions" (1947, 8); news can be an illusion with real connections to the place and time of its origin. So the historical-geographical "truth" conveyed by a newspaper is inevitably tainted and twisted, but nonetheless essential to geographers and other scholars.

Critical analyses of newspaper texts

If the previous section suggests that geographers must struggle to look *through* newspapers at the world, this section turns to a similar but distinct strand of geographic research—work that looks *at* newspapers. We shift the focus to relationships between power and knowledge. This approach is designed to expose and disrupt exclusionary notions of the public, and this critically informed approach is the most common way in which geographers have treated newspapers over the past two decades. This work benefits from engagement with various theoretical frameworks, including content analysis and frame theory (e.g. Entman 1991; 1993; Gamson et al. 1992), studies of discursive formations and governmentality (Foucault 1991; 2003), and geographical theories of power and representation (e.g. Ó Tuathail 1996a; 1996b; Rose 2001). Some critical news analysis involves measuring media content based on predetermined categories (content analysis), while some involves interpreting the meanings embedded in texts and images (semiotic theory, textual analysis and discourse analysis).

The most accessible research in this critical, theoretical vein focuses on newspaper representations of a particular place. For example, Potter studied representations of Haiti in five leading US newspapers over a 12-month period, showing that they constructed Haiti as a mélange of fantasy and absurdity: a realm of voodoo, zombies, mermaids and mud-eaters, employing stereotypes that have had "tremendous impact on the people of that particular location and can perpetuate or even legitimize social inequalities" (Potter 2009, 210). Likewise a study of Israeli newspaper coverage of peripheral towns in Israel found that newspapers viewed these towns as lawless, backwards, boring, abnormal and ugly, all of which succeeded in Othering both the towns and their inhabitants (Avraham & First 2006, 75, 81). Othering is a gaze that indulges in watching, is even captivated by the spectacle, but is diametrically opposed to a regard of mutual respect. The Othering gaze is desiring and condescending, infatuated and patronizing all at once. This gaze dominates portrayals of distant places (Duncan et al. 1999), for example typifying coverage of tourist destinations and war zones, but it can also be directed toward a newspaper's hometown or city. For example, 20th century Brazilian newspapers Othered favelas although these impoverished urban areas were close at hand: "including or excluding favelas from their maps, newspapers often represented these areas through divisions of 'modern' and 'backward,' 'legal' and 'illegal'" (Novaes 2014, 202). The dynamics of Othering have preoccupied geographers who view newspapers critically, and these dynamics position the audience in a way that draws a sharp line between "us" and "them," here and there, even when the linear distance between here and there is relatively short.

Another way in which newspapers engage in Othering is by promoting what Ó Tuathail (2002) called "geopolitical scripts." Geopolitical discourses in the news weave together stories of immigration, domestic politics, international conflict, trade and science, among other things. Geopolitical scripts guided coverage of the 1999 World Trade Organization (WTO) Ministerial Conference in Seattle by Australia's nationwide daily newspaper, *The Australian*, demonstrating the many ways "by which heterodox understandings are delegitimized" (McFarlane & Hay 2003, 228). Geopolitical uses of newspapers include visual components of the package. Cosgrove and della Dora (2005, 375) showed how maps published during World War II in the *Los Angeles Times* drew on the visual languages of various media to justify US expansionism. Conservative newspapers customarily frame immigrants as "terrorists," "criminals," and freeloaders "swamping" the nation, and even seemingly innocuous features like cartoons can work to "reinforce preconceived ideas and the editorial line of the newspaper" (Robson 2019, 119). The topics of critical news analysis can be retrospective, like a study of Cold War framing devices in newspaper coverage of Yucca Mountain, a Nevada site designated for storage of radioactive nuclear waste (Larsen & Brock 2005), or they can look to more recent events, like newspaper coverage of the Mars Pathfinder mission (Dittmer 2007). What runs through these various geographical investigations—with topics spanning most of a century, addressing diverse geopolitical situations, and focusing on everything from text to maps to photos to cartoons—is an interest in how journalism reinforces dichotomous ways of thinking, creating excuses, justifications and alibis for state projects. A subordinate theme is the disruption of such dominant discourses, for example articles and editorials that directly challenge dominant geopolitical frames or adopt alternative frames (Larsen & Brock 2005, 531–534). It is of geographical interest when articles and editorials give a sense that the established geopolitical order is breaking down (Farish 2001). In these various ways, newspapers are key players in the circulation of powerful geopolitical messages.

Newspapers are also studied as sources of environmental representations. For example Boykoff considers how newspapers represent and misrepresent the "complex and non-linear relationships between [climate] scientists, policy actors and the public" (Boykoff 2007, 478) by looking at "forms and content of the texts (such as headlines, framing techniques, salience of elements, ideological stances, tone and tenor, and relationships between clusters of messages)" (2008, 556). His samples include leading US newspapers and British tabloid papers from the last decade of the 20th century and the early 21st. Similarly combining content analysis with qualitative textual analysis, DiFrancesco and Young (2010) identify a "profound disjuncture between images and text" in the climate change coverage of Canada's major national newspapers. In this work, we can see that the relevant concerns for geographic analysis include representations of stakeholders and climate but also details of newspaper reporting and the social context of such reporting. They critically examine a social process through news stories, while simultaneously scrutinizing the role of the news. The fact that newspapers are archived and indexed more systematically than other media affords opportunities to address the historical unfolding of environmental issues in a systematic way by gathering many years of data. For example, Wakefield and Elliott (2003) studied news reports and public perceptions of a ten-year environmental assessment (EA) process for an industrial waste landfill in Ontario, Canada, and Larsen and Brock (2005) studied newspaper articles about the Yucca Mountain nuclear waste storage facility over a 14-year period. Comby et al. (2019) conducted an ambitious longitudinal study of *Le Monde*'s coverage of the Rhône River over a 68-year period. In other cases, a historical comparison of newspaper coverage at one or two selected times may be productive. An example is Rashid's (2011) comparison of flood reporting on a 1950 and a 1997 flood in Manitoba. The different infrastructural and institutional responses to flood risk on the two dates were reflected in the coverage, but in both cases the journalists voiced opposition to Canadian government responses.

While the majority of geographers critiquing newspapers focus on constructions of geopolitical and environmental phenomena, other issues have caught attention, such as the journalistic construction of women and their lives. There are many ways in which news narratives perpetuate patriarchal relations and gendered risks, including blaming women for the social and environmental risks they face (Faria 2008), or constructing women as emblems of both perfection and abjection (Walker 2005). Clearly more work could be done in this area. Geographical critiques of journalistic representations therefore provide an important supplement to other ways of studying social relations, geopolitics, human-environment relations and spatialized gender dynamics.

Newspapers as social actors

Newspapers also occupy the geographic literature as social actors or forces. For example, in a study of Ontario media representations of offshore farm labor, Bauder acknowledges that: "While my focus is on the discursive representations of offshore workers, I recognize that these representations are recursively linked to the material practices in the agricultural economy"; in this way news coverage contributes to the legitimation of Canada's offshore program (Bauder 2005, 41, 42). Such comments point to the need to consider contextual aspects of news production and consumption as parts of a dialectic or cycle. The questions that need to be asked are no less about what newspapers *are* than about what they *do*. "As powerful agents of social change, the news media exercise considerable influence in the construction of public understanding of political issues through their power to mediate societal discourses" (Ette 2017, 1481). For example, women politicians are marginalized by the amount of coverage and the framing of stories, and their concerns are sidelined by this lack of coverage because "having media presence… expedites publicity" and conversely being overlooked by the news media consigns one to obscurity (Ette 2017, 1481). Ette's use of the term "publicity" recalls the ideas of Lippmann (1965 [1922], 11) and Iggers (1999, 141), reinforcing the conclusion that news serves as the basis for democratic processes.

Finney and Robinson (2008) examine newspaper framing of asylum debates, and their critical approach aligns their work with the studies discussed above. There is an additional element to this study: consideration of how the local press relates to its community. This approach emphasizes the role of newspapers in defining a local community, as opposed to Anderson's (1991) "imagined community" of the nation. The ideal of community is mobilized as a way to marginalize and exclude, seeking to disperse and dilute immigrants. Other research shows how a newspaper can defend the interests of a marginalized group, like the newspaper *Xtra! West* which did not just try to speak to and for Vancover's gay and lesbian community, but also mobilized to defend the LGBTQ community's urban territory and economic vitality (Miller 2005, 69). This indicates how a newspaper can be a social actor, shaping lives and creating altered conditions for various groups, building or tearing down diverse publics and spatial justice. This points us back to the writings of Michael Warner, whose arguments helped to formalize the political role of a newspaper: "Our willingness to process a passing appeal determines which publics we belong to and performs their extension. [when we take part in publics the] direction of our glance can constitute our social world" (Warner 2002, 62). It is possible to infer the social role of a newspaper through the positions expressed in the paper relative to certain publics, and the way the paper becomes caught up in social dynamics around those publics.

To sum up the above arguments: the study of media geography brings focused attention to the newspaper as a vehicle for popularizing geographical insights, it relies on newspapers as sources of secondary data, it treats newspapers as samples of culturally and politically inflected representations, and it views newspapers as actors taking their place among other powerful social actors. We turn now to the transformation of the newspaper and how that change might be followed by media geographers.

What is "a paper" when it is no longer paper?

As the newspaper is undergoing a rapid transformation from paper to screen, from ink to bits, and from the one-way dissemination of information to audiences to a two-way information flow to and from audiences, the *affordances* of the medium have changed. It becomes possible for the owners of news companies, and third parties they contract with, to collect detailed information about newsreaders. The information constituting this reverse flow back from news consumer to news producer is not just information on newsreading habits, but also information about the tendencies of particular people (or types of people) to share information with others, what readers buy, who they know, how they spend their free time and even what they believe. In fact, not only are digital newspaper companies capable of gathering surveillance information from newsreaders directly, but "the news" provides a seductive and platform for others (advertisers and so-called "third parties") to gather rich surveillance information about newsreaders and share such data with affiliated companies and subsidiaries. This situation compromises the value of newspapers as vehicles for popularizing geographical insights, complicating newspapers' usefulness as sources of secondary data, and altering the relationships between news and democratic civil society.

Another way to say all of this is that journalism in the digital age is driven increasingly by *metrics*—measurements of how the audience engages with digital content: moment by moment, click by click, news story by news story, ad by ad. Masses of newsreader data are automatically generated once any piece of news is served online (Diakopoulos 2016), and this data is refined into metrics with names like "bounce rate," "click-through rate," "concurrent visits," "conversion rate," "engaged time," "entry rate," and "pageviews per visit." These metrics are subsequently assembled into analytics (Cherubini & Nielsen 2016), summary measures that have revolutionized the routines of news production. While the scrutiny of policy-makers and the public has raised concern thus far about privacy issues with social media sites, the attention on social media is starting to be a bit misplaced since "browsing news-related websites actually exposes you to over twice as much tracking as the rest of the web" (Libert & Pickard 2015). Much of this tracking arises from the "cookies" installed on news sites—small bits of code created by the designers of the site or by third parties (e.g. advertisers and special interest groups) to allow readers and their online activities to be tracked across various sites. More third party cookies are found on newspaper sites than on websites of any other type, including sports sites, game sites, shopping sites, recreation sites and adult sites (Englehardt & Narayanan 2016). A typical newspaper page designed for online reading contains about 40 third party cookies and, disturbingly, the majority of these cookies exchange data with Facebook (Libert et al. 2018, 3, 5). The line between social media and the digital reincarnation of "the newspaper" is exceedingly thin.

Various studies have confirmed that the shift to digital distribution is profoundly reshaping journalistic practices (Podger, 2009; Vu, 2014; Cherubini & Nielsen 2016; Ihlebaek & Larsson 2016; Larsson, 2017; Belair-Gagnon & Holton 2018; Zamith, 2018). It has also been clearly shown that journalists are coming to view the nature of the audience and their own roles in new ways (Anderson 2011; Hanusch & Tandoc 2017; Peruško et al. 2017; Ferrer-Conill & Tandoc 2018). At first glance, these changes seem to shift power away from journalists and editors to readers or audiences, since the vicissitudes and minutiae of audience attention are used to fine-tune the presentation and layout of news stories (Karlsson & Clerwall 2013, 65). The "people who not long ago were called an audience" (Pisani 2006, 44) now engage with news in ways that parallel social media use; they become, in a word, "produsers" (Bruns 2004; Pentina & Tarafdar 2014). This can seem like empowerment of the reader. But the reader is being read (surveilled, analyzed, followed, typed) and if news production and consumption are converging with social media production and consumption, then all of this prompts concern regarding who ultimately benefits from the supposed "democratization" of information flow.

As activity on a digital newspaper site is monitored, measured, and fed back into news production in a cycle reminiscent of social media development (Dwyer & Martin 2017), there are implications no less problematic than the problems already well known regarding the diffusion and popularization of social media. News metrics and analytics are created above all because they help recapture vanishing advertising revenues: "the most consequential audience construction takes place at the financing and distribution nodes" (Turow & Draper 2014, 643). This means that insofar as metrics create a picture of the audience, that picture is a projection of the needs of advertisers, and hence metrics "serve as currencies for most digital advertising—clicks, pageviews, and unique users" (Cherubini & Nielsen 2016, 38). Thus, while journalism professionals have always depended on simplified ideas of "the audience," metrics create a radically new dynamic insofar as audience members are reduced to hot buttons, and these profiles of manipulability are sold to the highest bidder as a way to micro-target ad appeals and political propaganda.

News metrics shore up the promise that digital newspapers can "navigate an ever-more competitive battle for attention" (Cherubini & Nielsen 2016, 7) in the world of convergent digital media. They are driven by the same logic that drives social media—the extraction of "behavioral surplus" (Zuboff 2019, 8), the latent value of surveilled and analyzed human actions to third parties who would seek to manipulate those actions. To subscribe to a digital newspaper, one must click "agree" on policies that allow the sharing of information about online activities with newspaper companies, their affiliates, advertisers and third parties of various sorts (Adams 2020). The pooling of newsreader data often results in the ability to intersect detailed information about the attention patterns of millions of readers. News media parent companies, including hedge funds such as Alden Global Capital, benefit in complex and largely unexplored ways from this ability to follow the process of news consumption and the formation of attitudes and beliefs, while locating individual readers within the spectrum of opinions. This affords these companies an opportunity to manipulate public attitudes via custom-tailored news. In short, newspapers are evolving into a social force that is almost entirely divorced from its original journalistic function of supporting democracy and informing readers about the world.

Geographers will continue to turn to newspapers as the go-to source of information about places both near and far. However, this research-based engagement with "the news" will be well advised to take into account the constructedness of news and the ways in which newspapers have long served as social actors. Beyond this business as usual, there is an urgent need for geographers to recognize that newspapers have undergone a radical transformation; they have new and largely unexplored capabilities that bring them in line with other media serving surveillance capitalism. This demands a careful re-thinking of the value of newspapers as venues for disseminating geographical findings, archives of historical-geographic data, samples of cultural-political representations, and actors within the social arena.

References

Adams, P. C. 2020. Agreeing to surveillance: Digital news privacy policies. *Journalism and Mass Communication Quarterly*, 97 (4), 868–889. https://doi.org/10.1177/1077699020934197.

Alderman, D. H. 2004. Toward a newsworthy cultural geography. *Journal of Cultural Geography*, 22 (1), 139–142.

Alderman, D. H. 2017. The serious business of public communication. *Newsletter of the American Association of Geographers*, 52 (August). http://news.aag.org/2017/08/the-serious-business-of-public-communication.

Alderman, D. H., and Inwood, J. F. J. 2019. The need for public intellectuals in the Trump Era and beyond: Strategies for communication, engagement, and advocacy. *The Professional Geographer*, 71 (1), 145–151.

Anderson, B. R. O'G. 1991. *Imagined Communities: Reflections on the Origin and Spread of Nationalism*. London: Verso.

Anderson, C. W. 2011. Between creative and quantified audiences: Web metrics and changing patterns of newswork in local US newsrooms. *Journalism*, 12, 550–566.

Avraham, E., and First, A. 2006. Media, power and space: Ways of constructing the periphery as the "other." *Social & Cultural Geography*, 7 (1), 71–86.

Bauder, H. 2005. Landscape and scale in media representations: The construction of offshore farm labour in Ontario, Canada. *Cultural Geographies*, 12 (1), 41–58.

Belair-Gagnon, V., and Holton, A. E. 2018. Boundary work, interloper media, and analytics in newsrooms. *Digital Journalism*, 6, 492–508.

Boykoff, M. 2007. From convergence to contention: United States mass media representations of anthropogenic climate change science. *Transactions of the Institute of British Geographers NS*, 32, 477–489.

Boykoff, M. 2008. The cultural politics of climate change discourse in UK tabloids. *Political Geography*, 27, 549–569.

Bruns, A. 2004. Reconfiguring journalism. In G. Goggin (ed.) *Virtual nation: The internet in Australia*, pp. 177–192. Sydney: UNSW Press.

Carlyle, T. 1993 [1841]. *On heroes, hero-worship and the heroic in history*. M. K. Goldberg, J. J. Brattin, and M. Engel (eds.). Berkeley, CA: University of California Press.

Carlyle, T. 1849. Occasional discourse on the negro question. *Fraser's Magazine for Town and Country*, Vol. XL, pp. 670–679. Collections Deposit Library Microforms, University of Texas at Austin.

Cherubini, F., and Nielsen, R. K. 2016. *Editorial analytics: How news media are developing and using audience data and metrics*. https://ssrn.com/abstract=2739328.

Coleman, J. 2014. Benevolent ladies and irrepressible busybodies: Contesting the bounds of "genuine" philanthropy. *Gender, Place & Culture*, 21 (9), 1071–1089. doi: 10.1080/0966369X.2012.731381.

Comby, E., Le Lay, Y.-F., and Piégay, H. 2019. Power and changing riverscapes: The socioecological fix and newspaper discourse concerning the Rhône River (France) since 1945. *Annals of the American Association of Geographers*, 109 (6), 1671–1690.

Cosgrove, D. E., and della Dora, V. 2005. Mapping global war: Los Angeles, the Pacific, and Charles Owens's pictorial cartography. *Annals of the Association of American Geographers*, 95 (2), 373–390.

De Haan, Y., Landman, A., and Boyles, J. L. 2014. Towards knowledge-centered newswork: The ethics of newsroom collaboration in the digital era. In W. N. Wyatt (ed.) *The ethics of journalism: Individual, institutional and cultural influences*, pp. 207–227. New York: I.B. Tauris in association with the Reuters Institute for the Study of Journalism, University of Oxford.

Diakopoulos, N. 2016. Computational journalism and the emergence of news platforms. In B. Franklin and S. A. Eldridge (eds.) *The Routledge companion to digital journalism studies*, pp. 176–184. London: Routledge.

Dittmer, J. 2007. Colonialism and place creation in Mars Pathfinder media coverage. *Geographical Review*, 97 (1), 112–130.

Duncan, J. S., Duncan, J., and Gregory, D. (eds.) 1999. *Writes of passage: Reading travel writing*. London and New York: Routledge.

Dwyer, T., and Martin, F. 2017. Sharing news online: Social media news analytics and their implications for media pluralism policies. *Digital Journalism*, 5, 1080–1100.

Englehardt, S., and Narayanan, A. 2016. Online tracking: A 1-million-site measurement and analysis. In *Proceedings of the 2016 ACM SIGSAC conference on computer and communications security*. New York: ACM.

Entman, R. 1991. Framing US coverage of international news: Contrasts in narratives of the KAL and Iran Air incidents. *Journal of Communication*, 41, 6–25.

Entman, R. 1993. Framing: Toward clarification of a fractured paradigm. *Journal of Communication*, 43, 51–58.

Ette, M. 2017. Where are the women? Evaluating visibility of Nigerian female politicians in news media space. *Gender, Place & Culture*, 24 (10), 1480–1497.

Faria, C. 2008. Privileging prevention, gendering responsibility: An analysis of the Ghanaian campaign against HIV/AIDS. *Social & Cultural Geography*, 9 (1), 41–73.

Farish, M. 2001. Modern witnesses: Foreign correspondents, geopolitical vision, and the First World War. *Transactions of the Institute of British Geographers*, 26 (3), 273–287.

Ferrer-Conill, R., and Tandoc, E. C. 2018. The audience-oriented editor. *Digital Journalism*, 6, 436–453.

Finney, N., and Robinson, V. 2008. Local press, dispersal and community in the construction of asylum debates. *Social & Cultural Geography*, 9 (4), 397–413.

Foucault, M. 1991. Politics and the study of discourse. In G. Burchell, C. Gordon and P. Miller (eds.) *The Foucault Effect*, pp. 53–72. Chicago, IL: University of Chicago Press.

Foucault, M. 2003 [1975/1976]. *Society must be defended: Lectures at the Collège de France 1975–1976.* London: Penguin.

Gamson, W. A., Croteau, D., Hoynes, W., and Sasson T. 1992. Media images and the social construction of reality. *Annual Review of Sociology*, 18, 373–393.

Griffin, C. J. 2006. Knowable geographies? The reporting of incendiarism in the eighteenth-and early nineteenth-century English provincial press. *Journal of Historical Geography*, 32 (1), 38–56.

Habermas, J. 1984. *The theory of communicative action 1: Reason and the rationalization of society.* Cambridge: Polity Press.

Habermas, J. 1985. *The theory of communicative action 2: Lifeworld and system: A Critique of functionalist reason.* Boston, MA: Beacon Press.

Hanusch, F., and Tandoc, E. 2017. Comments, analytics, and social media: The impact of audience feedback on journalists' market orientation. *Journalism*, 20, 695–713. doi:10.1177/1464884917720305.

Herman, E. S., and Chomsky, N. 1988. *Manufacturing consent: The political economy of the mass media*, 1st edn. New York: Pantheon Books.

Iggers, J. 1999. *Good news, bad news: Journalism ethics and the public interest.* Boulder, CO: Westview Press.

Ihlebæk, K. A., and Larsson, A. O. 2016. Learning by doing. *Journalism Studies*, 19, 905–920.

Karlsson, M., and Clerwall, C. 2013. Negotiating professional news judgment and "clicks": Comparing tabloid, broadsheet and public service traditions in Sweden. *Nordicom Review*, 34, 65–76.

Larsen, S. C., and Brock, T. J. 2005. Great Basin imagery in newspaper coverage of Yucca Mountain. *Geographical Review*, 95 (4), 517–536.

Larsson, A. O. 2017. The news user on social media: A comparative study of interacting with media organizations on Facebook and Instagram. *Journalism Studies*, 19, 2225–2242. doi:10.1080/1461670X.2017.1332957.

Libert, T., and Pickard, V. 2015. Think you're reading the news for free? New research shows you're likely paying with your privacy. *The Conversation* (November 6). http://theconversation.com/think-youre-rea ding-the-news-for-free-new-research-shows-youre-likely-paying-with-your-privacy-49694.

Libert, T., Graves, L., and Nielsen, R. K. 2018. *Changes in third-party content on European news websites after GDPR.* https://reutersinstitute.politics.ox.ac.uk/sites/default/files/2018-08/Changes%20in%20Third-Pa rty%20Content%20on%20European%20News%20Websites%20after%20GDPR_0.pdf.

Lippmann, W. 1965 [1922]. *Public opinion.* New York: The Free Press.

Miller, V. 2005. Intertextuality, the referential illusion and the production of a gay ghetto. *Social & Cultural Geography*, 6 (1), 61–79.

Moseley, W. G. 2010. Engaging the public imagination: Geographers in the op–ed pages. *Geographical Review*, 100 (1), 109–121. doi:10.1111/j.1931-0846.2010.00009.x.

Murphy, A. B. 2006. Enhancing geography's role in public debate. *Annals of the Association of American Geographers*, 96 (1), 1–13. doi:10.1111/j.1467-8306.2006.00495.x.

Novaes, A. R. 2014. Favelas and the divided city: Mapping silences and calculations in Rio de Janeiro's journalistic cartography. *Social & Cultural Geography* 15 (2), 201–225.

Ó Tuathail, G. 1996a. An anti-geopolitical eye: Maggie O'Kane in Bosnia 1992–93. *Gender, Place and Culture*, 3, 171–185.

Ó Tuathail, G. 1996b. *Critical geopolitics: The politics of writing global space.* Minneapolis, MN: University of Minnesota Press.

Ó Tuathail, G. 2002. Theorizing practical geopolitical reasoning: The case of the United States' response to the war in Bosnia. *Political Geography*, 21 (5), 601–628.

Pentina, I., and Tarafdar, M. 2014. From "information" to "knowing": Exploring the role of social media in contemporary news consumption. *Computers in Human Behavior*, 35, 211–223.

Peruško, Z., Čuvalo, A., and Vozab, D. 2017. Mediatization of journalism: Influence of the media system and media organization on journalistic practices in European digital mediascapes. *Journalism*, 1–25. doi:10.1177/1464884917743176.

Pisani, F. 2006. Journalism and Web 2.0. *Nieman Reports*, 60, 42–44. https://niemanreports.org/articles/journalism-and-web-2-0.

Podger, P. J. 2009. The limits of control. *American Journalism Review*, 31, 32–37.

Potter, A. E. 2009. Voodoo, zombies, and mermaids: US newspaper coverage of Haiti. *Geographical Review*, 99 (2), 208–230.

Rashid, H. 2011. Interpreting flood disasters and flood hazard perceptions from newspaper discourse: Tale of two floods in the Red River valley, Manitoba, Canada. *Applied Geography*, 31 (1), 35–45.

Robson, M. 2019. Metaphor and irony in the constitution of UK borders: An assessment of the "Mac" cartoons in the *Daily Mail* newspaper. *Political Geography*, 71, 115–125.

Rose, G. 2001. *Visual methodologies*. London: Sage.

Rycroft, S., and Jenness, R. 2012. J. B. Priestley: Bradford and a provincial narrative of England, 1913–1933. *Social & Cultural Geography* 13 (8), 957–976. doi:10.1080/14649365.2012.731700.

Shoemaker, P. J., Vos, T. P., and Reese, S. D. 2009. Journalists as gatekeepers. K. Wahl-Jorgensen and T. Hanitzsch (eds.) *The handbook of journalism studies*, pp. 73–87. London and New York: Routledge.

Turow, J., and Draper, N. 2014. Industry conceptions of audience in the digital space: A research agenda. *Cultural Studies*, 28, 643–656. doi:10.1080/09502386.2014.888929.

Vu, H. T. 2014. The online audience as gatekeeper: The influence of reader metrics on news editorial selection. *Journalism*, 15, 1094–1110.

Wakefield, S. E., and Elliott, S. J. 2003. Constructing the news: The role of local newspapers in environmental risk communication. *The Professional Geographer*, 55 (2), 216–226.

Walker, M. A. 2005. Guada-narco-lupe, Maquilarañas and the discursive construction of gender and difference on the US–Mexico border in Mexican media re-presentations. *Gender, Place & Culture*, 12 (1), 95–111.

Warner, M. 2002. *Publics and counterpublics*. Cambridge, MA: MIT Press.

Wren, K. 2001. Cultural racism: Something rotten in the state of Denmark? *Social & Cultural Geography*, 2 (2), 141–162.

Wright, J. K. 1947. Terrae incognitae: The place of the imagination in geography. *Annals of the Association of American Geographers*, 37 (1), 1–15.

Zamith, R. 2018. Quantified audiences in news production. *Digital Journalism*, 6 (4), 418–435.

Zuboff, S. 2019. *The age of surveillance capitalism: The fight for a human future at the new frontier of power*. New York: Hachette/PublicAffairs.

8

FAKE NEWS

Mapping the social relations of journalism's legitimation crisis

James Compton

What is the relationship of truth to news? This simple question is at the heart of much handwringing and debate among journalists, politicians and news consumers. We are a long way off from the casual faith audiences placed in Walter Cronkite's signature CBS Evening News sign off: "And that's the way it is." Cronkite was an iconic broadcaster who enjoyed widespread trust that no contemporary newsreader could reasonably claim. Today's polarized media landscape, by contrast, is fractured along deep political, economic and social cleavages. Since the 2016 American presidential election, much of the blame has been placed on so-called fake news. The initial spark for the debate was an influential report published by the online magazine Buzzfeed about how Macedonian teenagers in the small town of Veles were making money fabricating pro Donald Trump stories for pseudo-news websites with names such as USConservativeToday.com, and USADailyPolitics.com. The stories were created specifically to be shared by users of hyper-partisan Facebook pages (Silverman & Alexander 2016). The report dripped with irony, given that Buzzfeed had cut its teeth pioneering sharable "clickbait." And shortly following Trump's election victory concern spread after a North Carolina man was arrested for walking into a popular Washington, DC pizza restaurant carrying an assault rifle. The man told police he was investigating a false conspiracy theory circulating during the 2016 election that the restaurant was linked to a pedophile trafficking ring involving Democratic presidential candidate Hillary Clinton (Siddiqui & Svrluga 2016). The circulation of fake news was having dangerous real-world consequences.

Donald Trump famously embraced the term conferring the label on news organizations he disliked, such as CNN and the *New York Times*. Others would follow. The phrase became commonplace, used by authoritarian leaders, conspiracy theorists and satirical comedians around the world. The term's imprecision has lent itself to the description of a range of activities—from political propaganda, to misinformation and comedy (Tandoc Jr. et al. 2018). Perhaps most importantly, it is intricately connected to a moral panic concerning the legitimacy of journalism and liberal democracy (Walsh 2020). Anders Hofseth, writing for the Norwegian public broadcaster NRK, captured the concerns of many when he argued that fake news presents journalism and liberal society with a crisis of trust:

DOI: 10.4324/9781003039068-10

When the audience is trained to doubt everything they meet in the news, it may lead to devaluation and destabilization of society's system for information, and a vacuum might appear. This poses a threat not only to the media itself. It is challenging the entire structure of society.

(Hofseth 2017)

This chapter provides a critical social mapping of fake news, its historical context and implications for democracy. It situates current debates about fake news and concerns that we have entered a "Post-Truth" era (McIntyre 2018) within the broader political economy of accelerated commodity production and circulation. It discusses the flexible accumulation strategies of networked news organizations struggling to find new business models amidst audience fragmentation, advertising losses and massive job shedding. News media, formerly oriented to local geographic areas, are increasingly financialized institutions that now answer to hedge funds for whom loyalty to a community or place is downgraded or non-existent (Abernathy 2016). It argues that the fake news controversy is connected to a weakened social consensus and lack of trust in mainstream institutions, such as journalism, following the financial collapse of 2008. It details how the social space once occupied by mainstream institutions of government and media has been challenged and fragmented by promotionally interested social actors of various political stripes who use their knowledge of networked media and platforms to advance their interests via the circulation of misinformation, competing truth claims and "clickbait."

Fake news origin stories

While the term fake news gained global prominence following the 2016 American presidential campaign, its provenance can be traced back hundreds of years. In 1755 news of the Lisbon earthquake carried rumors the disaster was a form of divine retribution, prompting philosopher Voltaire to denounce religious explanations for natural disasters (Soll 2016). Tricks of the trade were refined in the early 19th century era of the penny press, in a transparent ploy to boost circulation. On August 25, 1835 the *New York Sun* newspaper published the first in a series of six sensational feature stories about the discovery of life on the moon. The articles were given the byline of Dr. Andrew Grant, said to be a colleague of a famous contemporary astronomer of the era Sir John Herschel. The stories featured graphic pictures of the discoveries made by Grant using a powerful telescope at his observatory on the Cape of Good Hope, South Africa, including large bi-pedal beavers and furry humanoid creatures with batwings. It was complete fantasy, but many readers, in an age of geographic exploration, were said to have been fooled as sales spiked. Close to a month later the *Sun* confessed to the hoax (Zielinski 2015), but sales of the paper did not suffer after the deception was revealed. Indeed, the sensationalist narrative style of the penny press was later extended into the infamous "yellow journalism" wars waged by Joseph Pulitzer's *New York World* and William Randolph Hearst's *New York Journal*. The term referred to the popular Yellow Kid comic strip that first appeared in the *New York World*, before its creator was wooed away to Hearst's *New York Journal*. It was then that the term "yellow journalism" became associated with their newspaper rivalry. The style, featuring bold typography, multicolumn headlines and dramatic illustrations has continued today in the pages of supermarket tabloids where sightings of space aliens and Elvis have become routine.

War reporting has often harboured false or misleading coverage. Hearst's *Morning Journal* notoriously stretched the truth to help launch the 1898 Spanish-American War. In 1964 news media repeated false claims by White House officials that a US destroyer had been attacked by North Vietnamese forces in the Gulf of Tonkin. The incident was used by American President Lyndon B. Johnson to ask Congress to authorize the use of force in the region, escalating the Vietnam War. Four decades later in 2003, the administration of President George W. Bush would tout the imminent threat from weapons of mass destruction held by Iraqi strongman Saddam Hussein to justify the invasion of Iraq. No weapons were found.

Errors in war reportage are often dismissed as products of "the fog of war." However, writing in 1925, senior Associated Press Editor Edward McKernan took a different view. He raised the well documented (Bloch 2013 [1921]) spectre of "fake news" circulating from the battlefields of Europe during World War I, blaming the heightened danger on the increased speed of news dissemination brought about by new global telegraph networks. "Advantage has been taken of every device of wit and science to speed up the report until the swift transmission of news is in itself a source of unprecedented danger" (McKernan 1925, 529). Because of the speed of these international networks, argued McKernan, news "explodes with a bang and its echoes are heard in the four corners of the earth" (p. 530). But what made McKernan particularly worried was the rise of individuals motivated to deliberately take advantage of the accelerated pace of the news file to misinform the public for the purposes of money, publicity or propaganda. He labelled these so-called "arch enemies of the Public and the Press... the Market Rigger, the News Faker and the Professional Propagandist" (p. 533).

All these examples are forms of what John B. Thompson calls "mediated quasi-interaction"—types of mass communication that, unlike face-to-face communication, or talking on the phone, are monological in character in that they are directed in one direction to an indefinite range of individuals "stretching social relations across space and time" (Thompson 2020, 5). Thompson, importantly, directs our attention to the social relations involved in different forms of mediated communication. Changes and innovations in printing technology in combination with large-scale, profit-driven national news agencies fundamentally reshaped 19th century social space. The accelerated flow of the news commodity made possible by electronic telegraphs profoundly remodeled the experience of space and time. As Anthony Giddens argues: "In conditions of modernity, place becomes increasingly *phantasmagoric*: that is to say, locales are thoroughly penetrated by and shaped in terms of social influences quite distant from them" (Giddens 1990, 19). The constraints of space that had historically separated people and events had been overcome through the swift delivery of news—whose value was intricately connected to time. Being first mattered. This process of "time-space compression" (Harvey 1989, 241) reconfigured the relationship of the local to the global. The production of this new mediated space was not neutral; it was from the beginning implicated in relations of control and power (Lefebvre 1991). "News agencies created the electronically mediated relationship between places in different parts of the world. However, these places had different exchange values depending on transmission time... and their location" (Rantanen 1997, 618). News agencies produced "phantasmagoric" stories of distant others from exotic locales favoring largely urban audiences located in cities such as London, Paris, Berlin and New York. This reshaping of space through mediated electronic communication was constitutive of the material interests of news agencies, reflecting hierarchical and colonial relationships, but they also created opportunities and constraints for action involving individuals who had the necessary skills and opportunity to exploit the accelerated global flow of information. These included McKernan's Market Rigger, News Faker and Professional Propagandist.

In this way, following Thompson, "the use of communication media involves the creation of new forms of action and interaction, new kinds of social relationships and new ways of relating to others and to oneself" (Thompson 2020, 4). This social dynamic continues to the present day, in the persons of Macedonian bloggers and far right online promoters of the QAnon conspiracy theory—a bizarre tale that falsely claims the world is run by a satanic pedophile ring led by Hillary Clinton and the so-called deep state. Memes such as this are examples of what Manuel Castells calls "mass self communication" (Castells 2009, 249), and have become constitutive of the production, circulation and consumption of the fake news commodity, by anonymous users on social media and President Trump himself who has retweeted QAnon many times (McIntire et al. 2019).

In what follows, we will contextualize the most recent iteration of fake news by situating the development and use of digital technology within a broader political economy that remains committed to the acceleration of commodity production, distribution and exchange. As David Harvey (1989) and Bob Jessop (2009) argue, the uneven spatio-temporal dynamics of capitalist globalization require spatio-temporal fixes to overcome tensions and crises in the flow of capital. A case in point is the overproduction crisis of the 1970s; it was overcome, in part, through the introduction of information and communication technologies that allowed for newly networked global supply chains. The elimination of space through the acceleration of time continues today with investments in internet and social media technologies—tools used to integrate world markets and accelerate the flow of commodities and financial capital. Importantly, this cycle of acceleration carries over into the cultural realm affecting the pace of everyday life and the rate of social change (Rosa 2009), "reshaping experience and perception" (Crary 2013, 39). The phenomenon of fake news, it will be argued, is one of the effects.

Journalism, change and uncertainty

To suggest journalism has a legitimation crisis is not novel. In fact, one might suggest the "craft" has long been dogged by uncertainty and criticism (Blumler & Gurevitch 1995; Tong 2018). To understand the historical conjuncture, Peter Dahlgren (2009) offers five key developments that have altered media geography creating uncertainty and change: proliferation, concentration, deregulation, globalization and digitization.

We are "awash" in media as content and format choices proliferate via cable, website and social media offerings (Dahlgren 2009, 35). Despite this proliferation of media, corporate mergers and acquisitions and other cooperative ventures have continued. Media behemoths, such as News Corporation, sit atop of integrated empires that encompass all forms of media, from print, to broadcast and internet. With this concentration comes a tension between the drive for profits and longstanding professional traditions of ethical behavior. Cyberspace is not immune from these forces. Google and Facebook now control more than 75 percent of the online advertising market (Abernathy 2018).

On the policy front, neoliberal deregulation and privatization have been ascendant. These policies have supported corporate concentration, but they have also enabled the financialization of the overall economy, including journalism. The repercussions have been enormous. The role of the finance sector has swelled around the world over the past couple of decades (Winseck 2010) with banks and hedge funds increasingly taking leadership roles.

Into this uncertainty is added the digitalization of media. Some critics (Benkler 2006; Shirky 2009) point to the rise of blogging and citizen journalism to suggest that the old hierarchies of so-called legacy media are giving way to a democratized mediascape of do-ityourself journalism. In the words of Shirky: "here comes everybody." Notwithstanding the welcome optimism, the utopian marketplace of ideas did not materialize, and instead we face the challenge of a moral panic over fake news.

Newsroom rationalization and the production of news deserts

The global proliferation of media, concentration of ownership, deregulation, globalization and digitalization have created many stress points and contradictions, perhaps none more visible than for local newspapers. Daily newspapers, boasting newsrooms significantly larger than their broadcast cousins, have traditionally been the bedrock of news production, with TV and radio following their lead. Newspapers had earned a reputation for their health and profitability, enjoying double-digit profit margins on the strength of being a core community hub for local events, classified and general print advertising. Those days are over, and the numbers have been dire. The trend has occurred worldwide, but the experience in the United States is instructive. Between 2004 and 2018 journalists employed by newspapers in the United States were slashed by half while print advertising, still the main source of revenue for newspapers, plummeted to record low levels with the Facebook/Google duopoly dominating online ad buys (Abernathy 2018). One should be careful not to blame the abstraction of the internet or some unholy alliance of citizen bloggers. The digitalization of media happened in conjunction with the forces of corporate concentration, deregulation and globalization. Nowhere is this more apparent than in the financialization of newspapers.

In the wake of the 2008 Great Recession, newspaper revenue plunged leaving many local papers in distress (Doctor 2020a). It was at this time that private equity hedge funds saw an opportunity and entered the newspaper market purchasing hundreds of papers. By 2018, researcher Penelope Muse Abernathy would report that:

> five of the 10 largest newspaper chains were owned by hedge funds, private equity firms and other types of investment groups, which have vast portfolios of unrelated holdings such as real estate, financial services, international debt and health care companies.
>
> *(Abernathy 2018)*

New Media/Gatehouse, itself a subsidiary of Fortress, and owned by a large Japanese telecommunications conglomerate, went on a debt-fueled purchasing binge to eventually become the owner of over 154 daily newspapers operating in 39 states. Digital First Media, owner of *The Denver Post, Salt Lake City Tribune* and *Boston Herald*, would become the property of New York-based hedge fund, Alden Capital, a company with a large international portfolio including holdings in a large Canadian pharmacy chain (Abernathy 2018). Consolidation continued with several significant leveraged deals, and bankruptcies in late 2019 (Doctor 2020b). The biggest transaction was the merger of Gatehouse with Gannett, the owner of *USA Today*. New Media Investment Group outbid Alden Capital in a $1.4 billion deal to seize control of Gannett creating America's largest newspaper chain—constituting a quarter of all newspapers—with 260 daily newspapers in the United States, along with more than 300 weekly publications, in 47 states (Tracy 2019). That same year, Alden Capital became the controlling shareholder of Tribune Publishing, owner of the *Chicago Tribune*.

Newspaper consolidation by private equity hedge funds fundamentally reshaped the geography of local news creating regional tensions and contradictions between local and national markets. Newspapers serving local and regional communities were now answerable to globally integrated transnational capital. The new business model did not involve building newsroom capacity and serving readers. Instead, these companies deployed the same strategy they had used when purchasing other distressed properties in different sectors of the economy. They drastically cut costs by slashing editorial staff, engaged in aggressive financial restructuring, including bankruptcy, and sold off fixed capital, such as printing facilities and buildings located on prime real estate. The moves left newsrooms gutted and unable to cover traditional news beats such as city hall or state and provincial legislatures with any rigor. Close to 1800 newspapers closed in the United States between 2004 and 2018, resulting in the production of "news deserts," most often affecting poorer and more vulnerable populations. "In an era of fake news," wrote Abernathy, "the diminishment of local newspapers poses yet another threat to the long-term vitality of communities. Many of our 7,100 surviving newspapers are mere shells, or 'ghosts,' of their former selves" (Abernathy 2018). According to Ken Doctor, one of the most dedicated chroniclers of the financialization trend:

> The impact is obvious. As America has moved from jokey indulgences in truthiness to a point where fact fights for its very life, it's the bankers who are deciding what will be defined as news, and who and how many people will be employed to report it.
>
> *(Doctor 2019)*

Contradictions in uneven media space

While finance capital was stripping value from newspapers to meet short-term debt payments, and monopoly online platforms were draining local newspapers of advertising revenue, a group of elite media were enjoying a so-called "Trump Bump." CNN rode the spectacle of the Trump White House to the highest ratings and revenues in the network's history—clearing $1 billion in annual profits in 2018 (Pompeo 2018)—while the *New York Times* and the *Washington Post* saw enormous increases in their online subscriptions (Molla 2017). The contradiction between local media and larger players able to tap into national and international markets was striking. While CNN, the *New York Times* and the *Washington Post* regularly bemoaned Trump's targeting of the media as "fake news," and accurately fact checked the president's many false statements through a series of well documented investigative scoops, their reporting on the never-ending Trump spectacle was also a boon for business. This reporting was a perfect fit for online engagement strategies of producing transposable, melodramatic stories designed to efficiently repurpose content and reaggregate fragmented audiences among a propriety network of print and online properties (Compton 2004). Two successful examples of this approach were "The Daily" and "Can He Do That?" podcasts produced by the *New York Times* and *Washington Post* respectively. Both podcasts were created to provide summary and analysis of the maelstrom of scandals and controversies connected to the Trump White House, including an FBI investigation into alleged Russian interference in the 2016 election, and the use of fake social media accounts to sow distrust and dissension in the political system (Lukito et al. 2020). As such, the podcasts were part of the boundary work and paradigm repair (Carlson & Lewis 2015) effort among elite news organizations, including CNN and MSNBC, aimed at blunting White House criticism of their reporting. These meta-discourses also served to contrast newsrooms with Fox News, whose primetime hosts were widely criticized for their embrace of unfounded conspiracies and pro-Trump apologias concerning the President's impeachment and subsequent Senate trial on charges of abuse of power and obstruction of Congress (O'Neil 2020). In turn, the acceleration in online news production more generally created pressure on editorial staff as work routines were altered and longstanding journalistic norms challenged.

The so-called "high modern" paradigm of journalism has been weakening for decades (Altheide & Snow 1991; Hallin 1994). Charges of tabloidization and hyper-commercialism are old hat, while the relationship between business interests and the newsroom has always been fraught. This history notwithstanding, the defense from within the craft has traditionally rested on the sincerely made claim that professional norms and standards of journalism worked to mitigate against the profane influences of advertising and politically motivated ownership (Alexander et al. 2016). In the wake of the Great Recession, and in conjunction with a search for a profitable online business model for newspapers, that changed. A once shameful thought was now advanced in the open—that the "Church and State" separation between the editorial and the business sides of news organizations should be rethought. When Joseph Ripp took the reins as Chief Executive Officer at Time Inc. in 2013, he quickly instituted a new policy to reorient editorial content to the company's multiple digital platforms and native advertising strategies—i.e. content paid for by clients, but given the appearance of traditional journalism (Ferrer Conill 2016). Media workers across the board were now being asked to embrace "a further intensification of the circulation of commodities" through social media and the encouragement of "promotional, auto-commodifying reputation-management"—now seen as a necessary "voluntary" contribution (Dyer-Witheford 2015, 92).

Major news organizations such as E.W. Scripps Company—later known as the Journal Media Group—teamed up with Knight Digital Media Center at USC/Annenberg to investigate the "transformation of newsrooms." According to Michelle McLellan, a co-author of a report for the Center: "It's a matter of re-engineering journalists' attitudes and their relationships with news consumers, as well as changing newsroom workflows and priorities" (McLellan 2015). This "re-engineering" was ostensibly in response to technological changes in the production and delivery of information brought about by digitalization. But such transformations entailed not only process changes, but fundamental alterations to the professional norms of editorial independence and integrity. The drive for structural and cultural re-engineering was so ubiquitous that even the Grey Lady herself, the *New York Times*, produced a widely discussed internal "Innovation Report" in 2014 calling for the abandonment of "our current metaphors of choice—'The Wall' and 'Church and State.'" The report called for the creation of "promotional teams" in the newsroom and for individual reporters and editors to better promote themselves and their work using social media (Benton 2014). These changes did not occur without resistance. Unionized reporters and editors working at Alden Capital owned Digital First newspapers were particularly vocal, organizing a large-scale media campaign to defend quality journalism amidst large-scale layoffs (Reynolds 2017). Resistance to poor working conditions spread, as union certification drives proliferated across traditional and online news organizations, such as the *Los Angeles Times*, Buzzfeed, Vice, Slate and the digital employees at NBC News (Greenhouse 2019).

What is clear from this overview is that the ground had already been prepared for a journalism legitimation crisis prior to the moral panic about the circulation of fake news in 2016. Financialized local newspapers as well as broadcasters had laid off tens of thousands of workers creating a large reserve army of skilled writers and editors as regional markets were rendered news deserts. As labor precarity in newsrooms increased, owners and management at both local and national news organizations openly pushed to "re-engineer" journalistic common sense to create a more flexible labor regime for online promotional strategies. Despite organized pushback from union activists, the "conversion of news companies into financial instruments" by "outside money men," according to Doctor, had stripped them "of civic responsibility" (Doctor 2019). This was the context of an internal struggle to impose a new form of discursive legitimation on the craft—one that sought to replace self-determined norms of public service with a new common sense that allowed for digital technical efficiency in the service of networked media companies.

Bad actors and the legitimation crisis of fake news

The proliferation of bad faith actors spouting misinformation in 2016—early 21st century versions of McKernan's Market Rigger, News Faker and Professional Propagandist—occurred within a broader legitimation crisis sweeping Western governments and civil-society institutions following the 2008 financial collapse and the state bailout of financialized capital—most prominently banks deemed "too big to fail." In the wake of this massive redistribution of wealth, public faith in government and civic institutions was tested, plunging social institutions into a broader social and political crisis, spawning the Occupy Wall Street movement on the left and various nationalist movements on the right, such as the Tea Party in the United States. "At the root of this crisis of political legitimacy," argues Manuel Castells, "was the crisis of a form of capitalism, global financial capitalism" (Castells 2019, 15).

As we have seen, these were the broader forces involved in the creation of news deserts, but they were also at play, argues Castells, in the related crises of democratic representation and identity. "The latent contradictions in the economy and society as transformed by globalization, the resistance of identity and the dissociation between state and nation, were all phenomena that became apparent in social practices during the economic crisis of 2008–10" (2019, p. 15). As Marc Bloch noted in his 1921 study of the false news emanating from the World War I battlefields of France and Belgium:

> False news is probably born of imprecise individual observations or imperfect eyewitness accounts, but the original accident is not everything: by itself, it really explains nothing. The error propagates itself, grows, and ultimately survives only on one condition—that it finds a favorable cultural broth in the society where it is spreading. Through it, people unconsciously express all their prejudices, hatreds, fears, all their strong emotions.
>
> *(Bloch 2013 [1921], 3)*

The complex social brew of the crisis of globalization, years of economic austerity, combined with a breakdown in political and institutional authority, supported the nationalist assertions of identity seen in the election of Donald Trump and the successful referendum bid for the United Kingdom to leave the European Union under the banner of "Brexit." This was the context in which "information systems" became vulnerable to "a mix of strategic disinformation from both national and foreign actors" (Bennett & Livingstone 2018, 127).

The remainder of this chapter will be given over to a brief mapping of the social relations at play within a new communication space in which traditional mass media networks are integrated with horizontal user-generated promotional commodity circulation. It is an uneven space, in which new companies, such as the Leaf Group, formerly Demand Media, and other so-called "content mills" (Frank 2010), take advantage of user-generated content (UGC) and the expanding reserve army of volunteer, or under-employed workers, who produce it. "What we are observing," argues Manuel Castells, "is the coexistence and interconnection of mainstream media, corporately owned new media, and autonomous Internet sites" (Castells 2007). The new social and political fields are "hybrid" spaces in which traditional media is integrated with the circulation logics of online commercially oriented platforms and online UGC (Chadwick 2017).

The low barriers to entry afforded by the proliferation of online media were a boon to both profit-oriented bloggers in Macedonia, eager to exploit the Facebook algorithm, and politically motivated conspiracy theorists, such as Alex Jones' Infowars website (Allcott & Gentzkow 2017). Both groups leveraged emotional content on social media "to generate attention and viewing time" (Bakir & McStay 2018). These actors then fed into a pre-existing moral panic over the deleterious effects of digital media on democracy (Wasserman 2020, 4). But while this hybrid media space made room for a variety of social actors, empirical research has found that far right activists had a disproportionate presence and influence during the 2016 American election, contributing to a more radical and polarized media system (Benkler et al. 2018; Bennett & Livingstone 2018). Fox News, along with other nodes on the network, played a prominent role, acting as amplifiers for disinformation circulating on sites such as Infowars, Breitbart, Truthfeed and Gateway Pundit, contributing to a "right-wing media ecosystem" that "represents a radicalization of roughly a third of the American media system" (Benkler et al. 2018, chap 1, loc. 338). Breitbart, funded by billionaire Trump supporter Robert Mercer, and The Daily Caller, backed by billionaire Charles Koch, both promoted a broad anti-government agenda (Bennett & Livingstone 2018). Another Mercer-funded organization that played a role in the dissemination of disinformation during the 2016 election was Cambridge Analytica. The voter-profiling firm took advantage of the Facebook algorithm using artificial intelligence technology in support of both the Trump campaign and the official "leave" side in the Brexit referendum (Benkler et al. 2018, chap 9, loc. 4828).

In addition to small-scale bloggers and well financed micro-targeting campaigns, the hybrid media space also includes state actors, most prominently in 2016 the Russian-sponsored Internet Research Agency (IRA). Special Counsel Robert Mueller's indictment of the IRA found that the agency "operated thousands of Twitter accounts posing as Americans to weigh in on US political discussions on social media between 2014 and 2017" (Kukito et al. 2020, 197). The IRA was particularly successful in tapping into traditional news media's desire to represent "vox popul," by having tweets from fake accounts appear in public opinion news round ups (Kukito et al. 2020).

Final thoughts

This chapter, in a preliminary way, has tried to map the broad changes to social space made possible by the forces of media proliferation, concentration, deregulation, globalization and digitization. In doing so, we have drawn attention to the wider social relations at play in the production, dissemination and consumption of the so-called fake news commodity. A popular notion has spread that so-called "filter bubbles" or "echo chambers" (Pariser 2011; Sunstein 2017) have contributed to selective exposure to news, thus deepening the polarization of media consumption and limiting exposure to information that may challenge ideological predispositions or assumptions. These fears have gained traction following the spread of conspiracy theories, such as QAnon, that appear immune to factual evidence. However, the filter bubble thesis has been challenged by research that suggests Facebook and Google had limited effect on the direct exposure to news (Cardenal et al. 2019), while other research has found evidence that social media also increased individual exposure to material from one's non-preferred ideological perspective (Flaxman et al. 2016). And while Benkler et al. (2018) report that right-wing networks are disproportionately responsible for the spread of disinformation, the authors shy away from blaming technology, or specifically social media, as a determinative driver. Their work, instead, suggests researchers should investigate institutional, political and cultural factors. In line with scholars such as Lance W. Bennett and Steven Livingstone, the chapter has focused less on the details of specific tales of disinformation, and instead attempted to broaden perspectives and "to resist easy efforts to make the problem go away by fact-checking initiatives and educating citizens about the perils of fake news" (Bennett & Livingstone 2018, 135). The roots of fake news, dating back as far as the *New York Sun*'s "moon hoax" of 1835, must be situated within the uncertainty of social, political and technological change and struggle.

References

Abernathy, P. M. 2016. *The rise of a new media baron and the emerging threat of news deserts*. Chapel Hill, NC: University of North Carolina Press.

Abernathy, P. M. 2018. *The expanding news desert*. www.usnewsdeserts.com/reports/expanding-news-desert/.

Alexander, J. C., Breese, E. B., and Luengo, M. 2016. *The crisis of journalism reconsidered: Democratic culture, professional codes, digital future*. New York: University of Cambridge Press.

Allcott, H., and Gentzkow, M. 2017. Social media and fake news in the 2016 election. *Journal of Economic Perspectives*, 31 (2), 211–236. https://doi.org/10.1257/jep.31.2.211.

Altheide, D., and Snow, R. 1991. *Media worlds in the post-journalism era*. New York: Aldine de Gruyter.

Bakir, V., and McStay, A. 2018. Fake news and the economy of emotions: Problems, causes, solutions. *Digital Journalism*, 6 (2), 154–175. https://doi.org/10.1080/21670811.2017.1345645.

Benkler, Y. 2006. *The wealth of networks: How social production transforms markets and freedom*. New Haven, CT: Yale University Press.

Benkler, Y., Faris, R., and Roberts, H. 2018. *Network propaganda: Manipulation, disinformation and radicalization in American politics*, Kindle edn. New York: Oxford University Press.

Bennett, W. L., and Livingston, S. 2018. The disinformation order: Disruptive communication and the decline of democratic institutions. *European Journal of Communication*, 33 (2), 122–139. http://doi.org/10.1177/0267323118760317.

Benton, J. 2014. *The leaked New York Times innovation report is one of the key documents of this media age*. www.niemanlab.org/2014/05/the-leaked-new-york-times-innovation-report-is-one-of-the-key-documents-of-this-media-age/.

Bloch, M. 2013 [1921]. *Reflections of a historian on the false news of the war*. http://miwsr.com/2013-051.aspx.

Blumler, J. G., and Gurevitch, M. 1995. *The crisis of public communication*. London: Routledge.

Cardenal, A., Aguilar-Paredes, C., Galais, C., and Pérez-Montoro, M. 2019. Digital technologies and selective exposure: How choice and filter bubbles shape news media exposure. *The International Journal of Press/Politics*, 24 (4), 465–486. http://doi.org10.1177/1940161219862988.

Carlson, M., and Lewis, S. (eds.) 2015. *Boundaries of journalism: Professionalism, practices and participation*. London and New York: Routledge.

Castells, M. 2007. Communication, power and counter-power in the network society. *International Journal of Communication*, 1, 238–266. https://ijoc.org/index.php/ijoc/article/view/46.

Castells, M. 2019. *Rupture: The crisis of liberal democracy*. Cambridge: Polity Press.

Chadwick, A. 2017. *The hybrid media system: Politics and power*. New York: Oxford University Press.

Compton, J. 2004. *The integrated news spectacle: A political economy of cultural performance*. New York: Peter Lang.

Crary, J. 2013. *24/7: Late capitalism and the ends of sleep*. London: Verso.

Dalhgren, P. 2009. *Media and political engagement: Citizens, communication, and democracy*. Cambridge: Cambridge University Press.

Doctor, K. 2019. *By selling to America's worst newspaper owners, Michael Ferro ushers the vultures into Tribune*. www.niemanlab.org/2019/11/newsonomics-by-selling-to-americas-worst-newspaper-owners-michael-ferro-ushers-the-vultures-into-tribune/.

Doctor, K. 2020a. *What was once unthinkable is quickly becoming reality in the destruction of local news*. www.niemanlab.org/2020/03/newsonomics-what-was-once-unthinkable-is-quickly-becoming-reality-in-the-destruction-of-local-news/.

Doctor, K. 2020b. *Six takeaways from McClatchy's bankruptcy*. www.niemanlab.org/2020/02/newsonom ics-six-takeaways-from-mcclatchys-bankruptcy/.

Dyer-Witheford, N. 2015. *Cyber-proletariat: Global labour in the digital vortex*. London: Pluto Press.

Ferrer Conill, R. 2016. Camouflaging church as state: An exploratory study of journalism's native advertising. *Journalism Studies*, 17 (7), 904–914. http://dx.doi.org/10.1080/1461670X.2016.1165138.

Flaxman, S., Goel, S., and Rao, J. M. 2016. Filter bubbles, echo chambers, and online news consumption, *Public Opinion Quarterly*, 80 (supp.1), 298–320. http://doi.org/10.1093/poq/nfw006.

Frank, T. 2010. Bright frenetic mills. *Harper's Magazine*, December. https://archive.harpers.org/2010/12/p df/HarpersMagazine-2010-12-0083200.pdf.

Giddens, A. 1990. *The consequences of modernity*. Cambridge: Polity Press.

Greenhouse, S. 2019. *Why newsrooms are unionizing now*. https://niemanreports.org/articles/why-news rooms-are-unionizing-now/.

Hallin, D. 1994. *We keep America on top of the world: Television journalism and the public sphere*. New York: Routledge.

Harvey, D. 1989. *The condition of postmodernity: An enquiry into the origins of cultural change*. Cambridge: Blackwell Publishers.

Hofseth, A. 2017. *Fake news, propaganda, and influence operations: A guide to journalism in a new, and more chaotic media environment*. http://reutersinstitute.politics.ox.ac.uk/news/fake-news-propaganda-and-in fluence-operations-%E2%80%93-guide-journalism-new-and-more-chaotic-media.

Jessop, B. 2009. The spatiotemporal dynamics of globalizing capital and their impact on state power and democracy. In H. Rosa and W. E. Scheuerman (eds.) *High speed society: Social acceleration, power, and modernity*, pp. 135–158. University Park, PA: Penn State Press.

Lefebvre, H. 1991. *The production of space*. Oxford: Blackwell Press.

Lukito, J., Suk, J., Zhang, Y., Doroshenko, L., Kim, S. J., Su, M. H., Xia, Y., Freelon, D., and Wells, C. 2020. The wolves in sheep's clothing: How Russia's internet research agency tweets appeared in US News as Vox Populi. *The International Journal of Press/Politics*, 25 (2), 196–216. https://doi.org/10.1177% 2F1940161219895215.

McIntire, M., Yourish, K., and Buchanan, L. 2019. In Trump's Twitter feed: Conspiracy-mongers, racists and spies. *New York Times* (November 2). www.nytimes.com/interactive/2019/11/02/us/politics/ trump-twitter-disinformation.html.

McIntyre, L. 2018. *Post-truth*. Cambridge, MA: MIT Press.

McKernan, E. 1925. Fake news and the public: How the press combats rumor, the market rigger, and the propagandist. *Harper's Magazine*, October. https://archive.harpers.org/1925/10/pdf/HarpersMaga zine-1925-10-0012896.pdf.

McLellan, M. 2015. *Leadership, culture are linchpins of digital transformation in the newsroom*. http://mediashift. org/2015/05/report-leadership-culture-are-linchpins-of-digital-transformation-in-the-newsroom/.

Molla, R. 2017. *The New York Times' digital business more than doubled in the past six years*. www.recode.net/ 2017/5/8/15578906/new-york-times-digital-business-doubled-subscription-advertising-revenue.

O'Neil, L. 2020. Fox News guests spread "disinformation"—says leaked internal memo. *The Guardian* (February 7). www.theguardian.com/media/2020/feb/07/fox-news-internal-briefing-disinformation-contributors.

Pariser, E. 2011. *Filter bubbles: What the internet is hiding from you*. London: Penguin.

Pompeo, J. 2018. CNN, despite Trump bump, prepares for dozens of layoffs. *Vanity Fair* (February 12). www.vanityfair.com/news/2018/02/cnn-prepares-for-dozens-of-layoffs.

Rantanen, T. 1997. The globalization of electronic news in the 19th century. *Media, Culture & Society*, 19, 605–620. https://doi.org/10.1177/016344397019004006.

Reynolds, J. 2017. *Untangling the web of Alden's dubious deals*. https://dfmworkers.org/untangling-th e-web-of-aldens-dubious-deals/.

Rosa, H. 2009. Social acceleration: Ethical and political consequence of a desynchronized high-speed society. In H. Rosa and W. E. Scheuerman (eds.) *High speed society: Social acceleration, power, and modernity*, pp. 77–112. University Park, PA: Penn State Press.

Shirky, C. 2008. *Here comes everybody: The power of organizing without organizing*. New York: Penguin.

Sididiqui, F and Svrluga, S. 2016. NC man told police he went to DC pizzeria with gun to investigate conspiracy theory. *The Washington Post* (December 5). www.washingtonpost.com/news/local/wp/ 2016/12/04/d-c-police-respond-to-report-of-a-man-with-a-gun-at-comet-ping-pong-restaurant/.

Silverman, C., and Alexander, L. 2016. How teens in the Balkans are duping Trump supporters with fake news. www.buzzfeednews.com/article/craigsilverman/how-macedonia-became-a-global-hub-for-pro-trump-misinfo.

Sol, J. 2016. The long and brutal history of fake news. *Politico* (December 18). www.politico.com/maga zine/story/2016/12/fake-news-history-long-violent-214535.

Sunstein, C. R. 2017. *#Republic: Divided democracy in the age of social media.* Princeton, NJ: Princeton University Press.

Tandoc, E. C.Jr, Lim, Z. W., and Ling, R. 2018. Defining "fake news." *Digital Journalism*, 6 (2), 137–153. https://doi.org/10.1080/21670811.2017.1360143.

Thompson, J. B. 2020. Mediated interaction in the digital age. *Theory, Culture & Society*, 37 (1), 3–28. http://doi.org/10.1177/0263276418808592.

Tong, J. 2018. Journalistic legitimacy revisited: Collapse or revival in the digital age? *Digital Journalism*, 6 (2), 256–273. https://doi.org/10.1080/21670811.2017.1360785.

Tracy, M. 2019. Acquisition of Gannett creates print goliath in a $1.4 billion deal. *New York Times* (August 6). www.nytimes.com/2019/08/05/business/media/gannett-acquired-gatehouse-media.html.

Walsh, J. P. 2020. Social media and moral panics: Assessing the effects of technological change on societal reaction. *International Journal of Cultural Studies*. https://doi.org/10.1177/1367877920912257.

Wasserman, H. 2020. Fake news from Africa: Panics, politics and paradigms. *Journalism*, 21 (1), 3–16. http://doi.org/10.1177/1464884917746861.

Winseck, D. 2010. Financialization and the "crisis of the media": The rise and fall of (some) media con-glomerates in Canada. *Canadian Journal of Communication*, 35 (3), 365–393. https://doi.org/10.22230/cjc.2010v35n3a2392.

Zielinski, S. 2015. The great moon hoax was simply a sign of its time. *Smithsonian Magazine* (July 2). www.smithsonianmag.com/smithsonian-institution/great-moon-hoax-was-simply-sign-its-time-180955761/.

9

FILM GEOGRAPHY

Elisabeth Sommerlad

"Films… often evoke a sense of place—a feeling that we, the… viewer, know what it's like to 'be there'" (Cresswell 2004, 7f). They introduce us to (un)known places and landscapes and allow us to experience them. The world we live in today consists more than ever of media imaginations that blend with our everyday experiences, conveying an overall sense of our world. The audio-visual images of motion pictures in particular contribute to this process. As cultural and social documents, films stage social realities and thus offer condensed and meaningful interpretations of everyday life (Schroer 2018). They shape and affect our perceptions and actions and contribute to the construction of world concepts (Gregory 1995, 474; Dimbarth 2018). Filmic geographies are not considered as mirrors of reality, but as medial constructions (Escher & Zimmermann 2001, 230). However, these imaginations create reality-generating meanings, and make them accessible for everyday discourses. They perpetuate, constitute and construct social realities in various ways and thereby influence, transform and manifest our knowledge of the world.

Film geography—sometimes also referred to as cinematic geography—is a research area that deals with the multi-layered, cinematographically generated geographical imaginations and their interconnections and effects manifested in the everyday world (Zimmermann 2007, 14). It integrates, both theoretically and methodologically, interdisciplinary approaches and deals with the relations between film, location, space and place (Roberts 2020). This chapter first gives an overview of the development of film geography. This is followed by an outline of selected topics that have been the focus of film geographical studies. The article concludes with an outlook on future trends in film geography.

Film geography: A brief overview

There is a long history of the discipline of film geography, which can be traced in several key scholarly publications. The analysis of filmic landscapes already appeared about 100 years ago in Balász's theoretical considerations of landscape photography, which, in his view, serves the scenographic spatial design of a feature film and in this role attributes high poetic qualities to it (Balász 1924). Another key publication examined the various aspects of cinematic presentation from a geographical perspective and focused on the juxtaposition of film and reality (Wirth 1952). Further precursors of film geography can be found in scholarly works which

DOI: 10.4324/9781003039068-11

addressed the interface between documentary films and regional geography (e.g. Dixon 2015, 40; Griffith 1953; Manvell 1956). It is noteworthy that the term *cinéma-géographie* was also mentioned for the first time by the French geographer Lacoste (1976). However, these publications are still isolated approaches that precede film geography as a discipline.

An important milestone in the development of film geography can be found with Burgess and Gold (1985), who argue for an engagement of geography with mass media content. A further landmark is an anthology published by Aitken and Zonn (1994), which compiled essays that dealt exclusively with film-related topics. At the beginning of the 2000s, Cresswell and Dixon (2002) pointed out that geography is required to take a critical look at cinematic realism. The concept of (cinematic) representation thus changes from an assumed process of mimetic depiction to a process of socially contested constructions (Lukinbeal 2010, 1110–1111). Subsequent publications have taken up the recognizable impetus of a critical debate within film geography and have continued to shape recent film geographical discourses (Aitken & Dixon 2006, 327). This is also reflected in the contributions by Escher (2006) or Lukinbeal and Zimmermann (2006; 2008), which explore different strands of research in film geography in a multi-perspective manner and highlight different thematic areas of study. The geographical examination of the subject of film has since developed steadily and differentiated considerably. Contemporary film geography addresses the medium of film from a variety of perspectives.

Topics of film geography

A review of the previously published introductory chapters on film geography shows that there are numerous possibilities for defining thematic areas of the subject. They arrange theoretical perspectives on film geography issues and research approaches in very different ways and thereby highlight different aspects (e.g. Aitken & Dixon 2006; Escher 2006; Lukinbeal & Zimmermann 2006; 2008; Lukinbeal 2009; Zimmermann 2009; Dixon 2014; Staszak 2014; Sharp & Lukinbeal 2015, 2017; Roberts 2020; Sommerlad forthcoming b). The contribution at hand offers a further possibility to explore the subject of film geography.

Film geographical studies can loosely be placed in a complex interplay of the following four perspectives: (1) the content conveyed on a *screen*, (2) the *locations* where filmic content is produced or located as well as the *places* charged with meaning by films, (3) the *reception* and *critical reflection* of filmic contents, practices and effects, and (4) film as a *tool* of academic research (see Figure 9.1).

In this context, film geographical studies approach an examination of film-related content, aspects of film productions and the (inter-)effects of film and life-world contexts. Without claiming to be exhaustive, the following six research areas are identifiable in the interplay of these relational perspectives: (1) geography in film, (2) geography of film, (3) screen tourism, (4) cinematic cartography, (5) didactics of geography and critical film geography, (6) film as a methodological instrument and medium of communication in research. This suggested classification, that serves to structure the chapter, focuses less on epistemological or theoretical perspectives underlying film geographical investigations; rather, it accentuates exemplary topics that delineate diverse fields of interest. In doing so, it is noteworthy that each of those thematic spheres is not merely anchored in a single film geographical perspective, but often combines different aspects—although one or other viewpoint is occasionally emphasized more intensively.

As diverse as the research interests in film geography are, so are the methodological approaches that tackle the discipline (Kennedy & Lukinbeal 1997). Studies that focus on films

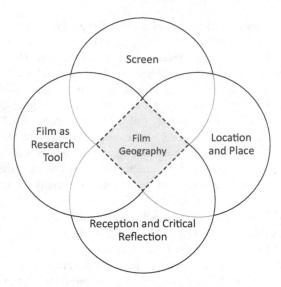

Figure 9.1 Relational perspectives on film geography: Overview

as text are often based on common methods of film studies and visual methodologies (e.g. Rose 2012). Analyses that go beyond textual film analysis draw on well established methods of quantitative and qualitative social research. Depending on research interests, cartographic methods and spatial analysis, approaches of cultural studies, media, communication and reception research, social media research or auto-ethnography are applied. The variety of methods is a creative challenge, as existing approaches require a meaningful combination (Zimmermann 2009, 307).

Geography in film

The basis of filmic geographies is a combination of camera perspectives, angles and viewpoints, "traditional visual guidelines and their inscribed stories" (Zimmermann 2009, 302, transl. by author). Filmmakers use these to engender spatial representations, which are examined under the term "cinematic space" (Heath 1981). Cinematic space is a central category of analysis in film studies and takes into account "the conceptual proximity to film technical and creative levels of film" (Zimmermann 2009: 302, transl. by author). From a geographical viewpoint, this involves theoretical examinations of forms of cinematographic design and questions of how an audiovisual continuity emerges, providing the spectator with a cohesive spatial impression of what is seen (Escher 2006; Clarke 2008). Anchored in cinematic space are narratively significant, cinematically conceived places. As an aesthetically and emotionally superimposed phenomenon, a cinematic landscape transports moods, atmospheric elements and myths (Higson 1987; Escher 2006, 309). A landscape can serve different roles in films and function as a framework for the plot, a setting, an actor or a symbol (Escher & Zimmermann 2001; Lukinbeal 2005). An overview of different approaches to the relationship between cinema and landscape can be found in Harper and Rayner (2010).

Furthermore, another and highly diverse key research topic in film geography is the analysis of cinematically imagined cities (Shiel & Fitzmaurice 2001; 2003; Lukinbeal & Sharp 2019). While recent approaches focus on the interactions between the urban and film and thereby transcend the boundaries of geography in film (e.g. Roberts 2012b), there is a long tradition in the textual analysis of the construct of a "Cinematic City" (Clarke 1997). As an aesthetic construction, a Cinematic City consists not only of the ensemble of filmed locations, architectures and staged society, but also of the interplay of spatial perspectives and the linking of urban sights attributed with semantic and spatial meanings. A Cinematic City is not only a location or setting for a film, but acts as a dramaturgically and dramatically influential actor across multiple movies (Escher & Zimmermann 2005). Corresponding studies deal with cities such as New York (Da Costa 2003; Fröhlich 2007; Sommerlad forthcoming b), Kuala Lumpur (Bunnell 2004), Marrakesh (Zimmermann & Escher 2005; Sommerlad 2019a), Cairo (Escher & Zimmermann 2005b), London (Brunsdon 2007), Tangier (Sommerlad 2019), Berlin (Natter 1994), and many more. Cinematic imaginations of landscapes or cities reveal a specific power of films: they do not simply depict a possible "reality," but reconfigure spatial views and thereby create new (illusionistic) places that would not exist without them (cf. Zimmermann 2007). They comprise a mixture of footage edited together from studio sets and real-life locations. Due to an increasing digitalization process, cinematic worlds are enhanced by digital sceneries (e.g. via computer generated imagery). For example, cityscapes that would not be possible in the "real world" are created through cinematographic techniques: locations that seem to be close to each other can actually be far apart, while other perspectives might be completely removed for the cinematic spatial image, whereas some others could be completely recreated by computer animation—thus creating new, ambiguous depictions of imaginary places (Zimmermann 2009, 302). A critical analysis of such spaces can uncover (representational) power relations that are embedded in the production processes as well as in related reception discourses, as Bunnell (2004) proves for example in his study on Kuala Lumpur.

Geography of film

Research in film geography is not limited to the film-diegetic world, but also aims to explore the interplay of film-related sites with everyday realities. Geography of film thereby draws attention to places and spaces as well as to industrial clusters and networks of media and film production, while also touching on issues related to (globalized) media economies and cultural politics. Research often takes place in conjunction with economic aspects, for example by taking into account approaches to economic geography or political economy (Sharp & Lukinbeal 2015, 22–23). Particular emphasis is placed on the formation, development and networking of local and national film industries, film studios and production clusters (Storper & Christopherson 1987; Storper 1989). A special interest lies in Hollywood's film industry (Scott 2005; McDonald 2008; Christopherson 2013; Gleich & Webb 2019), televisual and cinematic places (Fletchall et al. 2012), as well as the historical and contemporary development of film production sites and locations (Lukinbeal & Gleich 2018). On a global scale, increasing consideration is being given to examining global(ized) film industries. For instance the film business associated with Hollywood North in Vancouver and Toronto (Gasher 2002; Spencer & Ayscough 2003; Wilson 2016), Bollywood (Kavoori & Punathambekar 2008; Dagnaud & Feigelson 2012), Nollywood (Ogbeide 2012; Krings & Okome 2013; Miller 2016) or various East Asian film industries (Yau Shuk-ting 2009; Curtin 2015). Some studies (e.g. Govil 2015) address translocal and transnational perspectives on the film and television production industry.

The examination of film production sites is not necessarily limited to a (cultural) economic perspective. Lukinbeal (2006, 2012) argues for taking into account that cultural texts are inevitably embedded in both political and economic practices of their respective markets and that such conditions and production practices must also be considered within the framework of hermeneutic text analyses (Sharp & Lukinbeal 2015, 23–24). In taking into account the scope of cultural politics, questions arise as to what extent films produced in different regional contexts convey different cultural or political value patterns and interpretations of the world and society (e.g. geopolitical images, gender relations or concepts of identity). Morgan Parmett (2019) studies the cultural economy of film and television productions and explores the connection between geographies of race and class and television production practices. Therefore, she draws on the case study of *Treme*, a show filmed on location in the Tremé neighborhood of post-Katrina New Orleans. Furthermore, studies that look at global film cultures and border-crossing cinema from a postcolonial perspective are particularly interesting (e.g. Roy & Huat 2012)—although the deliberate examination of global film cultures within film geography is considered a desideratum.

Screen tourism

Film locations are exposed to tourist effects and are therefore a good example to trace the reciprocal effects between cinematic worlds and spheres of everyday life. There are numerous catchwords describing this phenomenon: film tourism, movie-induced tourism or set-jetting—to name just a few (Busby & Klug 2001; Tzanelli 2007; Roesch 2009; Zimmermann & Reeves 2009; Beeton 2016). Since touristic effects on places are no longer triggered solely by feature films, but also by TV shows and further formats, which are available on all types of devices, the term *screen tourism* is particularly well suited to describe the phenomenon and its accompanying practices (Kim 2010; Böcher 2018, 16). Academic research addresses the issue as a "complex social and tourism phenomenon with psychological, economic, geographical and political dimensions, among others" (Steinecke 2016, 21, transl. by author). As a tourism segment, screen tourism has become an established market phenomenon and is promoted by many countries and regions due to its high economic potential (Tooke & Baker 1996; Riley et al. 1998).

Screen tourism is not limited to the original locations of a film, but also affects regions, cities and even particular buildings that have served as the setting or place of inspiration. In addition, facilities of the cultural and leisure industry related to filmmaking are also addressed, such as cultural events, exhibitions, museums or theme parks (Steinecke 2016, 20). With regard to destinations and tourist practices at those very places, Escher, Sommerlad and Karner (2017, 157–161) suggest distinguishing between different types of screen tourism (see Figure 9.2).

Fan tourism (a) describes, for instance, the re-enactment of film characters or scenes at specific locations or events. *Set tourism* (b) takes place at locations created especially for a particular film or for film tourism, such as Hobbiton in New Zealand, a site that was in fact only reproduced and revitalized as a tourist destination after the filming of *The Lord of the Rings* trilogy had been completed. Entire landscapes are associated as a characteristic of a film genre in the case of *genre tourism* (c) and are staged in the context of touristic journeys. For example, Monument Valley is closely linked to the Western genre and has long been an established tourist magnet. In the case of *city tourism* (d), entire cities become tourist attractions because they have served as a film location/setting. A good example is Dubrovnik in Croatia, where tourists can trace *Game of Thrones* on specific tours. *Theme park tourism* (e) summarizes all tourism forms, which takes place in built recreational parks/studios, such as Universal Studios in Los Angeles. *Landscape tourism* (f) turns countries, regions or landscapes into destinations for screen tourists, as they are related to particular films. A prominent example is (once again) New Zealand, which has been promoting itself as Middle Earth for many years. A potential for future academic studies exists in investigating and analyzing the increasing interrelation between fictional film and factual everyday world(s). Such hyperreal

Figure 9.2 Types of screen-tourism: (a) fan-tourism (re-enacting *Mad Max* at Wasteland Weekend in Edwards, CA, Randy Lewis and Monti Sigg 2019); (b) set-tourism (Hobbiton in New Zealand, Marie Karner 2016); (c) genre-tourism (Monument Valley, USA, Claudia Finkler 2016); (d) city-tourism (*Game of Thrones* Tour in Dubrovnik, Croatia, Daniel Böcher 2018); (e) theme-park-tourism (Universal Studios in LA, Elisabeth Sommerlad 2016); (f) landscape-tourism (*Lord of the Rings* Tour in New Zealand, Marie Karner 2016)

sites continue to be re-invented through increasing marketing strategies, tourist activities and practices, and intermedia networks.

Cinematic cartography

Places featured in film are often illustrated with maps. An example are tourist maps for film sites in the context of screen tourism. Additionally, maps appear *in* films in manifold ways and perform a variety of narrative functions. They are used, for instance, to locate cinematic stories or to provide the spectator with an orientation within the fictitious cinematic world.

The interconnections between cartography, mapping practices, scale, maps and film are complex and have been analyzed under the heading of "Cinematic Cartography" (e.g. Caquard & Taylor 2009; Caquard & Cartwright 2014; Penz & Koeck 2017; Lukinbeal et al. 2019). This perspective integrates a diverse range of topics, involving more than just the question of what functionality maps have in film (Mauer & Sommerlad forthcoming). Roberts and Hallam (2014, 8) identify five relevant thematic fields of cinematic cartography: "(1) maps and mapping in films; (2) mapping of film production and consumption; (3) movie mapping and place marketing; (4) cognitive and emotional mapping; and (5) film as spatial critique." In order to understand why film itself is considered a "modern cartography" (Bruno 2002, 71) or a work of cartographic art (Conley 2007), it is necessary to expand the general idea of a map. From a film geographical perspective, emphasis is placed on the profound relations between the mediums of film and map, cartographic practices and cinematic narratives (Lukinbeal 2004; 2018; Caquard et al. 2009; Joliveau 2009; Sharp 2018; Lukinbeal & Sharp 2019). The concept of "Cinema's Mapping Impulse" is used to investigate the connections between cartographic and cinematographic practices (Castro 2009; 2011). Caquard examines the influence that cinematic maps have had on cartographic practices, for example, the very fact that current mapping practices and functions of digital cartography were conceptualized in much older cinematographic techniques (Caquard 2009; 2011). Roberts studies the extent to which films can be read as maps and asks about the spatial practices of cinematic cartography and how film maps and film mapping can be understood as geographical knowledge productions (Roberts 2012a).

Critical film geography and filmic geography didactics

Films never stand for their own sake, but are consumed and received as a product in a situational and situated manner and are thereby attributed with individual meanings (Sharp & Lukinbeal 2015, 27). Research on film reception and reflection investigates how the diversity of reception and interpretation can be grasped theoretically and empirically (Dixon et al. 2008). Aitken and Dixon (2006) criticize that the geographical interest in films often lacks a critical perspective. "Instead of focusing on the ways in which films generate meaning, film geographical research has often focused 'on the geographical realism of films'" (Cresswell & Dixon 2002, 7–8; Aitken & Dixon 2006, 326). Similarly, other scholars (Escher & Zimmermann 2001; Lukinbeal & Zimmermann 2006; Lukinbeal & Sharp 2015) problematize the Real/Reel Binary and call for the simplifying dual system of the real and the reel to be broken up in a critical perspective. Thus, the analysis of filmic texts should always take place in the awareness that films stage a pre-interpreted reality and that the resulting geographies must be understood as simulacra. Accordingly, films can only be discerned as socially constructed and discursive formations embedded in cultural and political global contexts that need to be deciphered. Therefore, a critically reflected film geography deals with topics such as popular geopolitics, cultural policies and practices, and complex globalized cultural industries. It calls for an anti-essentialist, critical examination of questions such as the power of spatial relations, how films generate socio-spatial meanings, and how they engender images that are materially reflected in everyday social practices (Aitken & Dixon 2006, Hughes 2007; Dodds 2008; Lukinbeal 2019; Sommerlad forthcoming a).

A critical perspective is also expressed in didactic contexts. Publications on film didactics include a critical perspective on the omnipresent use and methodological-didactic value of images, documentaries and feature films in teaching geography (Di Palma 2009; Plien 2017). Especially in educational contexts, a key aspect should be to critically examine the mediation of film narratives, their subtexts and apparent realisms, and how they influence the spectator's imagination (Aitken 1994). The use of "film as a critical pedagogic tool that confronts the viewers (students and researchers) with a whole series of questions about the ways in which spaces and identities are made up and dissolved within the structure of films," which Cresswell and Dixon (2002, 7) also advocate, should, in particular, question the relations between conveyed content and specific forms of filmic communication.

Film as a methodological instrument and communication medium of research

A further dimension of the geographic engagement with film consists in considering it not only as a subject of analysis, but also as a research tool, as a medium of communication, or as a result of academic research. Garrett (2010) discusses the advantages of including video production in the toolset of geographic research, as "[v]ideographic work gives researchers an avenue to depict place, culture, society, gesture, movement, rhythm and flow in new and exciting ways" (Garrett 2010, 536). The production of geographic knowledge could be achieved through acquisition and analysis of qualitative video data (*videography*, Knoblauch & Schnettler 2012), the production of films, or—in other words—the integration of film into the practice of geographic research. Possible ways include the integration of video footage for field documentation as well as including reflexive filmmaking and participatory videos in empirical research projects (Garrett 2010, 523–530). Similarly, Jacobs (2013; 2016) emphasizes that geography, as a visual discipline, must open itself more to the production of audiovisual materials and argues for a stronger engagement of geographers with filmmaking practices. This would allow geographers "to utilize new skill sets, reach new audiences and produce different forms of critically engaged audio-visualized knowledge" (Jacobs 2013, 724). Another benefit of integrating filmmaking into geographical practice is that cinematically mediated representations, power relations and spatial constructions can be better grasped if the process of filmmaking was actively experienced. This is especially possible through a self-determined and critical acquisition of methodical approaches to this medium (Thieme et al. 2019, 296). Knowledge of the possibilities of researching not only about but also with film thus relates not least to the incorporation of the practical production of media content such as (digital) films and videos into learner-centered educational environments, e.g. geography classes and training (Jacobs 2013). There are now numerous geographic institutes that operate media labs and integrate film-related training into their curricula (see Figure 9.3).

The background of such demands is also the call for a "Geographic Media Literacy" (Lukinbeal 2014)—i.e. the sensitization for a critical examination of (audio)visual media. This aspect is related to the previous paragraph, which focused on the didactic relevance of film geographical research. Furthermore, it is important to have a sound knowledge of how film content is produced in order to be able to analyze and understand it critically (Jacobs 2016, 453). Current explorations of filmic geographies also address questions about the dimension of the body and thus link the methodological debate with up-to-date approaches to embodiment research (Ernwein 2020).

Outlook

A high potential and key characteristic of film geography lies in its ability to focus on complex interconnections between cinematic and everyday life worlds. The field can be tackled best when theoretical and methodological approaches and perspectives are interdisciplinarily combined. Researchers in the field are invited to consider the presented perspectives

● 09.050.612 M5-MA Seminar: Film documentation

Veranstaltungsdetails

Lehrende/r: Julian Zschocke

Veranstaltungsart: online: Seminar

Anzeige im Stundenplan: M5-MA S: Fd

Semesterwochenstunden: 2

Credits: 5,0

Unterrichtssprache: Englisch

Min. | Max. Teilnehmerzahl: - | 34

Prioritätsschema: Senatsrichtlinie
Zulassung gemäß Richtlinie über den Zugang zu teilnahmebeschränkten Lehrveranstaltungen vom 07. März 2007.

Nähere Informationen hierzu entnehmen Sie bitte **www.info.jogustine.uni-mainz.de/senatsrichtlinie**

Inhalt:
This seminar aims to give students an insight into modern forms of documentaries that have emerged within the past 10 years. The rise of videoplatforms like YouTube and social media has paved the way for new forms of documentary filmmaking. Many of them focus on geographic phenomena. Examples include modern formats like

Vox: https://www.youtube.com/watch?v=cc0dqW2HCRc

Vice News: https://youtu.be/qi37th_N3Ck

Wendover productions: https://www.youtube.com/watch?v=MP1OAm7Pzps

and plenty of others.

The seminar is practical, meaning students will form groups to produce their own short documentary on a geographic topic within the semester. This starts with courses that teach knowledge on pre-production, how to form an idea, write a pitch, treatment and full script for a documentary. Productionwise students will be enabled to shoot their own movie with accessible means and the basics of post-production using DaVinci Resolve will be taught.
At the end students will present their final film to the class and have an in-depth discussion about their message, their filmmaking and how they can improve in the future.

Requirements to fulfill the course:
active participation, final documentary script, documentary film (5-10 min), presentation of the film

Figure 9.3 An example of the integration of filmic practice into a university curriculum in the MA program "Human Geography: Globalisation, Media, and Culture" at Johannes Gutenberg University Mainz (revised screenshot from the course catalogue, online: https://bit.ly/3kJW5Fn)

(screened content, location and place, reception and critical reflection, film as tool) and the linked six thematic fields not as separate approaches, but rather as interrelating spheres that can be creatively combined. Escher (2006; 2019), for example, reveals how cinematic and everyday cultures are interrelating through the decisive integration of various film geographical perspectives with historical and contemporary cultural phenomena. Another innovative approach for film geographic research is understanding film as spatial practice. Drawing on this perspective, Roberts (2018; 2020) advocates for an approach of "doing film geography," which incorporates aspects of performativity and haptics in order to creatively engage with the relation between place(s) and filmmaking practices.

A holistic film geography takes up the challenge of linking the complexities of meanings, materialities, places, spaces and scales and interpreting the patterns and effects underlying them:

A geography of film, however, is not limited to the cultural politics of place and space within film. It spans the spectrum from the individual cognitive realm to the socio-cultural level, as well as from the local to the global.

(Kennedy & Lukinbeal 1997, 47)

Film geography should therefore operate at a variety of different scales and levels and combine multiple perspectives, some of which have been introduced in this paper. It is desirable that the discipline continues to explore these complex and fascinating interconnections. Future film geographers should consequently encourage the emergence of critically reflective and creative approaches in order to foster the place of film geography in the edifice of media geography.

References

Aitken, S. C. 1994. I'd rather watch the movie than read the book. *Journal of Geography in Higher Education*, 18 (3), 291–307. https://doi.org/10.1080/03098269408709269.

Aitken, S. C., and Dixon, D. P. 2006. Imagining geographies of film. *Erdkunde*, 60 (4), 326–336. https://doi.org/10.3112/erdkunde.2006.04.03.

Aitken, S. C., and Zonn, L. E. (eds.) 1994. *Place, power, situation, and spectacle: A geography of film*. London: Rowman & Littlefield Publishers.

Balász, B. 1924. *Der sichtbare Mensch oder die Kultur des Films*. Wien: Deutsch-Österreichischer Verlag.

Beeton, S. 2016. *Film-induced tourism*, 2nd edn. Bristol, Buffalo, NY, Toronto: Channel View Publications.

Böcher, D. 2018. *Screentourismus in Dubrovnik: Vom Protocinema bis zur Social-Media-Kommunikation von Game-of-Thrones-Drehorten*. MA thesis, Johannes Gutenberg-University Mainz.

Bruno, G. 2002. *Atlas of emotion: Journeys in art, architecture and film*. New York: Verso.

Brunsdon, C. 2007. *London in cinema: The cinematic city since 1945*. London: British Film Institute.

Bunnell, T. 2004. Re-viewing the *Entrapment* controversy: Megaprojection, (mis)representation and post-colonial performance. *GeoJournal*, 59, 297–305. https://doi.org/10.1023/B:GEJO.0000026703.44576.6f.

Burgess, J., and Goldk, J. R. 1985. *Geography, the media and popular culture*. London and Sydney: Croom Helm.

Busby, G., and Klug, J. 2001. Movie-induced tourism: The challenge of measurement and other issues. *Journal of Vacation Marketing*, 7 (4), 316–332. https://doi.org/10.1177%2F135676670100700403.

Caquard, S. 2009. Foreshadowing contemporary digital cartography: A historical review of cinematic maps in films. *The Cartographic Journal*, 46 (1), 46–55. https://doi.org/10.1179/000870409X415589.

Caquard, S. 2011. Cartography I: Mapping narrative cartography. *Progress in Human Geography*, 37 (1), 135–144. https://doi.org/10.1177%2F0309132511423796.

Caquard, S., and Taylor, D. R. F. 2009. Editorial: What is cinematic cartography? *The Cartographic Journal*, 46 (1), 5–8. https://doi.org/10.1179/000870409X430951.

Caquard, S., and William Cartwright. 2014. Narrative cartography: From mapping stories to the narrative of maps and mapping. *The Cartographic Journal*, 51 (2), 101–106. https://doi.org/10.1179/0008704114Z.000000000130.

Castro, T. 2009. Cinema's mapping impulse: Questioning visual culture. *The Cartographic Journal*, 46 (1), 9–15. https://doi.org/10.1179/000870409X415598.

Castro, T. 2011. *La pensée cartographique des images: Cinéma et culture visuelle*. Lyon: Aléas.

Christopherson, S. 2013. Hollywood in decline? US film and television producers beyond the era of fiscal crisis. *Cambridge Journal of Regions, Economy and Society*, 6 (1), 141–157. https://doi.org/10.1093/cjres/rss024.

Clarke, D. B. (ed.) 1997. *The cinematic city*. New York: Routledge.

Clarke, D. B. 2008. Spaces of anonymity. In C. Lukinbeal and S. Zimmermann (eds.) *The geography of cinema: A cinematic world*, pp. 101–113. Stuttgart: Franz Steiner Verlag.

Conley, T. 2007. *Cartographic cinema*. Minneapolis, MN: University of Minnesota.

Conley, T. 2009. Locations of film noir. *The Cartographic Journal*, 46 (1), 16–23. https://doi.org/10.1179/000870409X430960.

Cresswell, T. 2004. *Place: A short introduction*. Malden, Oxford, Carlton: Wiley.

Cresswell, T., and Dixon, D. 2002. Introduction: Engaging film. In T. Cresswell and D. Dixon (eds.) *Engaging Film: Geographies of Mobility and Identity*, pp. 1–10. Oxford: Rowman & Littlefield.

Curtin, M. 2015. Chinese cinema cities: From the margins to the Middle Kingdom. In S. P. Mains, J. Cupples and C. Lukinbeal (eds.) *Mediated Geographies and Geographies of Media*, pp. 95–109. Dordrecht: Springer.

da Costa, M. H. B. V. 2003. Cinematic cities: Researching films as geographical texts. In A. Blunt, P. Gruffudd, J. May, M. Ogborn, and D. Pinder (eds.) *Cultural Geography in Practice*, pp. 191–201. London: Routledge.

Dagnaud, M., and Feigelson, K. (eds.) 2012. *Bollywood: Industrie des images*. Paris: Presses Sorbonne Nouvelle.

di Palma, M. T. 2009. Teaching geography using films: A proposal. *Journal of Geography*, 108 (2), 47–56. https://doi.org/10.1080/00221340902967325.

Dimbarth, O. 2017. Der Spielfilm als soziales Gedächtnis? In G. Sebald and M.-K. Döbler *(Digitale) Medien und soziale Gedächtnisse*, pp. 199–221. Wiesbaden: Springer.

Dixon, D. 2014. Film. In P. Adams, J. Craine and J. Dittmer (eds.) *The Ashgate research companion to media geography*, pp. 39–51. New York: Ashgate Publishing.

Dixon, D., Zonn, L., and Bascom, J. 2008. Post-ing the cinema: Reassessing analytical stances toward a geography of film. In C. Lukinbeal and S. Zimmermann (eds.) *The geography of cinema: A cinematic world*, pp. 25–47. Stuttgart: Franz Steiner Verlag.

Dodds, K. 2008. "Have you seen any good films lately?" Geopolitics, international relations and film. *Geography Compass*, 2 (2), 476–494. https://doi.org/10.1111/j.1749-8198.2008.00092.x.

Ernwein, M. 2020. Filmic geographies: Audio-visual, embodied-material. *Social & Cultural Geography*. doi:10.1080/14649365.2020.1821390.

Escher, A. 2006. The geography of cinema: A cinematic world. *Erdkunde*, 60 (4), 307–314. https://doi.org/10.3112/erdkunde.2006.04.01.

Escher, A. 2016. Driving the Cheyenne dream: The hero's journey dreamed by Philbert Bobo aka "Whirlwind Dreamer" in the feature film *Powwow Highway*. In M. Banerjee (eds.) *Comparative indigenous studies*, pp. 267–301. Heidelberg: Universitätsverlag Winter.

Escher, A. 2019. Rick's Café Américain, ein Nachtclub in New York City? In M. Stiglegger and A. Escher (eds.) *Mediale Topographien. Beiträge zur Medienkulturgeographie*, pp. 157–187. Wiesbaden: Springer VS.

Escher, A., and Zimmermann, S. 2001.Geography meets Hollywood: Die Rolle der Landschaft im Spielfilm. *Geographische Zeitschrift*, 89 (4), 227–236.

Escher, A., and Zimmermann, S. 2005. Drei Riten für Cairo: Wie Hollywood die Stadt Cairo erschafft. In A. Escher and T. Koebner (eds.) *Mythos Ägypten: West-Östliche Medienperspektiven II*, pp. 162–173. Remscheid: Gardez! Verlag.

Escher, A., Sommerlad, E., and Karner, M. 2017. "King's Landing gibt es wirklich!"—filminduzierte Reisen in imaginierte Welten. In Vorstand der Marburger Geographischen Gesellschaft e.V. in Verbindung mit dem Dekan des Fachbereichs Geographie der Philipps-Universität (ed.) *Jahrbuch 2016*, pp. 157–163. Marburg/Lahn: Selbstverlag der Marburger Geographischen Gesellschaft.

Fletchall, A., Lukinbeal, C., and McHugh, K. 2012. *Place, television, and the real Orange County*. Stuttgart: Franz Steiner Verlag.

Fröhlich, H. 2007. *Das neue Bild der Stadt. Filmische Stadtbilder und alltägliche Raumvorstellungen im Dialog*. Stuttgart: Franz Steiner Verlag.

Garrett, B. L. 2010. Videographic geographies: Using digital video for geographic research. *Progress in Human Geography*, 35 (4), 521–541. https://doi.org/10.1177%2F0309132510388337.

Gasher, M. 2002. *Hollywood North: The feature film industry in British Columbia*. Vancouver: UBC Press.

Gleich, J., and Webb, L. (eds.) 2019. *Hollywood on location: An industry history*. New Brunswick: Rutgers University Press.

Govil, N. 2015. *Orienting Hollywood: A century of film culture between Los Angeles and Bombay*. New York: New York University Press.

Gregory, D. 1995. Imaginative geographies. *Progress in Human Geography*, 19 (4), 447–485. https://doi.org/10.1177%2F030913259501900402.

Griffith, R. 1953. America on the screen. *The Geographical Magazine*, 26, 443–454.

Harper, G., and Rayner, J. 2010. *Cinema and Landscape*. Bristol: Intellect.

Heath, S. 1981. *Questions of Cinema*. Bloomington, IN: Indiana University Press.

Higson, A. 1987. The landscapes of television. *Landscape Research*, 12 (3), 8–13. https://doi.org/10.1080/01426398708706232.

Hughes, R. 2007. Through the looking blast: Geopolitics and visual culture. *Geography Compass*, 1 (5), 976–994. https://doi.org/10.1111/j.1749-8198.2007.00052.x.

Jacobs, J. 2013. Listen with your eyes: Towards a filmic geography. *Geography Compass*, 7 (10), 714–728. https://doi.org/10.1111/gec3.12073.

Jacobs, J. 2016. Filmic geographies: The rise of digital film as a research method and output. *Area*, 48 (4), 452–454. https://doi.org/10.1111/area.12309.

Joliveau, T. 2009. Connecting real and imaginary places through geospatial technologies: Examples from set-jetting and art-oriented tourism. *The Cartographic Journal*, 46 (1), 36–45. https://doi.org/10.1179/000870409X415570.

Kavoori, A. P., and Punathambekar, A. (eds.) 2008. *Global Bollywood*. New York: New York University Press.

Kennedy, C., and Lukinbeal, C. 1997. Towards a holistic approach to geographic research on film. *Progress in Human Geography*, 21 (1), 33–50. https://doi.org/10.1191%2F030913297673503066.

Kim, S. 2010. Extraordinary experience: Re-enacting and photographing at screen tourism locations. *Tourism and Hospitality Planning & Development*, 7 (1), 59–75. https://doi.org/10.1080/14790530903522630.

Knoblauch, H., and Schnettler, B. 2012. Videography: Analysing video data as a "focused" ethnographic and hermeneutical exercise. *Qualitative Research*, 12 (3), 334–356.

Krings, M., and Okome, O. (eds.) 2013. *Global Nollywood: The transnational dimensions of an African video film industry*. Bloomington, IN: Indiana University Press.

Lacoste, Y. 1976. Cinéma—géographie. *Hérodote*, 2, 153–168.

Lukinbeal, C. 2004. The map that precedes the territory: An introduction to essays in cinematic cartography. *GeoJournal*, 59 (4), 247–251. https://doi.org/10.1023/B:GEJO.0000026698.99658.53.

Lukinbeal, C. 2005. Cinematic landscapes. *Journal of Cultural Geography*, 23 (1), 3–22. https://doi.org/10.1080/08873630509478229.

Lukinbeal, C. 2006. Runaway Hollywood: Cold Mountain, Romania. *Erdkunde*, 60 (4), 337–345. https://doi.org/10.3112/erdkunde.2006.04.04.

Lukinbeal, C. 2009. Film. In R. Kitchin and N. Thrift (eds.) *International encyclopedia of human geography*, pp. 125–129. Amsterdam: Elsevier Science.

Lukinbeal, C. 2010. Film and geography. In B. Warf *Encyclopedia of Geography*, pp. 1110–1111. Thousand Oaks, CA: SAGE Publications.

Lukinbeal, C. 2012. "On location" filming in San Diego County from 1985–2005: How a cinematic landscape is formed through incorporative tasks and represented through mapped inscriptions. *Annals of the Association of American Geographers*, 102 (1), 171–190. https://doi.org/10.1080/00045608.2011.583574.

Lukinbeal, C. 2014. Geographic media literacy. *Journal of Geography*, 113 (2), 41–46. https://doi.org/10.1080/00221341.2013.846395.

Lukinbeal, C. 2018. *The mapping of "500 Days of Summer": A processual approach to cinematic cartography*. https://necsus-ejms.org/the-mapping-of-500-days-of-summer-a-processual-approach-to-cinematic-cartography/.

Lukinbeal, C. 2019. The Chinafication of Hollywood: Chinese consumption and the self-censorship of US films through a case study of *Transformers Age of Extinction*. *Erdkunde*, 73 (2), 97–110. https://doi.org/10.3112/erdkunde.2019.02.02.

Lukinbeal, C., and Gleich, J. 2018. Old Tucson: Studio and location, geography and film historiography. *Mediapolis: A Journal of Cities and Culture*, 3 (4). www.mediapolisjournal.com/2018/11/old-tucson-overview/.

Lukinbeal, C., and Sharp, L. 2019. Introducing media's mapping impulse. In C. Lukinbeal, L. Sharp, E. Sommerlad, and A. Escher (eds.) *Media's mapping impulse*, pp. 9–29. Stuttgart: Franz Steiner Verlag.

Lukinbeal, C., and Zimmermann, S. 2006. Film geography: A new subfield. *Erdkunde*, 60 (4), 315–326. https://doi.org/10.3112/erdkunde.2006.04.02.

Lukinbeal, C., and Zimmermann, S. 2008. *The geography of cinema: A cinematic world*. Stuttgart: Franz Steiner Verlag.

Lukinbeal, C., and Zonn, L. (eds.) 2004. Theme issue: Cinematic geographies. *GeoJournal*, 59 (4). https://link.springer.com/journal/10708/59/4/page/1.

Lukinbeal, C., Sharp, L., Sommerlad, E., and Escher, A. (eds.) 2019. *Media's mapping impulse.* Stuttgart: Franz Steiner Verlag.

Manvell, R. 1956. Geography and the documentary film. *The Geographical Magazine*, 29, 417–422.

Mauer, R., and Sommerlad, E. forthcoming. Die Karte im Film. In O. Bulgakowa and R. Mauer (eds.) *Dinge im Film.* Wiesbaden: Springer VS.

Miller, J. 2016. Global Nollywood: The Nigerian movie industry and alternative global networks in production and distribution. *Global Media and Communication*, 8 (2), 117–133. https://doi.org/10.1177%2F1742766512444340.

Morgan Parmett, H. 2019. *Down in Treme: Race, place, and New Orleans on television.* Stuttgart: Franz Steiner Verlag.

Natter, W. 1994. The city as cinematic space: Modernism and place in Berlin: Symphony of a great city. In S. Aitken and L. Zonn (eds.) *Place, power, situation, and spectacle: A geography of film*, pp. 203–228. New York: Rowman & Littlefield.

Ogbeide, B. R. 2012. *Nollywood. The Nigerian film industry.* Aachen: Shaker Verlag.

Penz, F., and Koeck, R. (eds.) 2017. *Cinematic urban geographies.* New York: Palgrave Macmillan.

Plien, M. 2017. *Filmisch imaginierte Geographien Jugendlicher: Der Einfluss von Spielfilmen auf die Wahrnehmung der Welt.* Stuttgart: Franz Steiner Verlag.

Riley, R., Baker, D., and van Doren, C. S. 1998. Movie induced tourism. *Annals of Tourism Research*, 25 (4), 919–935. https://doi.org/10.1016/S0160-7383(98)00045-0.

Roberts, L. 2012a. Cinematic cartography: Projecting place through film. In L. Roberts (ed.) *Mapping cultures: Place, practice, performance*, pp. 68–84. Basingstoke: Palgrave Macmillan.

Roberts, L. 2012b. *Film, mobility and urban space: A cinematic geography of Liverpool.* Liverpool: Liverpool University Press.

Roberts, L. 2020. Navigating cinematic geographies: Reflections in film as spatial practice. In T. Edensor, A. Kalandides and U. Kothari (eds.) *The Routledge Handbook of Place*, pp. 655–663. New York: Routledge.

Roberts, L., and Hallam, J. 2014. Film and spatiality: Outline of a new empiricism. In J. Hallam and L. Roberts (eds.) *Locating the moving image: New approaches to film and place*, pp. 1–30. Bloomington, IN: Indiana University Press.

Roesch, S. 2009. *The experience of film location tourists.* Bristol: Channel View Publications.

Rose, G. 2012. *Visual methodologies: An introduction to researching with visual materials*, 3rd edn. Los Angeles, CA: SAGE Publications.

Roy, A. G., and Huat, C. B. (eds.) 2012. *Travels of Bollywood Cinema: From Bombay to LA.* New Delhi: Oxford University Press.

Schroer, M. 2012. Gefilmte Gesellschaft: Beitrag zu einer Soziologie des Visuellen. In C. Heinze, S. Moebius and D. Reicher (eds.) *Perspektiven der Filmsoziologie*, pp. 15–40. Konstanz: UVK.

Scott, A. J. 2005. *On Hollywood: The place, the industry.* Princeton, NJ: Princeton University Press.

Sharp, L. 2018. *Embodied cartographies of the unscene: A feminist approach to (geo)visualizing film and television production.* https://necsus-ejms.org/embodied-cartographies-of-the-unscene-a-feminist-approach-to-geo visualising-film-and-television-production/.

Sharp, L., and Lukinbeal, C. 2015. Film geography: A review and prospectus. In S. P. Mains, J. Cupples and C. Lukinbeal (eds.) *Mediated Geographies and Geographies of Media*, pp. 21–35. Dordrecht: Springer.

Sharp, L., and Lukinbeal, C. 2017. Film. In D. Richardson, N. Castree, M. F. Goodchild, A. Kobayashi, W. Liu and R. A. Marston (eds.) *The international encyclopedia of geography*, pp. 1–4. New York: John Wiley & Sons.

Sharp, L., and Lukinbeal, C. 2019. Imagined cities (cinema). In A. M.Orum (ed.) *The Wiley-Blackwell encyclopedia of urban and regional studies*, pp. 1–4. New York: John Wiley & Sons.

Shiel, M., and Fitzmaurice, T. (eds.) 2001. *Cinema and the city. Film and urban societies in a global context.* Oxford: Blackwell Publishers.

Shiel, M., and Fitzmaurice, T. (eds.) 2003. *Screening the city.* London: Verso.

Sommerlad, E. 2019a. "Cinematic Faces" der Stadt Marrakech. In L. Pelizaeus (ed.) *Images du Patrimoine Mondial. Changement et persistance des images des sites du patrimoine mondial de l'UNESCO (du Maroc à la vallée du Danube—XVIIIe siècle à nos jours) / Welterbebilder. Veränderung und Persistenz der Bilder von UNESCO-Welterbestätten von Marokko bis in das Donautal vom 18. Jahrhundert bis heute*, pp. 321–339. Münster: Aschendorff Verlag.

Sommerlad, E. 2019b. Der Spielfilm Tangerine als Produktion eines (inter-)kulturell doppelt imaginierten Sehnsuchtsraums. In M. Stiglegger and A. Escher (eds.) *Mediale Topographien: Beiträge zur Medienkulturgeographie*, pp. 243–275. Wiesbaden: Springer VS.

Sommerlad, E.forthcoming a. Filmgeographie. In T. Thielmann and M. Kanderske (eds.) *Mediengeographie: Handbuch für Wissenschaft und Praxis*. Baden-Baden: Nomos.

Sommerlad, E. forthcoming b. *Interkulturelle Räume im Spielfilm*. Wiesbaden: Springer VS.

Spencer, M., and Ayscough, S. 2003. *Hollywood North: Creating the Canadian motion picture industry*. Montreal: Cantos International Publishing.

Staszak, J.-F. 2014. Géographie et cinéma: Modes d'emploi. *Annales de Géographie*, 123 (695/696), 595–604. https://doi.org/10.3917/ag.695.0595.

Steinecke, A. 2016. *Filmtourismus*. Konstanz: UVK Verlagsgesellschaft.

Storper, M. 1989. The transition to flexible specialisation in the US film industry: External economies, the division of labour, and the crossing of industrial divides. *Cambridge Journal of Economics*, 13 (2), 273–305. https://doi.org/10.1093/oxfordjournals.cje.a035094.

Storper, M., and Christopherson, S. 1987. Flexible specialization and regional industrial agglomerations: The case of the US motion picture industry. *Annals of the Association of American Geographers*, 77 (1), 104–117. https://doi.org/10.1111/j.1467-8306.1987.tb00148.x.

Thieme, S., Eyer, P., and Vorbrugg, A. 2019. Film VerORTen: Film als Forschungs und Kommunikationsmedium in der Geographie. *Geographica Helvetica*, 74 (4), 293–297. https://doi.org/10.5194/gh-74-293-2019.

Tooke, N., and Baker, M. 1996. Seeing is believing: The effect of film on visitor numbers to screened locations. *Tourism Management*, 17 (2), 87–94. https://doi.org/10.1016/0261-5177(95)00111–00115.

Tzanelli, R. 2007. *The cinematic tourist: Explorations in globalization, culture, and resistance*. London: Routledge.

Wilson, C. 2016. Welcome to Hollywood North, Canada: A world of stand-ins, tax breaks, studio expansion and cultural erasure/re-inscription. In J. Dobson and J. Rayner (eds.) *Mapping cinematic norths: International interpretations in film and television*, pp. 263–283. Oxford: Peter Lang.

Wirth, E. 1952. *Stoffprobleme des Films*. PhD dissertation, Albert-Ludwig-Universität Freiburg.

Yau Shuk-ting, K. 2009. *Japanese and Hong Kong film industries: Understanding the origins of East Asian film networks*. London: Routledge.

Zimmermann, S. 2007. *Wüsten, Palmen und Basare: Die Cineastische Geographie des imaginierten Orients*. PhD dissertation, Johannes Gutenberg-University Mainz.

Zimmermann, S. 2009. Filmgeographie: Die Welt in 24 Frames. In J. Döring and T. Thielmann (eds.) *Mediengeographie. Theorie—Analyse—Diskussion*, pp. 291–314. Bielefeld: Verlag.

Zimmermann, S., and Escher, A. 2005. "Cinematic Marrakech": Eine Cinematic City. In A. Escher and T. Koebner (eds.) *Mitteilungen über den Maghreb: West-Östliche Medienperspektiven I*, pp. 60–74. Remscheid: Gardez! Verlag.

Zimmermann, S., and Reeves, T. 2009. Film tourism: Locations are the new stars. In R. Conrady and M. Buck (eds.) *Trends and issues in global tourism 2009*, pp. 155–162. Heidelberg: Springer Verlag.

10

APPROACHES TO THE GEOGRAPHIES OF TELEVISION

James Craine

The importance of developing a better understanding of the television-geography relationship can perhaps be uniquely understood by revisiting the events of the 1952 election year. Rosser Reeves was an advertising executive with the famous Ted Bates Agency in the 1950s, when the industry of television was in its infant stages. He was most known at that time for creating the USP, or the "unique selling proposition," where one point is pounded home unrelentingly—it was annoying but effective as his Anacin ("fast *fast* **fast** relief") and his M&M ("melts in your mouth, not in your hand") commercials demonstrated. Always pushing the boundaries of advertising, Reeves uncovered the *value* of television's political landscape during the months ahead of the 1952 Eisenhower/Stevenson presidential election. Reeves was so good at selling M&Ms and Anacin that in late 1951 he was hired by the Republican party to see what he could do with their lackluster candidate, Dwight D. Eisenhower. Eisenhower may have been the hero of D-Day and the war in Europe, but when it came to public speaking, the General was hopeless and completely uninspiring. And he was up against a great public speaker, Adlai Stevenson. (Stevenson's best line, among many, was "I think the country is ready for more of the specific, and less of the General"). Eisenhower was at sea and so were Republican hopes for the White House unless Reeves could figure out a way to "package" the General… like M&Ms.

Reeves' great concept was "electorate penetration," his code for short, low-cost political spot ads—for television, not radio—that had an exposure potential into the many thousands as opposed to the hundreds garnered by in-person speech attendance. These ads could be focused on critical districts in critical states. They were too short to be tuned out unlike the long-winded speeches that were a Stevenson trademark. The ads were difficult to avoid—families only had one television set that was now the center of attention in a postwar homespace created and built throughout the United States to accommodate just such viewing engagements. The ads Reeves created were specifically designed to take advantage of this televisual landscape. They were *new*—no one had done this before—and people remembered them. Reeves estimated a content recall of nearly 90 percent for the Eisenhower ads while Stevenson's speeches registered at around a 10 percent retention rate.

By all accounts Reeves worked completely on the fly in a New York City studio on Broadway. He wrote a series of issue-based "answers" that were then transferred to large cue cards (Eisenhower needed glasses but refused to wear them in public). He put Eisenhower in

132

DOI: 10.4324/9781003039068-12

a studio, shot him from a low angle and had him repeat the series of "answers." When Reeves was satisfied with the "answers," passersby—"average people"—were literally grabbed off the streets of New York City. Reeves pulled them into the studio and asked a series of "questions," looking upward. He edited the two different parts, Eisenhower answering each question, first looking down in the direction of the speaker and then turning to talk directly into the camera and thus the television audience. Reeves thus created the first televised political commercials in America. Eisenhower handily won the election. Reeves understood the new medium of television better than anyone else at the time—no more criss-crossing the country in a train, the way Harry Truman had in 1948. Reeves understood what made people buy a product. And Eisenhower was a product to be sold just like M&Ms. And, thanks to Reeves, the image became more important than the issues (many of Reeves' commercials and all of the Eisenhower ads are easily available on YouTube).

With Reeves' ability to foresee the future of television, the landscape of American politics was instantly changed. The spatial aspect of how we, as a country, elected our political officials was now radically different from just a few months earlier. Television became an important geographical technology: homes were designed with an entertainment *space* and we became "viewers," consumers held in place by an instrument that now determined our very culture (my use of the term viewer/consumer is informed by Fiske (1987) who argues that the term suggests an *active* agent who contributes meaning to a media text). Americans moved from the interactions of the front porch to the "den." The industry of television rapidly took over American lives—it was how we learned of the world, how we consumed geography. It is this intersection that informs this chapter: the geography of television.

By way of an *entre* to this discussion of the geography of television we should first understand that one thing television does, in terms of *place*, is help foster a collective memory: television has made massive quantities of memory accessible to anyone with the means to own a receiver or simply have any type of viewing options anywhere. The way in which we experience memory of place has undergone a qualitative change—television and the ancilliary products it can distribute (i.e. DVDs, Blu Rays, streaming in all its forms) allow the viewer to relive whatever parts of the past the viewer chooses to experience. The past assumes a sensory presence, creating the illusion that the viewer is in the presence of a past reality. Televisual products, because of their narrative structure and temporal immersion in *place*, allow the viewer the opportunity of "being there"—of inhabiting that place. Thus, the geography of television can be very much phenomenological in the manner of Heidegger's conceptualization of *being-in-the-world*.

A problematic aspect of the geographical research related to television is, quite simply, the lack of such research. Much of the research related to the *geographical* and *spatial* aspects of television are found in communications studies and media studies and, therefore, geographers who do study television often utilize theory that is common to these fields and to geography. Media geographers have found Deleuzian theory quite useful, just as communications studies has, especially in the theories related to "new media" and beyond. A theme common across the disciplines is the concept that television operates as spatial form—it represents space, place and landscape within a series of electronic images and these spatial contexts have also shaped the practices of production and the meanings contained within televisual environments. Television also plays a central role in making social imagery concrete as part of the "real"— televisual content (in whatever format) has a material effect for those individuals and social groupings that construct and view them. Therefore, any analysis of the role of televison in the discipline of geography involves blurring the distinction between the real and the imagined. Geographers too often only consider space as the size of the geographical places and their associated processes. In other words, space is a macro-environment that exists in space-time, having complex processes and meaning. But space can also be form or structure, or pure space, or even space as geometry, or, importantly, in today's world, it can be a digital virtual space composed of information. Current research involves a theoretical and methodological approach that engages television as a specific material object existing in space and worth studying as a distinct geographical record within a broader set of practices and discourses.

As previously mentioned, and to put theory into a common context, one current engagement of media geography and the visual is grounded in the work of Gilles Deleuze and his concept of the movement-image—an actualization of the virtual in which images become embodied through an affective process. But the digital representation was unknown to Deleuze. Communication studies and some in geography therefore move beyond Deleuze to a geographic articulation of the concept of virtual *affectivity*. As one example in this movement, the work of Pierre Levy privileges the computational power that lies behind digital and virtual technologies, thereby promising an opportunity to more fully comprehend the geographic data coded as an array of iconic images and representations positioned within digital and virtual space.

As geographers interested in the role of television in space, we are beginning to explore these immersive virtual environments. The technologies have moved from basic telemantics (the synthesis of telephony and digital imagery) to virtual technologies that are digitized environments that rely upon a naturalized picture language that is more conducive to collapsing experiential differences between the virtual and the real than those previously available in the earlier analog applications of Reeves' time. Today's geographers can examine the intersection of geographic information conceived as a technology with the spatial relationships set up between the digital environments of the television and the body of the viewer/consumer in the hope that this new form of engagement can bridge the gap between geography and our understanding of the role of television in the creation and maintenance of space and place. To be relevant, geographical research must engage the representational spaces of the televisual medium and its effect (and *affect*) on modern spaces and identities, a methodology that incorporates new *geographic thought*. Media geographers have, therefore, created a critical geography of television to uncover patterns and relationships within the spaces of television.

One of the first articles to explore the relationship between humans and modern technologies was Horvath's 1974 "Machine space" piece in *The Geographical Review*. Horvath states:

> Now that the impact of humankind on the surface of the earth is appreciated in geographical thought, it is not premature to investigate the idea that the habitable area of the world itself may be in the process of being reduced quantitatively and qualitatively through the impact of modern technology… Until now, technology has been viewed largely as an a spatial phenomenon, and one of the major tasks here will be to translate technology into explicitly spatial terms… Machine space, or territory devoted primarily to the use of machines, shall be so designated when machines have priority over people in the use of territory.
>
> *(pp. 167–168)*

Horvath used the automobile as his illustration of a "machine space"—an example of how a technology was now not only capable of interacting with humans within space and time but could also supplant humans as the controlling force within that space. Horvath viewed this as a "problem" where "mounting evidence suggests more emphasis should be placed on seeking nontechnological solutions to problems of a technical origin" (1974, 185). His non-technological solution was to create "auto-free zones" and he offered a path for future research:

> The concept of machine space is offered as one way for geographers to participate in the wider questioning of human purpose and of the consequences of further tech-nological growth. Analyses of other types of machine space... need to be undertaken.
>
> *(1974, 188)*

Geographers did indeed study "technological growth" although it would be nearly 20 years before televisual technology was directly addressed.

In 1992, Paul Adams' "Television as gathering place" article was published in the *Annals of the Association of American Geographers*, marking the first true engagement of the "uniquely place-like" qualities of television in the discipline of geography. Adams was more interested in television as "center of meaning" as opposed to the "machine space" of Horvath, viewing television as an "environment" that allowed "people to gather and/or share experiences" (1992, 117). Adams discusses the prevailing theory at the time, one that looked at the "supposed psychological and social effects of television on society" and one that drew on "cultural materialism, linguistics, and semiotics to attempt to understand how television emerged from and works to reinforce existing structures and practices in society" (1992, 119). There is a reference to Raymond Williams' *Television: Technology and Cultural Form* (1974) and how class differences can determine cultural messaging from various forms of art. Williams, a cultural theorist, working during the birth of critical geography, explored how art is perceived by different social-class groups and, as Adams indicates, this book, along with Williams's later work, *Culture* (1981), is very relevant to media geographers, especially in making the connection between culture and class identity. But Adams' review of the current geographical literature demonstrates that while there was interest in television as part of the larger cultural practices of the communication landscape and industry, there was no exploration of televisual spatialities unto themselves. While Adams' main interest is his elaboration of television as "place" in the context of, say, Tuan and others of that school of place theory, Adams does, however, make the first linkages between television and affect, and by proxy, to the media theories discussed in more detail below.

Beyond the early work of Horvath and Adams, theories related to televisual processes, specifically those related to television spatialities, were more the domain of communication studies. Geographers began to bring into focus the subfield of *media geography* but the research was primarily devoted to cinema (a topic discussed in depth elsewhere in this book). While in principal one could indeed find some overlap in theories applied to cinema, televisual spaces found more purpose in research elsewhere, particularly in television's qualities of vir-tuality and affect, theories that became the foundation of "new media" research. Televisual spaces have become much more than simply a viewer consuming images imparted by a receiver—with the advent of immersive technologies such as DVDs, BDs and streaming, consumers now interact with the "real" spaces constructed around their home entertainment spaces *and* the virtual spaces created by these new televisual formats. Thus, beginning in the 1990s, new media-related theories were advanced, theories that attempted to incorporate the phenomenological aspects of both the concrete and the virtual, specifically how place, space and meaning are constructed in and through these new technologies. This research was multi-disciplinary, yet even those outside of geography addressed, directly or tangentially, the spatial aspects embodied within televisual technologies in their various formats.

Out of this research, the relationship between television and the body of the viewer in space became an important aspect of media theory. How do the visual electronic images produce affect? How does television create place via that affect? According to Weber:

> what distinguished television from other media is its power to combine such separation with the presentness associated with sense perception. What television transmits is not so much *images*, as is almost argued. It does not transmit representations but rather the *semblance of presentation as such*, understood as the power not just to see and to hear but *to place before us*. Television thus serves as a surrogate for the body in that it allows for a certain sense-perception to take place; but it does this in a way that no body can, for its perception takes place in more than one place at a time.
>
> *(1996, 166–167)*

Following Weber's work, Johnston (1999), drawing on Horvath's neologism, coined the term "machinic vision" that "presupposes not only an environment of interacting machines and human-machine systems but a field of decoded perceptions that, whether or not produced by or issuing from these machines, assume their full intelligibility only in relation to them" (Johnston 1999, 27). Integrated into geography, Johnston provides geographers with a discussion of the concerns of qualitative methodology and how the world is viewed, experienced and constructed by social actors. In Johnston's view, television becomes one of these "social actors" that codes and decodes space and place thus becoming invested in Deleuzian spatialities. This methodology provides the access to the motives, aspirations and power relationships that account for how places, people and events are made and represented, an interpretation of texts that can include landscapes, archival materials, maps, literature or visual images. A more recent example of how Deleuzian theory has found its way into the televisual discourse is how Salvado-Corretger and Benavente (2019) use the HBO series *Westworld* to explore questions of time and identity. They bring into the discussion the theories of Deleuze and Bergson and their process of "metaphysical reflection" to uncover the confusion between landscapes of reality of dreams and the role of memory in determining who exactly is "human" in the Westworld realm.

With the advent of interactive televisual media, new post-Deleuzian theories were needed to explore in more depth the relationship between the viewer and the body, the television, and the virtual interaction between those two entities. Media theorist Pierre Lévy (1998) moved visual theory into a modern neuroscience—there is now much more than a passive correlate of linkages between images—the body now has a creative capacity. Lévy was primarily interested in the affective qualities of the televised image, be it a series of images directed to the viewer (as from a television production) or via a computer monitor (for geographers, this would of course be geographic information systems (GIS) technology). By becoming *virtual*, as Lévy proposes, the viewer/consumer makes the connection between movement and sensation to the point that the slightest, most literal displacement invokes a qualitative difference—motion thus triggers affection as an active modality of bodily action. This began to find root in geographical research as media geographers have begun to privilege the power that lies behind digital and virtual technologies, thereby promising an opportunity to more fully comprehend geographic data coded as an array of iconic images and representations positioned within digital and virtual space.

More recent engagements with digital media, including the digital realms of television, such as Hansen (2003; 2006), facilitate the move into the realm of virtual affectivity. Whether virtual space is real or not, our experience of these spaces is a "real" experience. For Hansen, the viewer/consumer becomes the *virtual* subject. Thus, the virtual digital environment becomes a fundamental part of human experience. With this in mind, media geographers have now moved into discussions of specific modalities of the virtual (e.g. GIS, gaming, social media), utilizing a wide array of theory to more concretely place the virtual into the discipline. As an example of this research, Schneider (2018) uses Hansen's example of the CBS television show *Person of Interest* to discuss the role of television in new media theorizing. Drawing on Merleau-Ponty, Schneider uses Hansen's work to comment on the connective interactivity between digital devices, the virtual environment and the user of these technologies. Via television, both authors delve into how subjectivity is the capacity or power to sense and be sensed in today's media environment, a viewpoint certainly relevant to geography's technological turn. Hansen (2003; 2006; 2016) goes much further down this road with more extreme in-depth discussions of the development of media theory and its application to new forms of media, all of which is applicable to geography's theorization of virtual, televisual and cinematic space. Beyond Hansen's work in communication studies, Adams places television under the overarching umbrella of geographies of media and communication, stating: "the space of flows created by signals moving through the infrastructure are complementary parts of a spatial perspective we capture throughout this book with the term *media in space*" (Adams 2009, 1). Like Hansen and Lévy, Adams (109) recognizes the virtual qualities of television in his discussion of virtual centrality and how that concept applies to the manner in which television gratifies various desires. Further, Adams (p. 201) also explores "expressive being-in-place" via the phenomenological approach commonly found in qualitative geographical research, along with others discussions on the utilization of nonrepresentational geography and affect in the study of media and communications—theories that can be applied to specific research on television. Coleman and Oakley Brown (2017) conflate the concepts addressed in "new media" research such as how television is a virtual space/place, in this case a "surface" that visualizes and is also a surface in itself: television becomes one element in a "network imagination" of "teletechnology" that places television into contemporary network imaginations that are then thoroughly embedded in the flow of social life. In the end, the authors find that the linguistic model of understanding contemporary social life has been displaced by one of images, and how those images are produced, viewed and engaged (Coleman & Oakley-Brown 2017, 23).

The semiotic aspects of televisual space, first discussed by Adams in 1992, also advanced in conjunction with the development of new media theory. Dery (1999) postulated that a critical analysis of the content of popular media, like television, is necessary to understand modern culture because these representations map the material landscape by engaging audiences in the construction of new geographies that display the social and material world. Written pre-9/11, Dery nonetheless was able to foresee the dramatic impact the export of American culture would have on its creators. Media geographers will find this insightful for its discussion on the production of culture through media, and how that, invariably according to Dery, all goes wrong in the end.

Dery is also very aware of the role of television and its product-creating spaces of resistance and paranoia:

> The antigovernment sentiment that hangs menacingly over *The X-Files* first appeared on our mental horizons during Watergate (though it took Ronald

Reagan's covert policy of benign neglect toward a government he openly regarded as "not the solution to the problem" to whip the free-floating contempt into the angry thunderhead it is today). *The X-Files* is haunted by the restless ghosts of Watergate and Vietnam, with Richard Nixon, the patron saint of conspiratorial realpolitik and bunker paranoia, at their head.

(1999, 17)

Crang (2003a) makes the connection between semiotics and geography and visual media. This becomes important to geographers because it links visuality to the goal of obtaining geographical knowledge and discusses the views of most semioticians that meanings are relational rather than fixed, in that signs derive their meaning from other signs and from the wider system of signs, and not just from their actual form or content. Taken a step further, current theories of visualization surmise that the mental maps that compose the themes located within geographic space are mediated internally by the systems of signs, symbols and signals people have previously internalized through the experiential negotiation of constructed landscapes. Using Lévy to bring out the television/internet nexus, Craine (2009) explores how signifying elements work between actual places, television, and the internet by referencing the real landscapes of Los Angeles, the visualized landscapes of the Fox Network television show *The Shield*, and the virtual landscapes of http://theshieldrap.proboards45.com . Eva Kingsepp (2016) uses a form of semiotic analysis in her discussion of the portrayal of Egypt in World War II television documentary films, indicating how places are intimately connected to narratives, particularly how narratives transform place into space and space into place. Liz Roberts (2016) is among those who see the geography of visuality, and by extension, television, in terms of the idea of "image as text," thereby centering the study of televisual texts with the language of signs that are decoded by the consumer of the visual product. This approach echoes the work of Raymond Williams who believed that this "decoding" would vary from person to person depending on their specific social position.

Putting all of the above together, an example of how the research of a media theorist has a decidedly spatial context can be found in the work of Brian Massumi. Incorporating the phenomenological, virtual and semiotic aspects of televisual space, Massumi (writing about the Super Bowl) looks at how the interruption of home space by sports has an extreme affective quality:

> The televised game enters the home as a domestic player... The home entry of the game, at its crest of intensity, upsets the fragile equilibrium of the household. The patter of relations between house-held bodies is reproblematized. The game even momentarily interrupts the pattern of extrinsic relations generally obtaining between domestic bodies, as typed by gender. A struggle ensues: a gender struggle over clashing codes, rights of access to portions of the home and its contents, and rituals of servitude. The sociohistorical home place coverts into an event-space.
>
> *(2002, 80)*

The television then, for Massumi, "is more about delivery into a more-or-less open milieu than it is about the perspective of one closed space onto another, or of a closed space onto an open space" (Massumi 2002, 85). Television binds digital technologies to the "analog" process of consumption, thus cementing the relationship among a digital data source, actualized experience, embodied sensation and the virtual. Here, Massumi is one of the first to connect television's virtual space to movement and sensation through affectivity. His theories can be perceived as important to future discussions of GIS and other forms of virtual cartographies, a topic very underdeveloped (perhaps even misunderstood) in geography at this time.

There are numerous examples of how specific television productions *work* to create space, often referred to as "convergence." This can be the phenomenological sense related to the body's interaction with the show, the economic sense as a continuation of capitalist practices related to the production of value, or the production of cultural value. Jenkins explains "convergence" thusly: "the flow of content across multiple media platforms. The cooperation between multiple media industries, and the migratory behavior of media audiences who will go almost anywhere in search of the kinds of entertainment experiences they want" (Jenkins 2006, 2). Crothers (2010, 108) offers a similar definition of convergence, defining it as "the process by which synergy in the entertainment industry has been created... the term describes how the many companies and individuals who used to make popular culture have been reduced in number to the few corporations that control the trade today." As an example, Crothers noted the role of televisual familial relationships in decoding the space of the American home in his discussion of *The Cosby Show*:

> White Americans tolerated and even celebrated this vision of an employed, educated, and successful African-American family. The Huxtables reflected the broad patterns of American public culture in ways that reinforced the values and ideals Americans claim to value—such as hard work, tolerance, capitalist success, and so on... Put another way, it is hard to imagine a television show being as popular as *The Cosby Show* was if the family were dysfunctional and headed by a single woman working several jobs or receiving welfare who decried her fate as a black female in a racist America.
>
> *(2010, 35)*

Crothers also uses the original *Star Trek* and *The West Wing* to make the case for television as an expression of American political culture and ideals. He states (2010, 46–47) that:

> much of the continuing popularity of this science fiction program lies in its vision of a globalized, ethically driven, rights-respecting, democratic future—American culture manifest on a galactic scale... In many ways, then, the Federation manifests the ideals of American democracy in actual practice. The world of *Star Trek* expresses the ideals that are at the core of American public culture.

The West Wing "is the story of the best parts of American civic culture played out with sufficiently realistic touches to make the politics seem real" (Crothers 2010, 49). Jenkins uses two examples of American TV reality shows (*Survivor* and *American Idol*) to explain the convergence process—how those two shows "demonstrated the power that lies at the intersection between old and new media" (Jenkins 2006, 59). The growth of "reality tv," that intersection of old and new media, is shaped by what Jenkins terms "affective economics," a new configuration of marketing theory that "seeks to understand the emotional underpinnings of consumers' decision-making as a driving force behind viewing and purchasing decisions" (Jenkins 2006, 62). In terms of geography, understanding the consumption of media from the fan's point of view was now replaced by the need to understand how consumer desires shaped purchasing decisions. The spatialities of the simple consumption of the televisual product in the confines of the home, usually the domain of the "commercial," was replaced by a network product actively soliciting the viewer to make a purchase *outside* of the home. Jenkins recognizes the paradigm shift of "convergence"—what he sees as:

a move from medium-specific content toward content that flows across multiple media channels, toward increased interdependence of communications systems, toward multiple ways of accessing media content, and toward ever more complex relations between top-down corporate media and bottom-up participatory culture.

(Jenkins 2006, 243)

Applying the convergence theory globally, Zimmerman (2007) places the importance of televisual technology into the geographical discourse via his belief that "Film and television emotionalize space, place, movement, and identity thereby affecting the viewer's perception. Visual media have also become active agents in globalization in that they spread Western cultural beliefs and attitudes" (p. 60).

Researchers have looked at how gendered and LGBTQ spaces are created via television. Marcia England, in her chapter on "Visions of gender: Codings of televisual space," looks at the *Roseanne, The Cosby Show*, and *Murphy Brown* sitcoms to uncover the role of television in the formation of gendered social relations of (re)production by reinforcing traditional patriarchal frameworks. England explains how television gender roles are reinforced by the activities characters perform within their televisual spaces. Drawing on the feminist visual theories of Gillian Rose, England goes on to state that:

> Feminist geography needs to address the medium of television more directly due to the powerful force of reproduction and the spaces in which these processes take place... [and] the role of television in the formation of socio-spatial identity is important to analyze.

(2018, 106)

Similarly, Pinedo compares the Nordic crime drama *Forbrydelsen* to its American remake *The Killing* in an effort to draw out the significance of "postfeminism and neoliberalism as interpretive grids" (2019, 1) that limit feminist progress in the United States. The author also discusses the processes of transnational television distribution and how this affords the opportunity to experience cultural differences such as gender differences.

Herold looks at how the New York City LGBTQ community used public access television in the late 1970s and into the 1980s to provide a forum for their experiences, concerns, community and businesses. A special 1983 episode on the show *Our Time* looked specifically at the emerging AIDS crisis creating what the author termed a "televisual emotional pedagogy about AIDS" (2020, 25). The article explores the history of *Our Time* and its importance to the LGBTQ community, and how the affective qualities of urgency, fear, and anger reached communities underserved by broadcast networks and helped mobilize activists to begin producing more AIDS-related content for public access channels. Cavalcante (2018) studies affect and emotion via LGBTQ audience research, or what the author terms "resilient reception." How media representations "do things" to the cognitive and emotional life of audiences and how affect and emotion accumulate in individual bodies and in larger communities provide insight into the role of television in shaping LGBTQ identities. Interestingly, *Person of Interest* is referred to in the context of audience perception here, perhaps illustrating how productions that feature digital interactivity are important to shaping identity, a tenet advanced years earlier by Lévy.

There has also been interesting research uncovering television's role in creating political and economic spaces. Virino and Ortega discuss how the HBO show *Game of Thrones* paralleled the rise of the *Podemos* political party in Spain by way of forming a bind between the party and the participatory culture associated with the television show. The authors state:

> The Podemos-*Game of Thrones* case allows us to point out the key importance of TV fiction within the contemporary public sphere. Specifically, it allows us to establish the role of televisual fiction as a tool for political parties to communicate with their voters and sympathizers.
>
> *(2018, 4)*

Brunn et al. (2004) in their *Geography and Technology* volume understand television as a form of geographic technology and a form of geographical communication, with Wilbanks (2004, 10) exploring the role of television news in promoting public interest in geographic information and Rain and Brooker-Gross (2004, 315) exploring how the 24-hour global news media is seemingly placeless, just a stream of images and commentary "flowing *somewhere* into your living room." Rain and Brooker-Gross present the idea that geographers can move past traditional field and survey-type research by creating new methodologies that provide a better understanding of the geographical qualities of televisual news, especially the power of place depictions (p. 317). Within the broader discussion of global news, Rain and Brooker also discuss the landscape of news production and consumption and how that is a function of a geography of media ownership that is concentrated in just a few global cities (p. 319). Parmett and Rodgers look at the relocation of media production in the post-Fordist landscape. Instead of "flattening" as geographers have found with other capitalist economic processes, media production has become more local. We see the influence of "convergence" here as well in the way on-location filming practices have brought about rebranding and entrepreneurial competition in on-location urban spaces.

A more recent example of television's role in the creation of economic space is Mehta's (2020) discussion of the role of television in changes related to how Indians engage the screen. The over-reliance on soap operas that were often identity-based restricted viewers' ability to engage with the content. Corporate management stifled creativity and insisted on grueling production schedules thus creating an exodus of key production personnel who then began to create internet content. Based on interviews with actors from all aspects of the Indian television structure, the author found that the continuing reliance on the same content, coupled with low internet access prices in India, drove audiences away from television to internet streaming services in search of more diverse and varied programming, dramatically remaking the Indian media landscape.

Of particular note is Brett Christophers' 2010 book, *Envisioning Media Power: On Capital and Geographies of Television*. Christophers (2010a) devotes his book, the first book specific to television geographies, to uncovering the capitalist power relations that drive the television economy, including the "circuits of capital" that exist at every space and place within the local-global nexus. The work is important, even with its narrow focus, because it is a geography text that firmly connects television to space and place, predominantly in Australia/New Zealand, but also ranging to the major production centers of the television world. Christophers draws on the work of many prominent geographers, including David Harvey, Allen Scott, Neil Smith, among others, and presents his own analysis of how television is geographically located within capitalist place and space and how television can transform and maintain social and power relations within those spaces and places. While television-specific (and Australian television at that), the book is still a very insightful case study of how power is produced and maintained through televisual communication. Discussions of television are somewhat lacking in media geography, and the book does not ignore the place of television in the production of culture but instead concentrates on the results of that production. Christophers (2010b) later acknowledged that the study of television by geographers could be grouped into three "useful" categories: geographies *on* television, the way geography is represented on television; geographies *of* television, the organization of television as an industry and a technology; and "the myriad ways in which television shapes and reshapes the fundamental *experience* of space and place and of spatially mediated identities" (2010b, 2791).

While many of the above works are specific in their discussions related the application of visual theory, one should not overlook the contributions of Gillian Rose. Rose's (2007) *Visual Methodologies*, while perhaps more general in its approach, is often considered to be the key introductory text to visual methodologies. As a source for the ongoing development of the methodologies found in the study of media geography, it is useful for the different forms these methodologies can take. One of the keystone texts of media geography, it can be used particularly as a critical visual methodology. Rose explains how film manipulates the visual, the spatial and the temporal, in an attempt to structure "looking" and how that affects the gaze of the spectator, an insight that is applicable to the study of television geographies. Rose also maintains that there are important visual components to geography that should be further explored, and she proposes a variety of methodological approaches. One other text of interest is Steve Kosareff's *Window to the Future: The Golden Age of Television Marketing and Advertising*. The "window" is television, and the book follows two parallel paths: the history of the development of television as a technology and how television was advertised in various forms of American media (particularly printed media). The textual component is small because the bulk of the book consists of reproductions of the advertisements, but it nonetheless provides more insight into the growth of the televisual space in the American home. One last peripheral aspect of televisual space is that of sound. Coulthard uses affective theory to discuss how the music in three international televisual crime series is scored to "align with cinematic and international aesthetics" and reinforce the prominence of the female characters portrayed in these series (2018, 553).

Michele White makes an interesting observation regarding the value of cross-disciplinary studies of television:

> A number of academic disciplines, which include but are not limited to television studies, offer methods to consider systems of power and knowledge that have produced this economy. Feminist theories of spectatorship… present critical strategies to oppose the representations of the "other"… Postcolonial and critical race studies and a variety of other critical strategies can indicate how populations are made to seem dispensable because of their age, class, gender, global position, race, or sexuality. Engaging with the varied aspects of [electronic media] requires the adoption of critical strategies from different disciplines.
>
> *(2006, 196–197)*

Further, Warf and Arias (2009) have discussed how the "spatial turn" in geography has brought about collaborations between the disparate disciplines, especially in research that might be considered "nontraditional." Along that same thought, Hallam and Roberts (2014) reach the same conclusion after researching how the advent of television in Melbourne, Australia led to the closure of the city's cinemas—that there is geographical significance and benefit to be found in cross-disciplinary approaches. Advocating for geographers to undertake the study of video, Garrett (2016) refers to Crang (2003b) and the issues that stem "from the processual approach researchers have taken toward these methods as forms of "data collection," rather than making the opportunity to generate intellectually robust aesthetic materials" (2016, 685). Garrett further states that within geography there "remains a small minority of research in the discipline. There is a reluctance to undertake research with video that may stem from technological anxieties" (2016, 687). While Garrett is specifically referring to the use of video technologies to record "data," the inference still remains that those within geography who undertake research in video-related topics are a small minority.

The need for geography to successfully incorporate theories from other disciplines is perhaps the underlying theme of this chapter, especially in television-related discourse. While media geographers have done a masterful job interjecting cinematic space into the realm of geographic research, I would argue then that an ontology of television may be derived through an ontology of its geographical space. We can subscribe to the Heideggerian notion of technology as a mode of revealing: viewers/consumers coming in contact with televisual worlds in whatever format can now see something previously hidden. Televisual technology reveals something: viewers/consumers can now enjoy the capacity to control the presentation and performance of self in these contexts along with an increased ability to control the conditions of interaction. As White states above, engaging the "varied aspects" of television and its adjacent research offers the opportunity to better engage digitelevisual space and thus gain a deeper appreciation and wider comprehension of spatial information.

References

Adams, P. C. 1992. Television as gathering place. *Annals of the Association of American Geographers*, 82 (1), 117–135.

Adams, P. C. 2009. *Geographies of media and communication*. London: Wiley-Blackwell.

Brunn, S. D., Cutter, S., and Harrington, J. W. (eds.) 2004. *Geography and technology*. Berlin: Kluwer Academic Publishers.

Cavalcante, A. 2018. Affect, emotion, and media audiences: The case of resilient reception. *Media, Culture & Society*, 40 (8), 1186–1201. https://doi.org/10.1177/0163443718781991.

Christophers, B. 2006. Circuits of capital, genealogy, and television geographies. *Antipode*, 38 (5), 930–952. https://doi.org/10.1111/j.1467-8330.2006.00487.x.

Christophers, B. 2010a. *Envisioning media power: On capital and geographies of television*. Lanham, MD: Lexington Books.

Christophers, B. 2010b. Television and geography. In B. Warf (ed.) *The encyclopedia of geography*, pp. 2791–2793. London: Sage Publications. https://doi.org/10.4135/9781412939591.n1122.

Clough, P. T. 2000. *Autoaffection*. Minneapolis, MN: University of Minnesota Press.

Coleman, R., and Oakley-Brown, L. 2017. Visualizing surfaces, surfacing vision: Introduction. *Theory, Culture & Society*, 34 (7–8), 5–27. https://doi.org/10.1177/0263276417731811.

Connolly, W. E. 2002. *Neuropolitics*. Minneapolis, MN: University of Minnesota Press. https://doi.org/10.5749/j.cttts8p6.

Coulthard, L. 2018. The listening detective: Thinking music, gender, and transnational crime's affective turn. *Television & New Media*, 19 (6), 553–568. https://doi.org/10.1177/1527476418768008.

Craine, J. 2009. Virtualizing Los Angeles: Pierre Levy, "The Shield," and http://theshieldrap.proboards45.com. *GeoJournal*, 74 (3), 235–243.

Crang, M. 2003a. The hair in the gate: Visuality and geographical knowledge. *Antipode*, 35 (2), 238–243.

Crang, M. 2003b. Qualitative methods: Touchy, feely, look-see? *Progress in Human. Geography*, 27 (4), 494–504.

Crothers, L. 2010. *Globalization & American popular culture*. Lanham, MD: Rowman & Littlefield.

Deleuze, G. 1986. *Cinema I: The movement-image*. Minneapolis, MN: University of Minnesota Press.

Dery, M. 1999. *The pyrotechnic insanitarium: American culture on the brink*. New York: Grove.

England, M. R. 2018. *Public privates: Feminist geographies of mediated spaces*. Lincoln, NE: University of Nebraska Press.

Hallam, J., and Roberts, L. 2013. *Locating the moving image*. Indianapolis, IN: Indiana University Press.

Hansen, M. B. N. 2003. *New philosophy for new media*. Cambridge, MA: MIT Press.

Hansen, M. B. N. 2006. *Bodies in code: Interfaces with digital media*. London: Routledge.

Hansen, M. B. N. 2015. *Feed-forward: On the future of twenty-first century media.* Chicago, IL: University of Chicago Press.

Hansen, M. B. N. 2016. Media theory. *Theory, Culture & Society,* 23 (2–3), 297–306. https://doi.org/10.1177/026327640602300256.

Herold, L. 2018. Televisual emotional pedagogy: AIDS, affect, and activism on Vito Russo's *Our Time. Television and New Media,* 21 (1), 25–40.

Horvath, R. J. 1974. Machine space. *Geographical Review,* 64 (2), 167–188. https://doi.org/10.2307/213809.

Jenkins, H. 2006. *Convergence culture.* New York: New York University Press.

Johnston, J. 1999. Machinic vision. *Critical Inquiry,* 26 (1), 27–48.

Kaneva, N. 2018. Simulation nations: Nation brands and Baudrillard's theory of media. *European Journal of Cultural Studies,* 21 (5), 631–648. https://doi.org/10.1177/1367549417751149.

Kingsepp, E. 2016. Mythical space: Egypt in World War II TV documentary films. In C. Lukinbeal*et al.* (eds.) *Media's mapping impulse,* pp. 161–185. Berlin: Franz Steiner Verlag.

Kosareff, S. 2005. *Window to the future: The golden age of television marketing and advertising.* San Francisco, CA: Chronicle.

Lévy, P. 1997. *Collective intelligence: Mankind's emerging world in cyberspace.* R. Bononno (trans.). New York: Plenum Trade.

Lévy, P. 1998. *Becoming virtual: Reality in the digital age.* R. Bononno (trans.). New York: Plenum Trade.

Lotz, A. D. 2020. The future of televisions, a response. *Media, Culture & Society,* 42 (5), 800–802.

Massumi, B. 2002. *Parables of the virtual.* Durham, NC: Duke University Press.

Mehta, S. 2020. Television's role in Indian new screen ecology. *Media, Culture & Society,* 42 (7–8), 1226–1242. https://doi.org/10.1177/0163443719899804.

Parmett, H. M., and Rodgers, S. 2018. Re-locating media production. *International Journal of Cultural Studies,* 21 (1), 3–11. https://doi.org/10.1177/1367877917704479.

Pinedo, I. 2019. *The Killing*: The gender politics of the Nordic Noir crime drama and its American remake. *Television & New Media,* 22 (3), 299–316. https://doi.org/10.1177/1527476419875572.

Rain, D. R., and Brooker-Gross, S. R. 2004. A world on demand: Geography of the 24-hour global news. In S. D. Brunn, S. Cutter and J. W. Harrington (eds.) *Geography and technology,* pp. 315–337. Berlin: Kluwer Academic Publishers.

Roberts, L. 2016. Interpreting the visual. In N. Clifford*et al.* (eds.) *Key methods in geography,* pp. 233–247. London: Sage Publications.

Rose, G. 2007. *Visual methodologies: An introduction to the interpretation of visual materials,* 2nd edn. London: Sage Publications.

Salvadó-Corretger, G., and Benavente, F. 2019. Time to dream, time to remember: Patterns of time and metaphysics in *Westworld. Television & New Media,* 22 (3), 262–280. https://doi.org/10.1177/1527476419894947.

Schneider, J. 2019. New media pharmacology: Hansen, Whitehead, and worldly sensibility. *Theory, Culture & Society,* 36 (1), 133–154. https://doi.org/10.1177/0263276418806994.

Virino, C. C., and Ortega, V. R. 2019. Daenerys Targaryen will save Spain: *Game of Thrones,* politics, and the public sphere. *Television & New Media,* 20 (5), 423–442. https://doi.org/10.1177/1527476418770748.

Warf, B., and Arias, S. 2009. *The spatial turn: Interdisciplinary perspectives.* London and New York: Routledge.

Weber, S. 1996. *Mass mediauras: Form, technics, media.* Palo Alto, CA: Stanford University Press.

White, M. 2006. *The body and the screen: Theories of internet spectatorship.* Cambridge, MA: MIT Press.

Wilbanks, T. J. 2004. Geography and technology. In S. D. Brunn, S. Cutter and J. W. Harrington (eds.) *Geography and technology,* pp. 3–16. Berlin: Kluwer Academic Publishers.

Williams, R. 1974. *Television: Technology and cultural form.* New York: Schocken Books.

Williams, R. 1981. *Culture.* London: Fontana.

Zimmerman, S. 2007. Media geographies: Always part of the game. *Aether,* 1, 59–62.

11

GEOGRAPHICAL ANALYSIS OF STREAMING VIDEO'S POWER TO UNITE AND DIVIDE

Irina Kopteva

Online presence is a reality for billions of people who look for information, entertainment, shopping, avenues of self-expression, or social connections. China and the United States manage the 15 most popular global sites (Alexa Internet Inc. 2020). Two American sites lead the way: the search engine Google.com and the video sharing platform Youtube.com (YouTube). YouTube's global reach can be characterized by the following (Cooper 2019; YouTube 2020):

- two billion people use it monthly
- one billion hours watched daily
- over 100 countries stream its videos
- 80 languages used in video and interface
- 70% of videos viewed on mobile devices
- 15% of the global traffic comes from the United States
- 81% of 15–25 year-old Americans use YouTube

The list above demonstrates the multifaceted appeal of YouTube, which involves different generations in video production and consumption in numerous languages via multiple technologies. This chapter examines geographical dimensions of the YouTube phenomenon along with other video streaming services. A number of questions arise. What video streaming platforms exist in the world and what functions make the YouTube user experience special? How important is it for consumers to participate in the development and critique of video content? What creates YouTube's distinct culture of freewheeling opinions and like-minded followers? YouTube is widely popular in the United States, Brazil, Russia, Japan, the United Kingdom, Germany, France, Mexico, Turkey and India (Statista Research Department 2016). However, many countries in the world are not free democracies. What challenges do various political states present to local and global video streaming services and associated user experiences? Should streaming video be censored? Youth is exceptionally involved with YouTube. How can education employ this engagement? To address the questions, this chapter will review data and research on the use of video streaming technology across the world. Our geographical analysis reveals that a unique blend of creative

DOI: 10.4324/9781003039068-13

venues, diverse content, search capabilities, and social opportunities create powerful video sharing platforms, which unite and divide across cultural and geographical borders.

Spatial patterns of video streaming culture

The unifying power of video sharing community

YouTube attracts people at tremendous rates because it publishes user-generated content which facilitates creativity in ordinary people and reaches a diverse audience. New independent celebrities emerge, who do not associate themselves with any TV network and who publish their video commentaries about video games, makeup art, or share their thoughts or experiences on subjects of common interest. Some YouTube stars, also called influencers, have millions of subscribers who post comments on the videos and engage in the conversation with others across the globe. The biggest independent YouTube star, PewDiePie—a Swedish gamer—lives in the United Kingdom and his commentary on video games, viral videos and internet memes attracts 102 million subscribers (Leskin 2020). Figure 11.1 shows the world distribution of major YouTube influencers, whose fan bases vary from 20 to over 100 million.

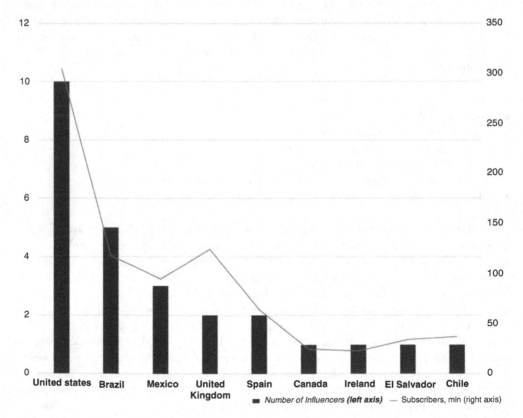

Figure 11.1 Countries with the top 26 independent YouTube influencers and their combined subscriber base

Out of the 26 YouTube influencers, ten stars broadcast from the US, five from Brazil, three from Mexico, two from the United Kingdom and two from Spain, while Canada, Ireland, El Salvador, and Chile have one star each (Leskin 2020). Europe hosts five influencers with 211.6 million subscribers. North and South America combined hold the most with 21 independent YouTube influencers, and the largest combined fan base of 613.5 million. The US has the largest number of active users and 90% of US internet users watched YouTube in 2018 (Clement 2019). The amount of traffic and broad participation demonstrate that YouTube, which started as an American phenomenon, has built a new global cultural community.

The opportunities of social interaction and quick fame offered by YouTube, Vimeo, Instagram, Twitter and other streaming video platforms attract people across the globe to such a degree that safety is often overlooked. Death stunts became a popular genre fancied by professional models or weathermen along with common folk just to acquire "likes" and monetary gain (Elgan 2019). People shoot pictures on high cliffs, on rooftops in Russia and China, in/on/near trains, when they dive in a dangerous pool, encounter wildlife, compete in the "fire" or "tide pod" challenges, stick their head in a microwave, or simulate gun play (the latter is often associated with US risk-takers). Research shows that deaths and injuries tripled from 2014 to 2015 due to risk-taking clips. Fatalities rose and 259 people died taking selfies in 2001–2017 (Bansal et al. 2018, 829). Most of the selfie deaths happened to occasional risk-takers, when they decided to picture an unusual moment like a close encounter with a bear. Some deaths went unreported, and others died when taped by someone else. Geographically, half of selfie deaths occur in India and many of them happen during "train surfing" (Cothier 2016). Russia, the United States and Pakistan follow India in the number of fatalities caused by drowning, experimenting with transportation and heights (Bansal et al. 2018, 830). Localities vary from dangerous shoots in public spaces like the Grand Canyon and Yosemite, a train in India or a Shanghai tower, to private places like yards and homes. YouTube and Twitter instituted policies against publishing videos and photos of harmful activities to curb dangerous incidents, however individual competition with others in the global risk-taking market continues to drive genre development.

The top ten countries with the most YouTube users relative to the number of internet users are presented in Figure 11.2. In 2016 in India, only 29% of people had access to the internet and 11% of the wired population watched YouTube (Figure 11.2). By 2019, the Indian online community had grown to 462 million people and roughly half of them watched YouTube (Diwangi 2019). The fivefold jump in YouTube consumption demonstrates a remarkable social change within four years.

The rise of YouTube can be attributed to content geared towards the popular love of music, comedies, beauty and how-to videos produced in vernacular languages. According to Poonam (2019), Indians can search for and view 95% of online videos in their native tongues in addition to English. This opens a window of opportunity for the marginalized population to connect with the world, form new communities, and learn new skills as well as enjoy entertainment and spiritual programming. YouTube is predominantly viewed on mobile devices in both urban and rural places varying from public to private areas that can include homes, workplaces, hotels or mass transit. YouTube's mobility contributes to its success in rural India where hard-wired connections are challenging to obtain. Following the astronomical growth of YouTube, the Indian entertainment model is changing. Big film corporations like Indian T-Series have become very successful and serve a huge market of more than 100 million subscribers by diversifying their content from devotional songs to popular music and production of video clips and films; 133 Indian independent content creators have

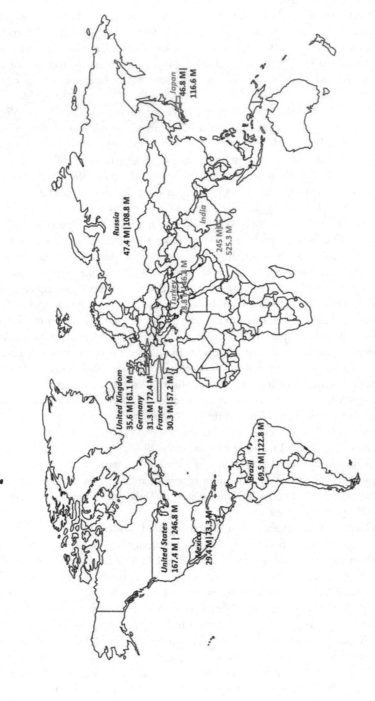

Figure 11.2 Top 10 countries with the most YouTube unique users and internet users, 2016

more than a million followers each and they often operate as a small business working with a team. Among video game commentary and beauty, money-making or purchasing advice, many Indian influencers deliver educational programs that schools may assign to students instead of a traditional class. Other programs teach how to improve life, to develop a career or just deliver motivational content. India is leading YouTube usage in Asia due to the affordable internet and mobile technology, diverse content produced in native tongues and the second largest population in the world.

Regional culture defines spatial patterns of video streaming platform development and usage. Though Japan is a less populous country than India, it is famous for technological innovations and for embracing high-tech tools and resources. A startling 92% of Japanese or about 116.6 million people were wired to the internet in 2016, which made it the fifth-largest online population in the world after China, India, the US and Brazil (Figure 11.2). More than 35% of the 47 million active Japanese YouTube users watch YouTube on their smartphones (Migiro 2018). A homegrown video sharing site NicoNico attracts 4 million subscribers and it started operations in 2006, one year after YouTube's launch. YouTube created a unique online culture of entertainment and education combined with an opportunity to form communities of special interest, and that culture manifests in Japan where users spend more time on video streaming platforms than on social websites (Spinks 2016). The flexibility of YouTube contributes to its success because subscribers can preserve their relative anonymity and they can modify content by displaying channels and clips that reflect their cultural preferences. YouTube communities allow Japanese fans to stay up-to-date and to express enthusiasm for news and events in real-time. NicoNico has advanced this flexibility to the next level by introducing new terminology, mascots and programs that are specific and dear to the local audience. NicoNico offers a variety of genres, including but not limited to anime, manga, idol groups, gaming, product reviews, how-to and music. These videos excite viewers who want to share their experiences with other followers. Technology recognizes the need for a high level of engagement and NicoNico's unique bullet chatting software displays user comments at corresponding points in the videos (Statista Research Department 2019a). An exceptionally personalized experience is the cornerstone of video streaming culture in Japan.

France created its own video sharing service and mobile app called Dailymotion in 2005, a month after YouTube was launched there. Dailymotion has 250 million monthly unique users who can view content in 18 languages, and it is the second-largest video sharing platform in the world for both professionals and amateurs (Mary 2016; Dailymotion 2020). Dailymotion organizes content in categories similar to YouTube, including News, Sports, Music and Entertainment, but it lacks Cooking, Health, Science and Education categories. YouTube has Google tools for creators to analyze the number of views and their origin, and that information helps further develop content and improve engagement. Dailymotion does not provide similar analytics. It has instead focused on software development to show high-quality videos shorter than 60 minutes. Content creators appreciate Dailymotion's attitude towards minorities, flexibility with copyright infringement and filtering options for adult content with Age Gate. YouTube restricts the upload of questionable videos to avoid legal battles. Dailymotion's most recent agenda is to expand their paid online TV service for the audience using smart TVs who can stream video programs and movies. This policy aims to attract advertisers looking to connect with "cord-cutters," the viewers who prefer to access on-demand digital content (Dailymotion 2020). Dailymotion's interface was updated to fit mobile screens and its design encompasses three clear sections: trending content with suggestions; a personalized collection of videos that the user follows; and a library where videos can be saved to watch later. Overall, Dailymotion is lagging in user-generated content and it is moving towards becoming an online extension of television services.

User-generated videos allow civil discussion to reach broad audiences, who otherwise would have to receive information from state-controlled media. Russian mainstream television rarely presents the views of political opposition or cultural antagonism and YouTube provides a platform for freewheeling opinions. YouTube in Russia is the second most visited online site after Google and it is more popular than homegrown sites such as Yandex, Vk. com, Whatsapp.com, Odnoklassniki.ru, and others (Elagina 2020). Nearly 79 million Russian users visited YouTube in July 2019, one and a half times more than in 2016 (Figure 11.2, MacFarquhar 2019). Of urban viewers aged 18 to 44, 82 percent search for news on YouTube, and young Russians, from 18 to 24 years old, demonstrate higher trust in internet news over conventional broadcasting. Mr. Dud [Dude], a Russian YouTube celebrity, has over 5.2 million subscribers who enjoy his provocative interviews with opposition and Kremlin favorites. Online Russian celebrities currently experience a fair amount of freedom of speech, but a new law in 2019 on "sovereign internet" aims to centralize information technology and censor online expression (Freedom House 2019a). The commercialization of YouTube presents another threat to homespun marginalized content, yet YouTube still provides a viable platform to publish an independent outlook and to engage the public in conversation.

Streaming video censure

The Communist government in China banned both Google and YouTube, which refused to censor their content under Communist Party guidelines. Local video sharing portals emerged such as iQiyi, PPTV, Sohu, LeTv, Tencent Video and the largest of these, Youku Tudou, Inc. (Davila 2019). In the most populous country in the world with roughly 1.4 billion people, 580 million users—or nearly 39% of the population—subscribed to Youku Tudou in 2018. Contrary to Youtube's strategy in facilitating user-generated content, Youku Tudou along with other nationally based video streaming sites focuses on delivering licensed exclusive content. Their interface does not aim to satisfy the unique taste of users; rather the sites facilitate online browsing for videos organized by category. The offered content reflects the leadership's opinion of what users should see, which transforms video streaming into an extension of online TV services. Any semiprofessional production undergoes Chinese government censorship and each video must be approved before its transfer to the internet, limiting freedom of expression (Curtin & Li 2018, 354; Wang & Lobato 2019, 361–362). Chinese amateur video content, grassroots activity and political critique ceased quickly after the state initiated copyright enforcement campaigns and required sites to partner with licensed TV networks owned by the state. The most popular video services were incorporated into technical giants such as Baidu, Alibaba and Tencent. Executives from these corporations participate in the National People's Congress, which is the supreme policy-making authority supervising communication culture. To operate, every video streaming website has to obtain multiple licenses, for example, iQiyi has 11 accreditations varying from a food trading permit to "China Internet Integrity" certification (Wang & Lobato 2019, 363). The government also participates in enlisting capital for internet project development, which further ties technical corporations to political structures. The close integration of the Chinese state and market leads to governmental control and that is the fundamental difference between the American and Chinese video streaming platforms. Wang and Lobato (2019, 367) argue that internet culture has historical and geographical roots, and the liberal values of

Californian cyber culture can determine the essence of digital services produced in the West. In contrast, Chinese video sites originate in perceptions formed by the government and are propelled by the "Chinese dream" and the "going out" policy. In Chinese ontology, "state and market are different facets of a common entity" and video streaming platforms are seen as participants in advancing state goals in the country and beyond (Wang & Lobato 2019, 364). Cultural and political contexts induce video production, which becomes an online extension of official mass media and is limited to enhancing the national agenda and collective work ideals.

Governmental censorship of online video culture is not unique in East Asia. The South Korean mobile messaging application KakaoTalk, with gaming and video sharing capabilities, has over 50 million monthly active users (Statista Research Department 2019b). In democratic South Korea, KakaoTalk receives thousands of executive requests to censor content, to provide users' information with a search warrant, and to wiretap under the pretense of preventing government insults and the spread of fake news (Yoon 2017). The Korea Communications Commission (KCC), established in 2008, manages internet content censorship, and the number of sites blocked and deleted has been growing ever since. The overall number of government intervention requests to online media tripled in 2018 compared to 2017, and cases varied from the investigation of pornography to cyber defamation of the South Korean and US governments (Freedom House 2019b). YouTube is gaining popularity and its number of Android users was 34 million in 2019, a 38% increase on 2016 (Crichton 2020). Most Koreans use a smartphone to watch YouTube, and time spent on the site exceeds time spent on the KakaoTalk app (Jin-Young 2018). Content hours uploaded to YouTube in South Korea increased by 50% from 2018 to 2019, and over 200 channels have over one million subscribers (Crichton 2020). A variety of global information deepens engagement on YouTube because the local apps focus on local content. In 2009, South Korea enacted laws that blocked the ability of Korean online users to upload material and publish comments, including anonymous trolling, on YouTube and approximately 150 other websites (Hoffberger 2012). That ban was active until 2012, and for three years, many Korean users could only watch videos on YouTube. To use the full set of features, many accessed YouTube illicitly by creating an account and logging in via another country's site. It is unclear if those users were prosecuted because the Korean administration must ask Google, an American company, to identify the content creators or commenters. For example, in 2018 the Deboreo Minju Party identified 104 YouTube programs as fake news and asked Google Korea to remove the content. The request was rejected because Google refrains from content credibility assessment to provide freedom of speech (Chavern 2019; O 2019). In 2019, South Korea introduced a system to temporarily block online services, including global operators like YouTube, Facebook and others, to strengthen "the Regulation of Illegal Information/Service" with a stated objective to inspect prohibited activities and protect users (Korea Communications Commission 2019). Many YouTubers and their followers think that the legislation suppresses freedom of speech and freedom of the press, and they point to a similarity with Chinese media control that restricts semiprofessional production. The government and professional media counter this with the argument that the new KCC plan is an attempt to provide quality journalism, reliable information, and protect the nation and individuals from the adverse effects of the internet.

Germany does not guard the government from criticism by requesting user data to initiate local investigation of such critics; rather the government focuses on hate speech (Pat 2010). Germany's past explains the strict rules prohibiting neo-Nazi or Holocaust deniers' content. All mass media, including global, local, online or conventional, are expected to follow the regulations. German authorities look broadly at the protection of minors and they delete or request the removal of media products that incite crime or have indecent, violent or racist content. Another serious offense is copyright violation. In 2009, the German music rights organization GEMA initiated a legal battle by requiring payments from YouTube for playing music that was created by artists registered with GEMA (Hill 2016). YouTube disagreed because it pays creators for advertisements which are displayed on channels that have at least 1,000 subscribers (Google 2020a). For seven years, from 2009 through 2016, thousands of music videos were blocked from view for German YouTube watchers. In 2016, YouTube announced that they had reached "a landmark agreement" with GEMA that enabled music artist members to receive compensation for future songs, videos published in 2009 through 2016, and advertising revenue. Videos of non-GEMA members may still be blocked. In addition to listening to music, German YouTube users follow gamers, and view how-to videos, educational and news programs. One of the young influencers, Rezo, delivered such well grounded and insightful polemics against the government that some establishment parties tried to mimic his style (Schuetze 2019). Though 80% of young Germans value news from newspapers over updates posted online, more and more millennials and conventional journalists recognize the importance of critiques posted by influencers on YouTube.

The increase in unregulated online content troubles governments regardless of their political structure. According to the Google Transparency Report (Google 2020b), 145 countries submitted requests to Google to remove online content because it may have violated a local law, or the country issued a court order to delete it. Additionally, authorities ask content to be reviewed if it violates Google's "product community guidelines and content policies" (Google 2020b). Expressed concerns may relate to national security, defamation, copyright, privacy and security, trademark, hate speech, obscenity and other issues. Google analyzes every incident, taking into account each country's respective regulations. The content may be blocked from view in the country that made the request or may be removed completely. Often the content stays or the owner makes changes or deletes it. In 2018, more than half of such requests, or 23,934 cases, were directed towards YouTube products. That trend represents a dramatic 28-fold increase compared to 2010, when Google started publishing these statistics. Online information is experiencing rising scrutiny, which increases both censorship and users' protection.

Personalization of the viewing experience defines the cultural divide

Video streaming platforms, especially YouTube and Netflix, focus on viewing experience personalization to increase viewership. The above analysis of streaming video popularity across the world demonstrates the commercial viability of the strategy, which attracts users of different cultural backgrounds. A high proportion—74% to 89%-—of viewers across generations use YouTube recommendations in selecting a new video (Smith et al. 2018). YouTube's search algorithm magnifies personal preferences by generating a sequence of choices. After the initial video search with keywords, the algorithm persuades the viewer to pick a video from a list alongside the current video they are viewing. These narrowed personal selections can lead to cultural fragmentation, which concerns scholars (Sunstein 2018, 77; Wang & Lobato 2019, 364). In our experience, a US geography professor who looks for a video about volcanic activity will probably select an English source from a reputable channel, like the National Geographic, USGS or BBC. Furthermore, the discussion of volcanic impact on society will mostly focus on negative destructive forces. On the contrary, the Japanese or Icelandic view of volcanoes would stress the beauty of the constructive forces that create and fertilize land, as well as providing sources of heat for life to thrive.

YouTube records and publishes user interactions with content, which can be used for research. The uploader's network, number of previous views and video age altogether determine the video's popularity (Bärtl 2018, 18). The fame of a very young video depends on the uploader's network size, and a broadly known creator receives the most views. For example, a new video created by the well known National Geographic channel can become viral in the United States faster than a video made by a foreign unfamiliar source. YouTube's search algorithm keeps recommending the often-viewed content, which proliferates older popular videos. Brodersen (2012, 249) points out that the algorithm creates barriers to the geographic diffusion of ideas because individuals make choices based on language, area of interest such as entertainment, news or sports, and the physical proximity of users. Overall, personal choices shape the online environment and most people trust widely promoted familiar media products.

YouTube aims to engage people across the world with a variety of interests. Tens of millions of channels with billions of videos are grouped into 18 categories such as Autos and Vehicles, Comedy, Education, Entertainment, Film and Animation, Gaming, How To and Style, Movies, Music, News and Politics, Nonprofits and Activism, People and Blogs, Pets and Animals, Science and Technology, Shows, Sports, Trailers, and Travel and Events (Bärtl 2018, 22). YouTube content has evolved over time and Google and YouTube regularly update stats that can be used by scholars to characterize the evolution (Google 2020b; You-Tube 2020). It is still technically impossible to analyze all of the videos hosted by YouTube (Wu et al. 2014, 98). To represent common trends, Bärtl (2018, 20) used the YouTube application programming interface (API) to gather and analyze statistically random data samples from 5,591,400 videos uploaded to 19,025 channels between 2006 and 2016. To minimize bias, the data collection tool was designed to randomly select up to five Latin letters and then duplicates were removed. The use of Latin letters discriminates against channels with titles in Arabic, Cyrillic, Greek, etc., though data on some number of such channels was retrieved when one or two letters matched.

The channel data samples from 2006 to 2009 show that YouTube creators uploaded mostly Music videos followed by categories such as Entertainment, People and Blogs, and Education (Bärtl 2018, 22). Then the People and Blogs category took the lead, eventually comprising about 75% of newly created channels by 2016. Gaming became the second most popular channel category in 2012 and has occupied that place since. Categories such as Pets and Animals, Autos and Vehicles, Nonprofits and Activism, and Travel and Events accounted for the least amount of new channels from 2006 to 2016. The most active channels are News and Politics, Entertainment, and People and Blogs, where new video uploads comprise 45%, 12% and 9% respectively of all content creation. Though the News and Politics category is very dynamic in contributing new videos, it accounts for just 3% of all YouTube channels because most news channels were established early in YouTube's existence and other categories have proliferated since then. For example, the People and Blogs category grows as more people create new channels. The dynamics of new content creation does not match the popularity of videos among users, who mostly view Entertainment, Music and Gaming programs with 24%, 17% and 13% of all views respectively. Throughout the years, popular preference has fluctuated between News and Politics, Music, Entertainment, Gaming, People and Blogs, Comedy, and Shows. To analyze the distribution of views among categories, Bärtl (2018, 26–27) divided annual data into the top 3% most viewed channels and the bottom 97%. There is a clear dominance of popular channels which is demonstrated by the long-term increase of the top 3% most viewed channels, both in the amount of uploads and number of views. Further analysis of success probability from 2011 to 2016 shows that a category influences whether a channel will break into the popular top 3%. During that period the best chance to become popular belonged to the Comedy, Entertainment, How To and Style, Gaming, and News and Politics categories. The People and Blogs category, which is very active, as well as the Sports, Education, Nonprofits and Activism channels had less than the average chance to enter the top 3%. Bärtl (2018, 30) concludes that a small percentage of channels dominate the rest of YouTube production due to (1) the common patterns in information development and dissemination, (2) a discrepancy in supply and demand of various content, and (3) YouTube's algorithms which guide video search and its advertisement policy which magnifies video popularity. An independent study by the Pew Research Center confirms that the YouTube API recommends increasingly popular and longer content regardless of the criteria used to select the initial video (Smith et al. 2018). In our opinion, these processes marginalize younger unfamiliar content, which deepens the cultural divide between individuals.

Application of YouTube in education

Many Americans use YouTube to expand their knowledge. The Pew Research Center conducted a survey among 4,594 US adults to evaluate if information on YouTube is very important, somewhat important, not very important, or not important (Smith et al. 2018). An astounding 87% of respondents indicate that YouTube content is important when figuring out how to do things they have not done before, and roughly half of 18 to 29 year olds say these videos are "very important." Every fifth user considers YouTube content to be "very important" when they want to understand things that are happening in the world or make purchasing decisions. One-third or 34% of surveyed parents look for children's content and allow their kids to watch YouTube regularly, while 81% of all parents do that sometimes. Nearly two-thirds or 61% of surveyed parents express concern that their children encountered inappropriate content. To protect children from disturbing videos, parents are advised to use a special product, YouTube Kids, that has the option to set up parental control and monitoring. The survey results confirm that people watch a variety of content and they are open to learning with streaming video.

Educators use audiovisual aids and various teaching methods to increase student engagement and to meet diverse learning needs. Anderson et al. (2001, 318–322) and Kopteva (2018, 85–86) argue that information delivery should support academic growth from the level of understanding through application and analysis to synthesis and evaluation levels described by Bloom's Taxonomy. Kopteva (2018, 88) points out that short instructional videos assist human geography students in learning complex spatial problems. For example, the study of Colorado Springs' urban sprawl requires the consideration of both physical geography such as the presence of the Front Range in the west, and human activities including but not limited to tourist and military support. Videos can demonstrate the development patterns throughout space and time in minutes, while a book typically has to present multiple color figures with pages of description that may be hard to process for novice geography learners. Berk (2009, 5) defines current students as the *Net Generation* used to receiving knowledge in video format which can be utilized to exploit students' multiple intelligences and learning styles. Even more, Berk (2009, 9–14) highlights 20 potential learning outcomes and 12 specific techniques for using video clips in the classroom. The integration of academic content with video material excites students and keeps them engaged, while associated discussion provides an active learning opportunity by correlating with a real-world scenario. Blended Learning Theory calls for a thoughtful integration of face-to-face and online learning experiences, which can include content on YouTube and Netflix or videos created by the instructor and hosted by a learning management system such as Blackboard, Canvas, Desire 2 Learn and others (Garrison & Kanuka 2004, 96; Fleck et al. 2014, 22; Kopteva 2018, 89). Instructors can find useful content in YouTube's educational hubs[1] or on individual channels. For example, GeographyNow,[2] Wendover,[3] National Geographic, PBS and others have various geography-related videos. Faculty from various disciplines employ YouTube in the classroom across the world, and the United States takes the lead (Snelson 2011, 163–165; Almobarraz 2017, 79).

The majority of YouTube's production is semiprofessional and one must carefully review an online video for content quality before showing it in the classroom. The selected video clips should be enhanced with questions to discuss in the course. Learning is a social process where dialogue helps shape thought and peer interaction is crucial for mind development (Vygotsky 1978, 137–140). Student-driven discussions, which are facilitated (not led) by the instructor, increase content comprehension, critical thinking, and student satisfaction. In the study of a hybrid psychology course, one or two video clips shorter than 5 minutes accompanied every textbook chapter, followed by a small group discussion; 85 students were surveyed at the beginning and end of the semester, where 80% reported positive perceptions of using YouTube to support learning and 97.3% enjoyed the experience (Fleck et al. 2014, 30). Kopteva (2018, 89) analyzed student feedback received during 2008–2016 before and after the integration of human geography online course content with streaming videos and correlated activities. The feedback showed that the integration provides opportunities for reflection and mutual sharing in discussion forums and it develops a learning community where students can exchange thoughts and have "aha" moments. The approach improves student comprehension of the concepts and their application in real-world situations, for example through the analysis of the spatial diffusion of AIDs or the impact of irredentism on the political situation in Iraq. In the fall of 2016, 88% of human geography students viewed videos created by the instructor, which increased their overall participation in the online course and satisfaction with the learning process. Kopteva (2018) calls the course improvements "the quality matters journey" and emphasizes that streaming video has a crucial role in the educational process.

Conclusion

Video streaming technology is a powerful communication tool that is widely used by billions of people across the world. The most popular platform YouTube pioneered the integration of video sharing and search capabilities with social opportunities. YouTube's focus on user-generated content and relative anonymity creates a new online culture that promotes personal creativity and attracts a diverse audience. The global spatial patterns demonstrate that individuals form video communities of like-minded followers regardless of cultural, geographic and political borders. New independent video creators engage broad audiences in conversation about video games, makeup art, shared experiences, social opportunities and such. The video streaming stars called "influencers" can attract over 20 million fans each. A clear geographic divide in independent broadcasting demonstrates that the biggest YouTube influencers video-stream from the Americas, where the largest fan base resides, and the heaviest traffic comes from the US. YouTube encourages subscription growth by paying popular channels to display advertisements. New cultural trends have emerged, such as a genre of "death stunts" where individuals compete with the global risk-taking market for quick fame by publishing online their dangerous activities that can lead to death.

Regional culture influences the video streaming products. Following the US, India is the second-largest YouTube market and its popularity can be attributed to content produced in vernacular languages in addition to English and the mobile capacity of video streaming. The latter contributes to the dominant use of YouTube over Netflix. YouTube offers entertainment as well as an opportunity to connect with and learn from the world, which attracted various demographic groups and led to gigantic growth for YouTube in India. The Japanese value YouTube's flexibility in displaying and updating the content of their choice, which allows them to stay current with other fans in respective online communities. Japan's homegrown NicoNico advanced video streaming by developing "bullet chatting software" to provide an exceptionally personalized experience. YouTube in Russia creates a platform for freewheeling opinion and conversation. A French YouTube rival Dailymotion focuses on producing smartphone-friendly and high-quality videos rather than on user-generated content. Chinese video streaming culture has drifted away from focusing on user-generated content and personal choices. YouTube was banned because it refused the Chinese censure, and local video streaming companies became an extension of official mass media that sends forth the "Chinese Dream" and the "going out" policy. Online broadcasting in South Korea is another example of heavily censored media in the world, where the options to upload material and comments can be turned on and off following governmental claims about "fake news" or defamation. Germany recognizes the importance of independent political criticism, though the authorities can ban YouTube programs because of copyright issues, hate speech or incriminating content.

A striking variety of video products exist in the world, however YouTube's search algorithm proliferates personal preferences and creates cultural divisions at the personal level. YouTube generates a list of suggestions to view alongside the current video and that obstructs geographic diffusion and can lead to cultural fragmentation. The algorithm persuades people to make choices among familiar content. There is a discrepancy in the video uploads and views over time. In 2011 through 2016, popular preference belonged to the categories of comedy, entertainment, how to and style, gaming, and news and politics. Other video categories and younger unfamiliar content from smaller networks have a small chance to become popular.

Younger Americans prefer watching streaming videos on their mobile devices. The surveys show that self-educating videos, as well as programs about products and world events, are important to viewers. These viewing preferences indicate that instructional videos should be short, and associated discussion should integrate academic content with real-world situations to provide an active learning opportunity. Using audiovisual tools in education increases student engagement and diversifies instruction to accommodate learning styles. YouTube's numerous educational hubs and channels can assist geography students in learning complex spatial problems.

Educators should employ public engagement with streaming video to enrich instruction and to build a learning community where cognitive development is reinforced by discussion of real-world examples, with sharing of ideas and "aha" moments.

Notes

1 www.youtube.com/edu
2 www.youtube.com/user/GeographyNow
3 www.youtube.com/user/Wendoverproductions

References

Alexa Internet Inc. 2020. *The top 500 sites on the web.* www.alexa.com/topsites.

Almobarraz, A. 2018. Utilization of YouTube as an information resource to support university courses. *The Electronic Library*, 36 (1), 71–81. http://dx.doi.org/10.1108/EL-04-2016-0087.

Anderson, L. W., and Krathwohl, D. R. (eds.) 2001. *A taxonomy for learning, teaching, and assessing: A revision of Bloom's Taxonomy of educational objectives.* New York: Longman.

Bansal, A., Garg, C., Pakhare, A., and Gupta, S. 2018. Selfies: A boon or bane? *Journal of Family Medicine and Primary Care*, 7 (4), 828–831. doi:10.4103/jfmpc.jfmpc_109_18.

Bärtl, M. 2018. YouTube channels, uploads and views: A statistical analysis of the past 10 years. *Convergence: The International Journal of Research into New Media Technologies*, 24 (1), 16–32.

Berk, R. A. 2009. Multimedia teaching with video clips: TV, movies, YouTube, and mtvU in the College Classroom. *International Journal of Technology in Teaching and Learning*, 5 (1), 1–21.

Brodersen A., Scellato, S., and Wattenhofer, M. 2012. YouTube around the world: Geographic popularity of videos. In *Proceedings of the 21st international conference on World Wide Web, Lyon, France, 16–20 April 2012.* New York: ACM. doi:10.1145/2187836.2187870.

Chavern, D. 2019. Want to stop fake news? Pay for the real thing: Google and Facebook should be allies of quality journalism, not its gravest threat. *New York Times* (January 31). www.nytimes.com/2019/01/31/opinion/google-facebook-fake-news-journalism.html.

CIA. 2020. *The world factbook.* www.cia.gov/library/publications/the-world-factbook/.

Clement J., 2019. *US user reach of leading video platforms 2018.* www.statista.com/statistics/266201/us-market-share-of-leading-internet-video-portals/.

Cooper, P. 2019. *23 YouTube statistics that matter to marketers in 2020.* https://blog.hootsuite.com/youtube-stats-marketers/.

Cothier, A. 2016. *Train surfing: One mistake and this illegal "sport" might kill you.* Video, 14:48. www.nationalgeographic.com/video/shorts/885887555575/.

Crichton D. 2020. *YouTube has seen soaring growth in South Korea.* https://techcrunch.com/2020/02/05/youtube-has-seen-soaring-growth-in-south-korea/.

Curtin, M., and Li, Y. 2018. iQiyi: China's internet tigers take television. In D. Johnson (ed.) *From networks to Netflix: A guide to changing channels*, pp. 343–354. New York and London: Routledge.

Dailymotion. 2020. *Press and media.* https://press.ondailymotion.com/en.

Davila, D. 2019. *Youku Tudou vs. YouTube: A financial comparison (YOKU, GOOG).* www.investopedia.com/articles/insights/053016/youku-tudou-vs-youtube-financial-comparison-yoku-goog.asp.

Diwanji, S. 2019. *India: Number of internet users 2015–2023.* www.statista.com/statistics/255146/number-of-internet-users-in-india/.

Elagina, D. 2020. *Most popular online resources in Russia in 2019, by audience.* www.statista.com/statistics/1051512/russia-most-popular-online-resources-by-audience/.

Elgan, M. 2019. *People are falling off buildings in search of the perfect Instagram shot.* www.fastcompany.com/90287323/people-are-falling-off-buildings-in-search-of-the-perfect-instagram-shot.

Fleck, B. K. B., Beckman, L. M., Sterns, J. L., and Hussey, H. D. 2014. YouTube in the classroom: Helpful tips and student perceptions. *Journal of Effective Teaching*, 14 (3), 21–37.

Freedom House. 2019a. *Russia.* https://freedomhouse.org/country/russia/freedom-net/2019.

Freedom House. 2019b. *The crisis of social media.* www.freedomonthenet.org/country/south-korea/freedom-on-the-net/2019.

Garrison, R. D., and Kanuka, H. 2004. Blended learning: Uncovering its transformative potential in higher education. *The Internet and Higher Education*, 7, 95–105. doi:10.1016/j.iheduc.2004.02.001.

Google. 2020a. *YouTube partner program overview & eligibility.* https://support.google.com/youtube/answer/72851?hl=en.

Google. 2020b. *Google transparency report.* https://transparencyreport.google.com/government-removals/overview?removal_requests=group_by:products;period:&lu=removal_items&removal_items=group_by:products;period:

Google. 2020c. *YouTube data API.* https://developers.google.com/youtube/v3.

Hill, P. 2016. *YouTube and GEMA reach agreement over blocked videos in Germany.* www.neowin.net/news/youtube-and-gema-reach-agreement-over-blocked-videos-in-germany/.

Hoffberger, C. 2012. *South Korea lifts ban on Youtube uploading and commenting.* www.dailydot.com/news/south-korea-youtube-upload-ban-legal/.

Jin-Young, C. 2018. *Enjoyment without responsibility: YouTube ranks 1st in S. Korea for number of users.* www.businesskorea.co.kr/news/articleView.html?idxno=20222.

Kopteva, I. 2018. Humanizing human geography online: The quality matters journey. *The Geography Teacher*, 15 (2), 80–91. https://doi.org/10.1080/19338341.2017.1417880.

Korea Communications Commission. 2019. *The Korea Communications Commission announces key plans for 2019.* https://eng.kcc.go.kr/user/ehpMain.do.

Leskin, P. 2020. From PewDiePie to Shane Dawson, These are the 26 most popular YouTube stars in the world. *Business Insider* (February 7). www.businessinsider.com/most-popular-youtubers-with-most-subscribers-2018-2.

MacFarquhar, N. 2019. Looking for free speech in Russia? Try YouTube. *New York Times* (June 9). www.nytimes.com/2019/06/09/world/europe/youtube-russia-putin-state-tv.html.

Mary. 2016. *Video site comparison: YouTube vs Dailymotion.* https://acethinker.com/youtube-downloader/youtube-vs-dailymotion.html.

Migiro, G. 2018. *Which countries watch the most YouTube?* www.worldatlas.com/articles/which-countries-watch-the-most-youtube.html.

O, T. 2019. *South Korea introduces measures to block YouTube, Google, Facebook, Twitter, Instagram, Apple, Netflix.* https://eastasiaresearch.org/2019/03/14/south-korea-introduces-measures-to-block-youtube-google-facebook-twitter-instagram-apple-netflix/.

Pat. 2010. Policing the WebGermany No. 2 in Google "censorship" rankings. *Spiegel* (April 22). www.spiegel.de/international/germany/policing-the-web-germany-no-2-in-google-censorship-rankings-a-690569.html.

Poonam, S. 2019. How India conquered YouTube. *Financial Times* (March 13). www.ft.com/content/c0b08a8e-4527-11e9-b168-96a37d002cd3.

Schuetze, C. F. 2019. The German YouTuber emerging as the voice of a generation. *New York Times* (October 18). www.nytimes.com/2019/10/18/world/europe/germany-rezo-youtube.html.

Smith A., Skye, T., and Van Kessel, P. 2018. *Many turn to YouTube for children's content, news, how-to lessons.* www.pewresearch.org/internet/2018/11/07/many-turn-to-youtube-for-childrens-content-news-how-to-lessons/.

Snelson, C. 2011. YouTube across the disciplines: A review of the literature. *MERLOT Journal of Online Learning and Teaching*, 7 (1), 159–169.

Spinks, R. 2016. How YouTube and NicoNico fuel online fan culture in Japan. *The Guardian* (April 21). www.theguardian.com/media-network/2016/apr/21/youtube-niconico-video-fuels-fan-culture-japan.

Statista Research Department. 2016. *Countries with the most YouTube users as of May 2016.* www.statista.com/statistics/280685/number-of-monthly-unique-youtube-users/.

Statista Research Department. 2019a. *Nico Nico Douga's penetration rate in Japan 2019, by age group.* www.statista.com/statistics/1077384/japan-nico-nico-douga-penetration-rate-by-age-group/.

Statista Research Department. 2019b. *Kakaotalk: Number of monthly active users worldwide 2013–2019.* www.statista.com/statistics/278846/kakaotalk-monthly-active-users-mau/.

Sunstein, C. R. 2018. Polarization. In *#Republic: Divided Democracy in the Age of Social Media*, pp. 59–97. Princeton, NJ: Princeton University Press.

Vygotsky, L. S. 1978. *Mind in society.* Cambridge, MA: Harvard University Press.

Wang, W. Y., and Lobato, R. 2019. Chinese video streaming services in the context of global platform studies. *Chinese Journal of Communication*, 12 (3), 356–371.

Wu, X., Zhu, X., Wu, G. Q., *et al.* 2014. Data mining with big data. *IEEE Transactions on Knowledge and Data Engineering*, 26 (1), 97–107.

Yoon, J. 2017. *South Korea and internet censorship.* https://jsis.washington.edu/news/south-korea-internet-censorship/.

YouTube. 2020. *YouTube in numbers.* www.youtube.com/intl/en-GB/about/press/.

PART III

Mobile media and surveillance

12

EVOLVING GEOGRAPHIES OF MOBILE COMMUNICATION

Ragan Glover-Rijkse and Adriana de Souza e Silva

Mobile communication has long played a part within popular culture with people fascinated by the ability to communicate anytime, anywhere. As early as 1946, the comic character Dick Tracy used a "watch phone" to call others. And, in 1966, the television show Star Trek featured the "Star Trek" communicator, a handheld phone-like device said to have inspired the design of the first mobile (cellular) phone. Yet, the first mobile phone was not introduced until 1973, and mobile phones did not become a part of many people's everyday communications until the turn of the 21st century. During the past 70 years, what it means to engage in mobile communication has changed significantly, shaping how we relate to one another and the spaces around us.

When mobile phones became common at the turn of the century, people felt the need to talk about their (mobile) locations. A frequent question whenever someone received a call was: "Where are you?" This happened because telephones were now detached from physical locations and, as a result, users could make and receive calls while on the go. During these "old" days, we also would frequently hear about mobile phones disconnecting people from their surrounding spaces by connecting them to remote others. As such, early mobile communication was embedded into a rhetoric of disconnection between people and spaces, overlooking its value for producing "co-presence," coordinating face-to-face encounters, and connecting users to places and locations (Hjorth 2007; 2008; de Souza e Silva 2013). Although some of this rhetoric of disconnection still persists, the emergence of smartphones in the late 2000s normalized the constant tracking of location and the mapping of spaces with an array of location-based apps. As a result, carrying a mobile technology not only allowed communicating with remote others, but also embedding digital information into physical spaces through which people moved. During this time, navigation apps, location-based games and mobile social networks gained popularity.

Today, smartphones are just one of the nodes—albeit a very important one—of the immense network of interconnected devices (aka "the internet of things") that populate our spaces. A constellation of mobile infrastructures, such as sensors, watches, beacons, cars and radio frequency identification (RFID) tags, constantly track our location as well as personal and physiological data. They also allow for mobile networks to extend into new spaces where network connectivity might be infeasible or inconsistent, such as rural areas. In doing so, mobile infrastructures allow new networked spaces to emerge; they also change our

DOI: 10.4324/9781003039068-15

mobilities within, and relationship to, these spaces by reshaping interactions and ways of knowing about these spaces. This progression of mobile communication over the past 20 years begs the question: What are the evolving geographies of mobile communication? In this chapter, we look at the evolving ways in which mobile phones in particular, and mobile infrastructures in general, are intrinsically connected to the production and experience of spaces (Lefebvre 1991). We analyze the progression of relationships between people and spaces mediated by mobile infrastructures, focusing on three key moments: (1) the late 1990s and early 2000s, when mobile phones were widely diffused and adopted into new spaces, (2) the emergence of smartphones in the late 2000s, along with the normalization of locations embedded with digital information, and (3) the spread of mobile infrastructures in the second decade of the 21st century, focusing on how they relate to communication and mobility.

Extending phones into new spaces

Closer to the turn of the 21st century, "mobile phone adoption exploded, with subscriptions reaching a half billion worldwide and well into the billions in the following decade" (Campbell 2007, 34). The release of second generation (2G) networks in the early 1990s provided digitally encrypted and consequently more reliable phone calls. It also brought with it the ability to send SMS (short messaging service)—known colloquially as text messaging. At the same time, mobile phones became much smaller, allowing users to carry these devices with ease into a variety of spaces; 2G phones also included new features, such as display screens, access to the mobile web (with the development of the Wireless Application Protocol—WAP—in 1996), and the integration of ringtones and address books. These new features, along with dropping prices, helped to integrate mobile phones into quotidian life. Mobile phones quickly went from a luxury item to a social device used by teenagers and younger adults to coordinate daily life (Kasesniemi & Rautiainen 2002; Ling & Yttri 2002; Matsuda 2005). As Ito (2005) described, time became more flexible because mobile users could negotiate meeting arrangements in real-time, and forgetting the grocery list at home could be solved with a quick call. Micro-coordination shaped everyday life spaces as flexible spaces. In addition, younger people started to use mobile phones as expressions of their self and identity by customizing their phones and interacting with friends (Weilenmann & Larsson 2001; Fortunati & Cianchi 2006).

Public spaces also became places for communication with remote others. The introduction of the mobile phone into new social contexts required renegotiating social norms within public spaces, as the mobile phone allowed for bringing formerly private conversations into public spaces. A particularly heated debate around this time revolved around using mobile phones on public transportation and in restaurants, as these spaces were traditionally only composed of interactions among co-present people. Bringing outside contexts into a local physical space created all sorts of uncomfortable situations for people who did not participate in those phone conversations (de Souza e Silva & Frith 2012). Likewise, many expressed concerns about how mobile phones impacted a person's relationships with others occupying the same space. Some scholars began to study how mobile phones could help create intimacy between people by connecting distant spaces. Licoppe (2004), for instance, discussed the idea of "connected presence" to refer to the mediated presence of a person who is geographically distant; Habuchi (2005) offered the term "telecocoon" to discuss the maintenance of private relationships while using the mobile phone in public spaces; and Fujimoto (2005) described the mobile phone as a "territory machine," capable of inscribing a personal boundary for communication in a public space. Collectively, these scholars describe a qualitative change to the experience of space, brought forth by the mobile phone's capacity to merge otherwise distinct contexts. As a result, people had to learn to negotiate a new reality that included the merging of formerly defined private and public spaces.

Mobile phones were also often blamed for disconnecting users from their physical surroundings in favor of connecting with distant social networks (Gergen 2002; Puro 2002). Gergen (2002), for instance, argued that mobile phone users physically inhabited a space but dedicated their attention elsewhere, thereby creating an "absent presence." This argument has gained some traction over the years (Turkle 2011); however, many scholars have rejected it, arguing instead for a more nuanced understanding of how the mobile phone is embedded into society and social spaces. Yes, mobile phones can indeed distract us from the surrounding spaces, but they also become a part of those spaces. They help to facilitate encounters, find venues and inscribe locations. They increase our connections to nearby spaces. This became increasingly evident with the popularity of GPS-enabled phones in the mid-2000s.

Smartphones and digitally inscribing place / space

The mid-2000s reflected a major shift in mobile communication practices with the introduction of smartphones. Smartphones converged multiple, distinct media into a single portable device—a process referred to as "technological convergence" (Humphreys et al. 2013). They also normalized the use of mobile internet and location-based services on the phone, which have been key to the ongoing process of digitally inscribing information onto space and place. According to Hjorth (2008), with these developments, "one could almost forget that mobile media arose from an extension of the landline telephony" (p. 143). Such a statement reflects a change in tenor from the discourse of the late-1990s and early-2000s, which primarily emphasized the mobile phone as an extension of the landline. Mobile phones were still called "phones" but, in fact, they became mini-computers—able to run applications, access the internet and, most importantly, display their locations (de Souza e Silva 2006).

In the early 2000s, United States President Bill Clinton removed the degradation of civilian-use GPS signals, allowing the public to use GPS-enabled devices to pinpoint a geographic location to within a couple of feet. However, it was not until 2004 that Qualcomm developed a software—called gpsOne assisted GPS—that enabled mobile phones to display their precise geographic location. This development was key to the later creation of location-based services for smartphones, allowing users (and their mobile apps) to identify their location in space through an interface. Also, in the early 2000s, cellular networks were upgraded to 3G connections (3rd generation cellular networks), allowing for mobile broadband internet. This infrastructural change enabled high-speed internet connections for smartphones—critical to many of the data-consumptive and location-specific uses of the smartphone today.

Among the early popular location-based services were navigation systems, such as Google Maps and Waze, and location-based social networks, such as Loopt, Brightkite and, eventually, Foursquare (de Souza e Silva & Frith 2010). These apps allowed users to "check-in" to locations, inscribing digital information onto spaces and broadcasting their locations to a social network of friends. For example, in a fashion that later led to apps such as Yelp, people could use Foursquare to write a review about a restaurant or bar. That review was tagged with latitude/longitude coordinates, and anyone within a radius of that location was able to read the review when using the app. Although restaurant reviews might seem like an irrelevant example, the ability to inscribe locations with digital information represents a significant shift in the meaning of locations. For the first time, locations could be embedded with digital information, which then became a part of that location in what Zook and Graham (2007) refer to as a "DigiPlace." While DigiPlace accounts for the "range of political, economic, and cultural considerations" of how digital code represents and interacts with physical spaces (p. 466), we also offer de Souza e Silva's (2006b) concept of hybrid spaces to account for the social changes to spaces as a result of this process. This embedding of digital information meant that our social places were no longer purely physical spaces, with physically co-present people and things. Any serious attempt to understand communication and social interaction from then on, needed to take into consideration the confluence of physical and digital spaces—along with both physically co-present and distantly connected people. Hybrid spaces are essentially mobile communication spaces, because they arise from the social use of mobile technologies. Understanding that we live in hybrid spaces means that we cannot address social interactions by looking exclusively to their happening in physical or digital spaces, because they occur in the merging of both.

Hybrid spaces not only denote a different way of inscribing spaces with information; they represent a new logic of social conduct in these spaces. For example, "checking-in" came to characterize part of early 2010s culture, with mobile users sharing their location when they thought it might reflect something positive about their identity—that is, a user would check-in if they were at a "cool" concert, but not at the cheap neighborhood market. Sharing locations also served as ways of inscribing places with identities. Places' identities are shaped not only by descriptive information about those places (such as the reviews), but also by the people in them. Judging from the profile of people who are in certain locations (even those not in one's social network), one could potentially be more inclined to visit that place (de Souza e Silva & Frith 2012; 2013). Alternatively, sorting through this information could lead to racist, classist or even gendered connotations of spaces based on those who occupy them (Leszczynski 2016, 1697). From this, we note that inscribing spaces with digital information at once offers opportunities to better connect with surroundings, but at the same time exacerbates existing inequalities and prejudices that shape how many people experience particular spaces.

Of course, location-based services such as these raised many issues about our social constructions of space and the creation of differential spaces—that is, spaces which people experience differently depending on their access to a mobile interface and their ability to actually interact with the information that is embedded into those spaces. Differential experiences of space and ways of moving through spaces, however, are not only created by mobile technologies; they have always existed (Graham 2005). According to Wood and Graham, "from the moment some people rode or were carried while others walked, there have existed differences in mobility which reflect and reinforce existing social structures" (Wood & Graham 2005, 177). Nevertheless, we must still look at how mobile technologies contribute to current experiences of inequality and prejudice. Sheller's (2018) call for mobilities justice has offered an important intervention for interrogating how differences in identity "interact with mobility regimes and control systems that reproduce uneven mobilities" (p. 18). We therefore situate access to and experiences of hybrid spaces as an important site for understanding these contemporary social structures, particularly as the ways that people interact with hybrid spaces increase.

We argue for using hybrid space to theorize the changes introduced by the convergence of digital and physical spaces. Hybrid spaces are different from other understandings of the mixing of digital and physical information, such as code/space (Kitchin & Dodge 2011), augmented space (Manovich 2007), mixed reality (Milgram & Colquhoun 1999; Benford & Giannachi 2011), DigiPlace (Zook & Graham 2007), and Hertzian spaces (i.e. a space characterized by signals from our various electronic devices) (Varnelis & Yoshida 2008). Although all of these perspectives point to some kind of mix between digital information and physical spaces, hybrid spaces refer specifically to the *social transformations* of space that occur as a result of interacting with mobile technologies. The emphasis on the social aspect is important to understanding the dimensions of power and control that are embedded in our interactions with mobile technologies. For instance, hybrid spaces take into consideration people's ability to inscribe and "read" locations, as part of ongoing interactions with a mobile phone. de Souza e Silva and Frith (2013) show how digital information can be inscribed onto physical spaces in a practice they name "writing space." People can write space when they are able to add digital information to places, adding longitude/latitude coordinates to text, audio and video, and as a result, transforming places into locations. Inversely, smartphone users can "read" spaces for digitally inscribed information. We do that basically every day, whenever reading information from a location-based application, such as Google Maps or Yelp.

Nevertheless, much of the digital information inscribed onto places often reproduces dominant views of space and accentuates the biases of social groups. For example, Leszcyznski (2016) and Thatcher (2013) note how safety apps represent spaces as safe/unsafe based on factors like crime statistics and street lighting—factors which disproportionately represent traditionally white and affluent neighborhoods as "safe," while representing traditionally BIPOC and impoverished neighborhoods as "unsafe." Moreover, the inscription of digital information onto physical spaces reflects a preference toward urban rather than rural spaces and commercial rather than non-commercial spaces. Together, these examples demonstrate that hybrid spaces do not inherently reflect homogenizing or democratizing potential, whereby the experience of the digital would mitigate inequalities within the physical. Instead, hybrid spaces can compound those inequalities—reflecting and reinforcing them to mobile users. One way of addressing these inequalities comes from "participatory mapping," which happens when everyday users create their own maps of the world reflective of their experiences in spaces (Kabisch 2008, 225). An example of participatory mapping is the project WikiMapas, developed by the Rede Jovem program in Brazil. Though the WikiMapas platform, slum dwellers can plot points of interest on Google Maps, such as the locations of hospitals, schools, stores and streets in their low-income communities. This is particularly important, as these locations are not represented in traditional online satellite maps. As such, WikiMapas helps to integrate these communities into official city maps and geography (de Souza e Silva et al. 2017). Hybrid spaces, therefore, have the potential to engender creativity, as everyday mobile users can contribute to how spaces are experienced.

Likewise, hybrid spaces are also ludic spaces, experienced by playfully interacting with the surrounding world. Location-based mobile games, for example, turn urban spaces into games boards. Using a mobile or smartphone as an interface to play the game, people can find digital objects embedded into physical locations and interact with other people depending on their physical distance (de Souza e Silva & Hjorth 2009; de Souza e Silva & Sutko 2009). Location-based games gained popularity in 2016 with Pokémon Go, but in fact their history traces back to the beginning of the 2000s (de Souza e Silva 2013). For example, Mogi—a popular game in Tokyo, Japan—allowed players to see the location of virtual creatures and other players on their mobile phone screen. Players had to physically walk to the locations of creatures and people to interact with them; that is, they needed to be nearby in physical space to be able to play the game. Other location-based mobile games have even transformed the physical properties of spaces. Spallazzo and Mariani, for instance, explain how interacting with physical objects helps to "achieve results in the digital world" (Spallazzo & Mariani

2020, 41). In some of their location-based games, players need to literally destroy physical objects to retrieve codes, which then need to be typed into the mobile phone game interface to advance in the game world. In all these examples, the interplay between physical and digital spaces is what defines location-based games—the game would not exist had one of these spaces been removed from game play. Moreover, both spaces become transformed through their interaction with the other. Importantly, though, the mobility required by these games has exposed how inequalities and prejudices become structured into spaces and experiences via the interactions in hybrid space. Aspects of gender, race and class impact how players move through spaces to achieve game objectives; access to infrastructures controls when and where players can play the game. Thus, even in such playful moments, we must attend to those limitations that shape our encounters with spaces.

The idea that our interactions with digital location-based information shape our movements in and through space is reflected in the new mobilities paradigm (Hannam et al. 2006; Sheller & Urry 2006; 2016). According to Sheller (2016), examining mobilities includes looking to how mobile technologies and infrastructures enable or constrain movement. It also addresses the politics of movement, also considering how mobile technologies and infrastructures might contribute to the production of mobilities by enacting force or friction, controlling velocity, dictating routes, and shaping the experience (Cresswell 2010). From this perspective, we can at once see how mobile technologies can limit our mobilities: shaping if and how spaces are represented to us, enabling or prohibiting our access to them. On the other hand, mobile technologies can also enhance our mobilities and increase our control over spaces. Sutko and de Souza e Silva (2011) build on this idea of coordination of time to argue that mobile phones equipped with GPS also influence how people coordinate space, such as when people move differently or for specific purposes based on their knowledge of space(s). They describe two kinds of interfaces by which users might gain knowledge about a space: eponymous and anonymous. Anonymous interfaces, such as the crowdsourcing app Waze, afford coordination with strangers and places. For instance, if a person sees there is a car crash in a specific location, they will likely avoid that route, even though they do not know who inscribed that location with the crash information. Differently, eponymous interfaces identify users, including information about them like their name, interests and location. For instance, many social platforms like Instagram and Facebook allow for adding geotags to photos and status updates. As such, if a person sees that several friends are having coffee nearby after work, they might be tempted to join them on their way home. These apps foreground connections between the social and the spatial as they facilitate spatial coordination.

Embedding mobile and networked infrastructures

Increasingly, our spaces are not only embedded with digital information, but with all sorts of mobile infrastructures connected to digital networks, such as sensor networks, CCTV cameras and smartcards for urban transit systems. McQuire (2017) suggests these urban infrastructures are constructing "increasingly rapid and precise feedback loops between mobile subjects and their particular urban routines" that actively transform spaces as well as mobilities through them (McQuire 2017, 4). For instance, Georgia Tech's Living Lab has partnered with the city of Atlanta to create "smart corridors." In its first implementation, 100 sensors were integrated into North Avenue to monitor and regulate all forms of traffic. Accordingly:

> The technology provides emergency vehicle pre-emption through traffic signalized corridors (where signals change to green for faster response times), provides drivers

with signal phasing and timing data to their phones and cars, alerts drivers when they are speeding through a school zone or sharp curve, and alerts cyclists and pedestrians of vehicles approaching too fast or too close. All of the communication from the infrastructure and different mobility users is disseminated to a smartphone app called Travel Safely that improves safety for the overall traveling public.

(Levine 2017)

Like in this example, increasingly our smartphones interact with more than just cell towers and GPS satellites, but also a number of mobile infrastructures spread throughout (often urban) spaces that can determine how we move through spaces and our ways of knowing about spaces. We should therefore consider our smartphones' interactions with these infrastructures as part of the changing geographies of mobile communication.

Our smartphones have embedded within them a number of sensors and other hardware that allow them to interact with mobile infrastructures that populate our surroundings, including QR code and barcode readers, WiFi routers/dongles, Bluetooth beacons, and near field communication (NFC) devices. While these mobile infrastructures allow access to digital inscriptions of spaces, they also track many of our behaviors as we move through space. For instance, major retailer Walmart introduced Bluetooth beacons to their stores, allowing many customers access to relevant coupons. However, these Bluetooth beacons also track the mobilities of customers who carry Bluetooth-enabled mobile devices. These devices are small and inconspicuous, leaving most customers entirely unaware of their intervention. This relates to the premise of the ubiquitous computing paradigm, which suggests that sensors and computational devices should "weave themselves into the fabric of everyday life, freeing users to focus on other tasks (Weiser 1991, 94). While the concept of ubiquitous computing dates to the early 1990s, it was just in the past decade that ubiquitous computing—now commonly understood as the "internet of things"—has gained traction with the widespread adoption of smartphones and other "smart" devices (e.g. sensors, RFID tags, Bluetooth beacons, etc.) that are embedded into the infrastructure of everyday spaces, thereby creating "smart spaces." We can understand "smart" spaces as networked spaces that attempt to seamlessly connect our surroundings, transforming whole cities into hybrid spaces. Such devices do not permeate just urban public spaces; domestic spaces are also embedded with smart technologies creating what has been termed the "smart home" (Dourish 2016, 27; Fortunati 2018).

The integration of these devices into various arenas of life raises questions about how to interpret these changes. Some scholars understand these devices as forming a massive complex of networked infrastructures. For instance, Andrejevic (2007) offers the concept of digital enclosures—an ever-expanding (surveillant) space constituted by networked devices. Although digital enclosures do not have visible borders, we cannot escape them, because it is almost impossible to be disconnected—we carry networked devices on our person, we interact with them in the public urban spaces of everyday life, and we are connected to them in our homes. Likewise, Bratton (2016) uses the term "the stack" to refer to this proliferation of devices that span the globe. He suggests that instead of considering these devices as different and independent of each other, we should pay attention to their interconnectivity, "forming a coherent and independent whole" (Bratton 2016, 5). By contrast, Dourish and Bell argue that networked spaces are "inherently heterogeneous," messy and assembled for various purposes (Dourish & Bell 2011, 43). They point out the fact that often "smart" spaces and cities are not planned spaces; they develop organically and without a central purpose. Because of this, smart infrastructures are often distributed in an uneven manner—with affluent regions exhibiting "smart" infrastructures, while poverty-stricken and rural regions lack these networked resources. Importantly, there is methodological gain from both perspectives, but also political stakes for how we frame individuals' relationships to these "smart" infrastructures.

For example, a consequent debate concerning these networked infrastructures relates to how it shapes urban mobility. As we have shown, early mobile phones allowed for micro-coordination and smartphones facilitated spatial coordination. Now, networked infrastructures, when embedded in urban and domestic spaces, heavily determine how people can move—or cannot move—through these spaces. Take the case of the shared bikes in São Paulo, Brazil. In 2018 the start-up company Yellow introduced dockless location-based bikes in the city. In order to use a bike, people had to install an app, unlock the bike through the app and pay for the bike at an hourly rate. At first, it seemed that these bikes helped mobility in the city, because users could pick them up and drop them off wherever they wanted. However, the bikes had geofencing technology which uses GPS tracking to disable any bike that is taken outside of a designated area. Yellow placed the bikes in the Jardins neighborhood (an affluent region of São Paulo) and programmed the bikes to disable the moment they moved outside of this geofenced area. Thus, programmed bias ensured that only some regions of the city had access to these "smart" infrastructures and reinforced power relations between affluent and poverty-stricken spaces.

On the other hand, smart infrastructures do not just shape people's movements but additionally track them in nearly all arenas of life, placing them in asymmetric power relationship to those who track and collect their data. For this reason, it is useful to understand the interconnectivity between these devices that enables data to be fed from one network to another. For instance, CCTV cameras in conjunction with face recognition software can identify where people are; proximity sensors and cell phone signals can detect movements and concentrations of people in urban spaces, and location-based smartphones are constantly transmitting individuals' locations to cell phone towers. When people have their location history turned on, companies, such as Google, can tap into those location reports and build very precise tracks for each smartphone user on Earth. In 2020, Google's COVID-19 Mobility Reports[1] have crossed personal locational histories with mapping geographic information to produce detailed reports about where people are moving or not moving. Such reports help us to produce new associations and understandings of spaces, and the people inhabiting them, as well as to facilitate action that transforms these spaces (e.g. limiting who can inhabit them, how people move through them). While this information is mostly anonymized, as Andrejevic and Burdon suggest, the "emerging practices of data collection and use… complicate and reconfigure received categories of privacy, surveillance, and sense making" (2014, 20). As such, we need a framework for rethinking these categories amid the changing geographies of mobile communication.

Towards mobile communication futures

As we have seen, our mobile devices have transitioned from a device that initially only supported voice communication, to a mobile technology that allows communication with voice and text, and then to a device that communicates our location. Now, our mobile devices have undergone yet another transformation as they are integrated into the fabric of urban infrastructures. Through these various functions, mobile devices shape our perceptions of spaces, coordinate our mobilities and render digital information about these spaces. Because these mobile devices are so important to shaping our perceptions of space, it is important to consider how power and control are enacted through their use. As such, scholars might attend not only to the diffusion of mobile devices within society, but additionally to *how* they are integrated into particular spaces, *where* they are placed and *who* has access to them. Such work is important to understanding the production of differential spaces.

Additionally, scholars might account for the kinds of information gathered from these infrastructures and how it becomes represented to inform our understandings of space. We argue that, at a minimum, these activities should be transparent to the end-user. Solove (2008) and de Souza e Silva and Frith (2012) suggest that often tracking activities are either undisclosed or that everyday mobile users lack the necessary digital literacy to understand how to control their privacy settings. The fact that tracking apps normally run in the background on these devices without the user's knowledge actually follows an early premise of ubiquitous computing, called "calm computing"—the notion that our networked devices should not draw attention to themselves and fade into the background of our everyday lives (Weiser & Brown 1996). Since then, some scholars have addressed the political and social stakes of calm and transparent technology. For instance, Takayama (2017) points to the political stakes of devices fading from our awareness and thus operating and enacting consequences beyond a user's control. Likewise, Galloway (2004) and Dourish and Bell (2011) have called for seam*ful* design of networked technologies—those that readily violate the principles of calm computing and, instead, make their mediation evident to users. The purpose behind such calls is that when technologies fade into the background of spaces, we become less conscious of their operations—the means through which they enact power and control in and over our lives. Importantly, we also become less conscious of how they shape our understandings and experiences of geographies.

There is an opportunity for creative interventions that challenge the uneven distribution of networked technologies in urban spaces. For instance, in Rio de Janeiro, some residents have stolen GPS-enabled scooters, removed their trackers and moved them outside of the geofenced region for "free" daily use (Guimarães et al. 2019). There is also opportunity to push back against asymmetric data-collecting measures, such as when artist Simon Weckert brought 99 smartphones into a city street, creating a fake traffic jam on Google Maps (Shammas 2020). Although seemingly trivial, such interventions resist the top-down constructions of our spaces and expose the inequalities that shape them.

Taking these points together, we suggest that one of the new challenges for mobile communication and mobilities studies is to understand the competing geographies that arise with the widespread integration of mobile and networked infrastructures throughout many urban spaces (Freudendal-Pedersen & Kesselring 2017). This perspective recognizes that infrastructures take part in the production of space and the inscription of borders (Tahwil-Souri 2015). It considers that mobile infrastructures, like Bluetooth beacons, RFID chips and WiFi routers, actively inscribe new territories for the governance, modulation and surveillance of action, including mobilities and communication, in everyday places and often without notice. Increasingly, though, these territories are inscribed not just by hegemonic forces, but also by everyday individuals aiming to produce their own territories. We must look to how the subsequent contestations over space play out through the implementation of mobile infrastructures, paying particular attention to the impact on those interacting within those spaces. At stake is an understanding of the geographies of mobile communication which accounts for the inherent dynamism of spaces and the ongoing ways that mobile technologies shape our relations to them.

Note

1 www.google.com/covid19/mobility/

References

Andrejevic, M. 2007. Surveillance in the digital enclosure. *The Communication Review*, 10, 295–317.

Andrejevic, M., and Burdon, M. 2014. Defining the sensor society. *Television & New Media*, 16 (1), 19–36.

Benford, S., and Giannachi, G. 2011. *Performing mixed reality*. Cambridge, MA: MIT Press.

Bratton, B. H. 2016. *The stack: On software and sovereignty*. Cambridge, MA: MIT Press.

de Souza e Silva, A. 2006. From cyber to hybrid: Mobile technologies as interfaces of hybrid spaces. *Space and Culture*, 3, 261–278.

de Souza e Silva, A. 2013. Location-aware mobile technologies: Historical, social and spatial approaches. *Mobile Media & Communication*, 1 (1), 116–121. doi:10.1177/2050157912459492.

de Souza e Silva, A., and Frith, J. 2010. Locative mobile social networks: Mapping communication and location in urban spaces. *Mobilities*, 5 (4), 485–506.

de Souza e Silva, A., and Frith, J. 2012. *Mobile interfaces in public spaces: Locational privacy, control, and urban sociability*. New York: Routledge.

de Souza e Silva, A., and Frith, J. 2013. Re-narrating the city through the presentation of location. In J. Farman (ed.) *The mobile story*, pp. 46–62. New York: Routledge.

de Souza e Silva, A., and Hjorth, L. 2009. Playful urban spaces: A historical approach to mobile games. *Simulation and Gaming*. doi:10.1177/1046878109333723.

de Souza e Silva, A., and Sutko, D. M. (eds.) 2009. *Digital cityscapes: Merging digital and urban playspaces*. New York: Peter Lang.

de Souza e Silva, A., Duarte, F., and Damasceno, C. S. 2017. Creative appropriations in hybrid spaces: Mobile interfaces in art and games in Brazil. *International Journal of Communication*, 11, 1705–1728.

Dourish, P. 2016. The internet of urban things. In R. Kitchin and S. Y. Perng (eds.) *Code and the city*, pp. 27–48. London and New York: Routledge.

Dourish, P., and Bell, G. 2011. *Divining a digital future: Mess and mythology in ubiquitous computing*. Boston, MA: MIT Press.

Fortunati, L. 2018. Robotization and the domestic sphere. *New Media & Society*, 20 (8), 2673–2690. doi:10.1177/1461444817729366.

Fortunati, L., and Cianchi, A. 2006. Fashion and technology in the presentation of self. In J. Höfflich and M. Hartmann (eds.) *Mobile communication in everyday life: Ethnographic views, observations and reflections*, p. 203. Berlin: Frank and Timme.

Freudendal-Pedersen, M., and Kesselring, S. 2017. Networked urban mobilities. In *Exploring networked urban mobilities*, pp. 1–18. New York: Routledge.

Fujimoto, K. 2005. The third-stage paradigm: Territory machines from the girls' pager revolution to mobile aesthetics. In M. Ito, D. Okabe and M. Matsuda (eds.) *Personal, portable, pedestrian: Mobile phones in Japanese life*, pp. 77–102. Cambridge, MA: MIT Press.

Galloway, A. 2004. Intimations of everyday life: Ubiquitous computing and the city. *Cultural Studies*, 18 (2/3), 384–408.

Gergen, K. 2002. The challenge of absent presence. In J. Katz and M. Aakhus (eds.) *Perpetual contact: Mobile communication, private talk, public performance*, pp. 227–241. New York: Cambridge University Press.

Graham, S. 2005. Software-sorted geographies. *Progress in Human Geography*, 29 (5), 562–580. doi:10.1191/0309132505ph568oa.

Greenfield, A. 2006. *Everyware: The dawning age of ubiquitous computing*. London: New Riders.

Guimarães, A., Leitão, L., and Soares, P. R. 2019. *Polícia recebe relatos de roubos de patinetes no Rio mesmo com tecnologia de rastreamento*. https://g1.globo.com/rj/rio-de-janeiro/noticia/2019/06/14/policia-recebe-r elatos-de-roubos-de-patinetes-mesmo-com-tecnologia-de-rastreamento.ghtml.

Habuchi, I. 2005. Accelerating reflexivity. In M. Ito, D. Okabe and M. Matsuda (eds.) *Personal, portable, pedestrian: Mobile phones in Japanese life*, pp. 165–182. Cambridge, MA: MIT Press.

Hjorth, L. 2007. The game of being mobile: One media history of gaming and mobile technologies in Asia-Pacific. *Convergence*, 13 (4), 369–381.

Hjorth, L. 2008. Being real in the mobile reel. *Convergence: The International Journal of Research into New Media Technologies*, 14 (1), 91–104. doi:10.1177/1354856507084421.

Humphreys, L., Von Pape, T., and Karnowski, V. 2013. Evolving mobile media: Uses and conceptualizations of the mobile internet. *Journal of Computer-Mediated Communication*, 18 (4), 491–507. doi:10.1111/jcc4.12019.

Ito, M., Okabe, D., and Matsuda, M. (eds.) 2005. *Personal, portable, pedestrian: Mobile phones in Japanese life*. Cambridge, MA: MIT Press.

Kabisch, E. 2008. Datascape: A synthesis of digital and embodied worlds. *Space and Culture*, 11 (3), 222–238.

Kasesniemi, E., and Rautiainen, P. 2002. Mobile culture of children and teenagers in Finland. In J. Katz and M. Aakhus (eds.) *Perpetual contact: Mobile communication, private talk, and public performance*, p. 170. Cambridge: Cambridge University Press.

Kitchin, R., and Dodge, M. 2011. *Code/space: Software and everyday life*. Cambridge, MA: MIT Press.

Lefebvre, H. 1991. *The production of space*. Malden, MA: Blackwell Publishers.

Licoppe, C. 2004. Connected presence: The emergence of a new repertoire for managing social relationships in a changing communication technoscape. *Environment and Planning D: Society and Space*, 22, 135–156.

Ling, R., and Yttri, B. 2002. Hyper-coordination via mobile phones in Norway. In J. Katz and M. Aakhus (eds.) *Perpetual contact: Mobile communication, private talk, public performance*, pp. 139–169. New York: Cambridge University Press.

Manovich, L. 2007. *The poetics of augmented space: Learning from Prada*. www.manovich.net/DOCS/Augmented_2005.doc.

Matsuda, M. 2005. Mobile communication and selective sociality. In M. Ito, D. Okabe and M. Matsuda (eds.) *Personal, portable, pedestrian: Mobile phones in Japanese life*, pp. 123–142. Cambridge, MA: MIT Press.

McQuire, S. 2017. *Geomedia: Networked cities and the future of public space*. New York: John Wiley & Sons.

Milgram, P., and Colquhoun, H. 1999. A taxonomy of real and virtual world display integration. In Y. Ohta & H. Tamura (eds.) *Mixed reality: Merging real and virtual worlds*, pp. 5–30. New York: Springer.

Puro, J. P. 2002. Finland, a mobile culture. In J. Katz and M. Aakhus (eds.) *Perpetual contact: Mobile communication, private talk, public performance*, pp. 19–29. Cambridge, MA: Cambridge University Press.

Shammas, B. 2020. A man walked down a street with 99 phones in a wagon: Google Maps thought it was a traffic jam. *The Washington Post* (4 February). www.washingtonpost.com/technology/2020/02/04/google-maps-simon-weckert/.

Solove, D. 2004. *The digital person: Technology and privacy in the information age*. New York: New York University Press.

Solove, D. 2008. *Understanding privacy*. Cambridge, MA: Harvard University Press.

Spallazzo, D., and Mariani, I. 2020. Keeping coherence across thresholds: A narrative perspective on hybrid games. In A. de Souza e Silva and R. Glover-Rijkse (eds.) *Hybrid play: Crossing boundaries in game play, player's identities and play spaces*, pp. 30–45. New York: Routledge.

Takayama, L. 2017. The motivations of ubiquitous computing: Revisiting the ideas behind and beyond the prototypes. *Personal and Ubiquitous Computing*, 21 (3), 557–569.

Turkle, S. 2011. *Alone together: Why we expect more from technology and less from each other*. New York: Basic Books.

Varnelis, K., and Yoshida, N. 2008. Architecture for Hertzian space. *Architecture and Urbanism*, 5 (452), 14–19.

Weilenmann, A., and Larsson, C. 2001. Local use and sharing of mobile phones. In B. Brown, N. Green and R. Harper (eds.) *Wireless world: Social and interactional aspects of the mobile age*, pp. 99–115. Godalming and Heidelberg: Springer-Verlag.

Weiser, M. 1991. The computer for the 21st century. *Scientific American*, 265 (3), 94–104.

Weiser, M., and Brown, J. S. 1996. Designing calm technology. *PowerGrid Journal*, 1 (1), 75–85.

Wood, D., and Graham, S. 2005. Permeable boundaries in the software-sorted society: Surveillance and the differentiation of mobility. In M. Sheller and J. Urry (eds.) *Mobile technologies of the city*, pp. 177–191. London: Routledge.

13

MOVING

Mediated mobility and placemaking

Roger Norum and Erika Polson

In a suburb of Helsinki, a driverless, electric bus picks up passengers for free to ferry them to the city's conference center. A Zoom conference call hosted in Singapore has on it nine remote attendees from across Asia, Europe and the Canadian Arctic, two of whom are sitting on transport (train and car) and one of whom is cycling to work, her camera off and microphone muted so that her colleagues do not notice the fact that she is not at her desk. On Apple iPhones around the world, the "Significant Locations" feature records and can broadcast the device's location triangulated through GPS, Bluetooth, crowdsourced WiFi and cell tower signals. A school librarian in Montgomery, Alabama packs up books which local kids have reserved online, then delivers them to Wing, a Google-owned drone service that uses GPS satellite tracking to subsequently drop off the books in front yards all across her community. Driving global society, movements such as these transport people, workforces, money, products, customs and ideas from one corner of the planet to another. These diverse movements, and their crossings and collisions, result in multiplicities and relations that are co-produced alongside new types of spatial configurations (Massey 2012, 91). Through many encounters and points of connection, such circulations maintain, enrich and transform economies and societies, while the unequal distribution of these infrastructures both reflects and reinforces growing gaps in economic and social equity. The prevalence of the term "mobile media technologies" suggests not just that this mobility is mediated or that media are mobile, but that technologically mediated mobile practices are increasingly at the core of how humans live their everyday lives.

As we began to write this chapter (in Spring 2020), much of the world's movements had come to a halt due to a contagious virus that immobilized everything in its highly mobile path. Never has the omnipresence and power of both mobility and immobility, on both collective and individual scales, been made more clear than during the COVID-19 pandemic. It was in all likelihood the hypermobile practices of the global travel industry that initially facilitated COVID-19 spreading so quickly and so pervasively across the planet. Within days of the virus being declared a pandemic, international research conferences and business meetings went online; AirBnB's Experiences product, which once promised travelers the chance to "hang out with a local," went fully digital by launching Online Experiences, offering a "window to" another world; Instagrammers who once posted #travelgram photos began sharing dreams of where they would #travelsomeday. Digital imagery from satellites

DOI: 10.4324/9781003039068-16

reinforces anecdotal evidence about how air quality was impacted by the mobility-break. At the local, analog level, drive-in movie theaters became popular once again. The ways in which mobility is mediated have also been transformed, crystalized through signifiers that demonstrate viral spread: transport routes alongside statistics of infection; cell phone movements surveilled to track social distancing; a contact tracing app that lets people know when they have spent 15 minutes or more in close proximity of somebody who has tested positive.

The importance of mobility also becomes clear if we simply consider the mobile media technologies on which many have relied to cope with the societal immobilities during COVID-19. Even before this, sea changes in digital technology had led to massive shifts in the ways in which people exist, move and communicate across space and time. Technological advances such as GPS-linked devices, smartphone apps driven by real-time location data, web-based video calls, networked social media, internet-linked public transportation, and the many new business and social practices that accompanied them, had already enabled an unprecedented shift in close connections between location and social (and economic) relations. This familiarity is perhaps what made the changes necessitated by the quarantine at once more painful to accept and easier to adapt to.

As suggested in the introduction to this volume, this mobility/media nexus has been at the forefront of scholarship in multiple disciplines. Across the social sciences, in what has become known as the "mobility turn" (e.g. Urry 2003), researchers have begun to center mobilities in relation to social and economic relations, infrastructures and governance—attending not only to historical and contemporary practices but increasingly bringing a normative lens to focus on mobility justice and access to mobility (Sheller 2014). This turn has intensified discussions about relationships between mobility and media (see e.g. Falkheimer & Jansson 2006) where a growing interest in places and spaces has grown along with developments in locative, locating or geo-media. Shifting tensions such as those between mobility and location, engagement and escape, place and "emplacement," are foundational to the new subjectivities emerging in an era of mediated mobility. In this chapter, we investigate the linkages of media and mobility through a critical analysis of the dynamic relationships created by various forms and instantiations of "mediated mobility."

More and more, digital media technologies are used to inspire, enable, guide and trace mobility, both in the context of trajectories of human movement (e.g. migration, travel/tourism) and in the machinations and practices of everyday life. Locative media platforms, such as real-time mapping apps that suggest and recommend proximate activities or social connections, can influence not just how people move through space, but also whom they connect with, what they experience and how they remember their experiences. Such affordances of connective media foster new forms of mobility and, through processes of "digital placemaking" (e.g. Polson 2016; Halegoua 2019; Norum & Polson 2021), facilitate engagements with new environments. This chapter explores the co-construction of media and mobilities, engaging key conceptual issues surrounding their cultural, social and spatial implications.

Defining "mediated mobilities"

To develop a framework for working with the concept of "mediated mobility," we unpack the multidimensionality of its two constituent terms before then re-linking them through their connections and mutually constitutive components. First, to mobility. Though many have offered particular approaches to what mobility is and how it can be wielded theoretically (see e.g. Urry 2003; Hannam et al. 2006; Salazar 2014), Ziegler and Schwanen's frame of the concept is particularly fruitful for this article's discussion of how mobility is embedded within processes of mediation. They describe mobility as "the overcoming of any type of distance between a here and a there, which can be situated in physical, electronic, social, psychological or other kinds of space" (Ziegler & Schwanen 2011, cited in Salazar et al. 2017, 1). The authors deliberately emphasize distance as a challenge or obstacle, while stressing that mobility is itself a process rather than a state.

Second, to mediation, a fundamental attribute of digital media and communications. In addition to its lay definition as a form of arbitration, an "intervention in a dispute in order to resolve it" (OED), mediation also refers to media representations of phenomena across time and space, and in particular the ways in which such representations may give life to new phenomena (Chouliaraki 2008; Tsatsou 2009). However, here we wish to focus more on Silverstone's (1999) broad definition of mediation that considers at once what media do to us and what we do with media. He describes mediation as "the fundamentally uneven, dialectical process in which institutionalized media are involved in the general circulation of symbols in social life" (Silverstone 2005, 189). From this, we build on Grossberg et al.'s broad analysis suggesting that three overarching aspects of media—technology, institutions and cultural forms—are held together by the concept of mediation, which might include any of the following: a formal relationship necessary to connect previously unconnected activities or people; an intervening entity through which a force acts or an effect is produced; a contrast with "the immediate or the real"; and finally, the "space between individual subjects and reality as a space of experience, interpretation and meaning" (Grossberg et al. 1998, 14–15).

Having joined these two understandings of mobility and mediation, we now use these definitions to outline several of the forms which mediated mobility take, and to explore how these forms may function culturally, socially and economically in the contemporary world. The cases we use to draw out these forms below are not meant to be an exhaustive review of literature or cases relevant to each category, but rather are used to illustrate a variety of differences, similarities and interconnections in how mediated mobilities might be explored heuristically. Although there are many areas of crossover in these definitional categories below, this organization serves as a useful heuristic for unpacking phenomena of mobility and mediation in interaction.

Connecting mobile people, actions or things

To begin, we consider mediated mobility to be a constitutive process that links previously unconnected or disconnected mobile activities, entities or people. This can be observed most clearly in the capacity of infrastructural entities that create new connections between actors and/or things that inhabit both a "here" and a "there" (Ziegler & Schwanen 2011). As transitive media, infrastructural forms enable "the possibility of exchange over space... physical networks through which goods, ideas, waste, power, people, and finance are trafficked" (Larkin 2013, 328). Examples might include infrastructures which carry, permit or thwart the movement or transport of things—vessels such as roads (Harvey & Knox 2015) or shipping containers (Morley 2017), or processes such as the managed operations of autonomous trucking (Dickerson 2018) or electricity distribution (Gupta 2015). Such media enable or disable mobility through the application or removal of various frictions (Tsing 2014) on external bodies. Digital entities, too—most notably digital platforms—as mediants of communicative or economic exchange, are recognized to be key social and material sites of infrastructuring processes (Gillespie 2010; Plantin et al. 2018). Such entities, whether digital, material or a combination of both, are also intricately linked to global supply chains that connect multiple communities, across cities and continents with mobile (or motile) entities in other parts of the world.

Technological innovations in this sphere have transformed the way many services are provisioned and experienced, and indeed, this is one of the ways through which mediation most impacts mobility in the digital era. For example, smartphone-enabled transport apps such as Uber or Lyft connect riders with drivers, just as they provide visualizations of the car's location before and during the provided service, which can increase senses of safety on the part of the rider (Pink et al. 2019), but also industrial-state surveillance of the driver (Rabih 2020). The same mobile technologies powering on-demand labor platforms such as Taskrabbit and MTurk, which have been shown to be potentially exploitative (Graham & Shaw 2017), are also used to organize cooperative, employee-owned businesses of drivers, cleaners, dog walkers or even journalists (Scholz 2017) geared towards mobilizing a nomadic workforce under more fair and just conditions.

Among the new social relationships that may germinate from the connecting of multiple communicators through digital media, many digital affordances rely on location data to bring mobile people together in co-located space. Such is the case of location-based dating apps, such as Tinder or Grindr, in which the physical location of participants is used to "sort potential matches by proximity, with the aim of expediting [a] localized encounter" (Miles 2017, 1596). Such apps also enable connections beyond their purported purpose of facilitating a sexual or romantic tryst; for example, Leurs and Hardy (2019) find increasing evidence of tourists using dating apps to meet people who may provide "instant access to local knowledge" and "company for solo travelers" (p. 323). The ability of smartphones to keep one connected interpersonally as well as to provide access to banks of information and resources has proven essential for mobile individuals reliant on transience as livelihood or for survival, such as refugees (Waltorp 2020) or people experiencing homelessness (Humphry 2019). Future research in this area will doubtless consider the growing use of socially mediated mobile apps that connect users based on proximity, yet in online spaces. Consider, for example, the German "hyperlocal community app," Jodel, which enables people who are physically near to one another to engage in on screen chats, opening up new opportunities for people in motion to obtain information about or find social connections in (and about) the places they roam. Other examples include the short-range signaling technologies that connect proximate *offline* users, such as the Bluetooth-enabled Zapya app for file-sharing and chatting when they have moved away from WiFi hotspots (Polson 2019).

While most scholarly attention on the convergence of communication technologies and mobility focuses on smartphone-enabled interpersonal and commercial connections, in fact radio frequency identification (RFID) tags are much more commonly used in mobile communication than smartphones—at least in terms of pure number of daily connections (Frith 2015). For example, the Bluetooth-powered Tile device helps users locate keys, wallets and other small items without expensive, data-intensive GPS signaling. Across processes of tracking people, activities or things, Frith (2015) demonstrates how the integration of various levels of RFID technologies into everything from passports to food items on a warehouse shelf has ushered in a vast, virtually undetectable integrated network of devices that "talk" to each other, to users, or to a given authority or central network.

Forces acting on mobile bodies through intervening entities

One of the oldest definitions of mediation refers to the occupation of a middle position, or the process of acting as an intermediary (Grossberg et al. 1998), to produce an outcome or effect. If we consider how mobility is mediated through some intervening entity, we might think of mediated mobility in relation to how real-time digital stimuli act on the movement of bodies, objects or data across space, direction and time. This can take place across more interactive and dynamic scales that may involve mobility on the part of the media that are being controlled, the media that are doing the controlling and the controller itself. Marketing teams recognized early on the ability to affect decision-making in real-time based on the location information of mobile people (c.f. Hopkins & Turner 2012). The growing embeddedness of media into mobile practices intervenes in consumption practices, for example, as people are targeted by digital advertising from proximate retailers while strolling through a shopping district (e.g. Wilken 2019).

Perhaps the most ubiquitous example of how digital media might produce an effect on mobile bodies is in the communication of live, location-based information (such as through GPS-enabled orientation media), which have created a range of new social practices and consumption practices, particularly in urban space (McQuire 2016). As Halegoua points out, technologies of wayfinding and mapping can "alter the context and purpose of movement, the knowledge of surroundings, the organization of spatial and social relations, and the sense of self within urban space" (2019, 108). Indeed, rather than simply influencing the street a driver might take or the direction someone might head in while walking in a city, the "effect" of locating and orienting technologies might work even at the ontological level. Parks (2001) suggests that the GPS map enables users to recognize themselves as "located," and from this position (as part of a "dynamic geographical interface"), they become "subject[s] produced through a series of movements and encounters" (pp. 213–214). The ability of location-aware media to place the subject in space has corresponded to a trendiness in performative mobilities, seen for example in users consciously "checking in" on Foursquare or Facebook to show they are present in a location or "safe" following a disaster (Humphreys & Liao 2013) or using geocoded social media posts that transmit what was being done where (Schwartz & Halegoua 2015). Such practices can be seen to be means of performing the self spatially to others.

One of the most direct contemporary examples in this category is the growing popularity of wearable fitness technologies—activity trackers, smartwatches and other wireless-enabled wearable technology devices that sense and store personal corporeal metrics, such as the number of paces walked or steps climbed, considered in relation to fitness data such as heart rate, calories burned and even quality of sleep. Top sellers include the Apple Watch, Fitbit and Samsung Galaxy Fit, which work in communication with various smartphone applications to track these activities through biometrics. Beyond tracking, however, these devices nudge users into action through reminders (such as a message from Apple Watch informing you it's time to stand up) and visualizations that demonstrate a goal is as yet incomplete (such as Apple Watch's urging to "close your rings" each day, as a visual marker of having burned enough calories, walked briskly and stood up enough). Gilmore (2016) suggests such devices, which he calls "everywear" to account for their pervasiveness, produce new forms of everyday practice as they are "tethered to bodies and, through habitualization, designed to add value to everyday life in the form of physical wellbeing" (p. 2525). The "effects" of everywear relate both to individual health and social belonging, as Gilmore points out that alongside "structures of motivation" for users to become more fit and healthy, the quantification of their physique and movements also reshapes how they think about their bodies and place in the world.

Mediation distinct from "the immediate" or "the real" in mobility

We now consider how the idea of "mediation" is used in contrast to notions of the "immediate" or the "real" in discourses surrounding mobile practice. When considering mediated mobility, the acronym "IRL" comes to mind—a term developed from online chat culture to specify that the "place" where something actually happens is "in real life," which is considered somehow distinct from the virtual realm where the actors are typically engaged in conversation. However, the mediation of mobilities has also altered how so-called IRL moments might be understood.

This can be seen emblematically in the way technologies of immediacy have changed the experience of mobility in travel or tourism, in which digital communication technologies can, on the one hand, produce a sense of immediacy as people can feel their social networks are actively traveling with them, while, on the other hand, these multiple and active virtual connections can be seen as an "obstacle to 'real' connections" (Molz & Paris 2015, 187). In interviews with digitally networked backpackers, Molz and Paris find many people believing that digital connectivity takes away from any "real" engagement with local cultures. One such couple described feeling overly focused on shooting video for their blog when at a gorilla refuge before deciding to "put their cameras away so they could 'just [take] it all in and *really* experience the situation'" (2015, 186 emphasis added). Ironically, now that travelers can share information about their trips in "real-time," they may feel less like they are having a "real" travel experience. Such experiences are mediated by the real-time circulation of data in and around a moment (Simanowski 2017), which can similarly lead to people, who are otherwise trying to not just document an experience but respond to live comments real-time, potentially altering the experience itself, feeling less connected to the moment (Norum 2021).

Mobile communication technologies also increasingly overlay, augment and perhaps even reconstruct the "realness" of "everyday life," for example through location-based mobile games. As Hjorth and Richardson (2017) describe in the introduction to a special issue about the game Pokémon Go, players progress through physical space while battling for digital accolades, "accessing a microworld through their smartphone via the digital overlay of game objects and virtual locations across the actual environment" (p. 4). They point out that to fully engage, players must adopt an "as if" attitude, "moving through the environment 'as if' it were game terrain or an urban playground" (p 5). This online-to-offline mobile overlay entails what de Souza e Silva (2006) calls a "hybrid reality" and Licoppe (2017) refers to as "hyper-realistic." The hybridization of reality might also be seen as a temporal hybridity, for example in projects such as the StreetMuseum app that uses archival photographs called up through location-based data to overlay the past and the present in augmented reality (see Verhoeff 2016).

Along these lines, advancements in extended reality (XR)—a term capturing all real-and-virtual combined environments and human-machine interactions generated by computer technology and wearables—allow users to explore and interact in real-time in artificial environments that are designed to look, sound and "feel" real. In that XR designers and researchers are increasingly able to provide a "comprehensive simulation experience" (Murray & Sixsmith 1999, 317), they grapple with IRL-embodied challenges such as motion sickness (Shaw et al. 2015). The compulsion for programmers to design out of XR experiences the physiological trauma that stems from either simulating space via real motion (VR, or virtual reality) or simulating motion in real space (AR, or augmented reality) suggests that there are important considerations at play here in planning what the "real" is that is being simulated, and how. Finally, the interactivity of navigation technologies underlying so many mediated mobilities has called attention to the flexibility and performativity of maps. Once seen as direct representations of spatial reality or location, as maps are increasingly built through geo-coded texts, images or artifacts, they are understood as representative of "ways of being and representing the world" rather than as objective reality (Farman 2010 in Schwartz & Halegoua 2015, 1657).

Mobilizing new interpretations between subjects and "reality"

Finally, we consider how mobile media facilitate the creation of new spaces of experience, interpretation and meaning between the individual subject and "reality." This definition acknowledges that the perception or experience of reality is always mediated by any given subject's social, historical and economic positionality—embodied experiences that shape how situations are made sense of and understood. This differs slightly from the previous section in that, rather than asking how media are seen as somehow separate from "reality," this begins with the assumption that experience with "reality" is always *already* mediated. Media studies have long drawn from this understanding of mediation in relation to communication technologies, and how such technologies may help construct realities in relation to other people, places, objects or events.

When we add mobility to this consideration, we find the newly produced interpretations are frequently connected to "place." For example, Williams explores how a growth in broadcasting technologies serving populations that had become both increasingly mobile and home-centered, or individualized, were a form of "mobile privatization," responding to growing fragmentation which created a need for new forms to help with understanding the world "out there" (2004 [1974], 21). Interpretations of the "out there" that produced the world as an understandable place are crucial to mediation at this level. The earliest radio broadcasts were described by Paddy Scannell as "place doubling" processes (1996, 91), during which listeners could feel proximity to distant phenomena and events. Moores' (2011) intervention builds on this to suggest that in the context of mobility, places are not made through or in physical geographic locations, but rather are generated experientially, which Hill (2018) argues occurs through affective processes that involve bodily experiences. Now, such experiences are even more closely approximated through the technologies of extended, virtual and augmented reality (Murray & Sixsmith 1999; Backe 2016).

Considerations of how digital media can foster new interpretations of reality by mobile subjects can be exemplified through a number of locative art and digital storytelling projects that serve to produce new understandings of spaces as they are encountered. Galloway (2006) traces a series of interactive projects that use the invisible nature of "ubiquitous computing" (known as ubicomp) to augment experiences of the city, whether through mapping and organizing space in alternative ways or through unique experience with sound in relation to movement. Farman (2015) explores a number of mobile storytelling projects that endeavor to make a given site's forgotten (or invisible) histories visible, suggesting these uses of mobile media may produce a meaningful, embodied experience in place. By using mobile media to tell counter-narratives of place, he argues, "data and the material world inform each other" (Farman 2015, 107).

Mediated mobilities can also work to interpellate (Althusser 1971) individual subjects. Althusser's earlier context speaks to how subjects are always already being constituted by apparatuses of state ideologies. In the current age of ubiquitous digital computing and surveillance, this might speak to the iPhone's use of its always-on Location Services function to generate data about one's movements for interested advertisers, or to the use of GPS technologies to track and render citizens as targets, serving to normalize perpetual militarization and states of war (Kaplan 2006).

Conclusion: Mobility, mediation and digital placemaking

The manifold, diverse contexts and configurations which we have outlined in this chapter point to the broad range of applications, experiences and impacts that mediated mobilities maintain in relation to cultural, social and spatial practices across various spheres of everyday life. One of the central tenets running through the categories we have outlined is the tension between mobility and place, and the roles of media technologies in either producing, mollifying or in some form mediating these tensions. Recent exponential growth in the ubiquity of mobile media has shone a light on the mobilities (and on the uneven distribution of mobilities) of people, things and ideas, connecting people to location in new ways. It has also created new ways of experiencing in-situ engagement, as the very nature of geographic location now contends with the mediated presence of multiple (IRL, virtual, augmented) places in any given moment. While mediated mobilities influence how actors position themselves in time and space, the negotiation of spatio-temporal boundaries via mobile mediated practices enables not merely a reconceptualization of time and space but in fact a re-shaping and re-structuring of both (Tsatsou 2009; see also Ling & Campbell 2009).

By way of concluding thoughts on how mediations of mobility operate in regard to both place and person, we want to consider briefly how place, movement and translocality are formed by various media and through various placemaking practices (Bork-Hüffer 2012). Place is a necessary element for how mediated mobile practices are conceptualized (Wilken & Goggin 2012), as are various processes of placemaking embedded in such practices. In the early days of mobile media, scholars showed concern that the increasing use of networked, mobile devices corresponded to a growing individualization—such as Wellman's (2001) "networked individualism" and Willson's (2006) concern that the singular nature of device interaction disconnected individuals from their geographic surroundings. However, many researchers have critiqued the notion that the individual nature of use might encourage people to focus more completely on their private interests (e.g. Nip 2004; Miller 2013), and some have pointed out that such privacy can facilitate comfort in public places, such as Ito, Okabe and Anderson's (2009) finding that travelers on public transportation could use their media to "avoid contact with others in the shared physical space" and escape into "private cocoons" (p. 74). Halegoua (2019) demonstrates that people can actually use mobile media to themselves become placemakers, and Polson (2016) has shown that through a combination of geo-social practices, people can create a "mobile sense of place" (p. 51).

In travel and tourism studies, access to location-based and digital media has impacted how travelers engage with and impact foreign places. For example, although networked media can allow travelers to remain connected to dispersed networks by posting to social media (Molz 2012) and sharing photos while on the go (Rubinstein & Sluis 2008), such networked travel photos broadcast place to wider audiences, communicating "not just 'I was here'; but 'I am here right now, having this experience in real time, and here is the evidence that this is the case'" (Dinhopl & Gretzel 2016, 127). The navigation support in networked devices empowers travelers by providing access to new terrain (Jansson 2007) and helps connect tourists to local activities through location-based travel apps (Polson 2018; Jansson 2019). As Jansson (2019) points out, while enabling tourist mobilities, such devices also foster imbalances in local economies and can be a factor in gentrification.

In fact, mobility has long been seen as a threat to particularities of locality: mobility and location have been understood by many to be mutually exclusive phenomena (Reig & Norum 2019). And yet, social scientists and humanities scholars often seek to bring place into conversation with mobility. What is in place and what is in a state of motion—mobility and fixity, movement and location—should together be the point of departure for any consideration of space. Indeed, for Ingold (2011) places are themselves made *through* movement, or as Bunge (1966, xvi) put it even some half a century earlier, "any explanation of a location involves the notion of movement." If places are found in the complex, interstitial encounter between emplacement and displacement (Lems 2016), then the forms and formations of human engagement with media that we have considered throughout this chapter importantly draw our attention to the necessary processes that enable people to negotiate and make meaning out of a world forged through the mediation of a multitude of mobilities. In these senses, place is neither found nor revealed but *generated* out of variegated phenomena—between entities and actions that are mobile and fixed, moved and located, motile and potentially sedentary.

References

Althusser, L. 1971. Ideology and the state. In *Lenin and philosophy and other essays*. New York: Monthly Review Press.

Backe, E. L. 2016. A review of virtual reality ethnographic film, or: How we've always been creating virtual reality. *The Geek Anthropologist* (November 17). Available at: https://thegeekanthropologist.com/2016/11/17/a-review-of-virtual-reality-ethnographic-film-or-how-weve-always-been-creating-virtual-reality.

Bork-Hüffer, T. 2016. Mediated sense of place: Effects of mediation and mobility on the place perception of German professionals in Singapore. *New Media & Society*, 18 (10), 2155–2170.

Bunge, W. 1966. *Theoretical geography*. Lund: Gleerup.

Chouliaraki, L. 2008. The media as moral education: Mediation and action. *Media, Culture & Society*, 30 (6), 831–852.

de Souza e Silva, A. 2006. From cyber to hybrid: Mobile technologies as interfaces of hybrid spaces. *Space and Culture*, 3, 261–278.

Dickerson, D. 2018. *No hands: The autonomous future of trucking*. www.cognizant.com/whitepapers/no-hands-the-autonomous-future-of-trucking-codex3867.pdf.

Dinhopl, A., and Gretzel, U. 2016. Selfie-taking as touristic looking. *Annals of Tourism Research*, 57 (C), 126.

Falkheimer, J., and Jansson, A. (eds.) 2006. *Geographies of communication: The spatial turn in media studies*. Gothenburg: Nordicom.

Farman, J. 2015. Stores, spaces, and bodies: The production of embodied space through mobile media storytelling. *Communication Research and Practice*, 1 (2), 101–116.

Frith, J. 2015. Communicating behind the scenes: A primer on radio frequency identification (RFID). *Mobile Media & Communication*, 3 (1), 91–105.

Galloway, A. 2006. Intimations of every life: Ubiquitous computing and the city. *Cultural Studies*, 18 (2–3), 384–408.

Gillespie, T. 2010. The politics of "platforms." *New Media & Society*, 12 (3), 347–364.

Gilmore, J. 2015. Everywear: The quantified self and wearable fitness technologies. *New Media & Society*, 18 (11), 2524–2539.

Graham, M., and Shaw, J. (eds.) 2017. *Towards a fairer gig economy*. London: Meatspace Press.

Grossberg, L., Wartella, E., and Whitney, D. C. 1998. *Media making: Mass media in popular culture*. Thousand Oaks, CA: Sage Publications.

Gupta, A. 2015. An anthropology of electricity from the Global South. *Cultural Anthropology*, 30 (4), 555–568.

Halegoua, G. 2019. *The digital city: Media and the social production of place*. New York: New York University Press.

Hannam, K., Sheller, M., and Urry, J. 2006. Editorial: Mobilities, immobilities and moorings. *Mobilities*, 1 (1), 1–22.

Harvey, P., and Knox, H. 2015. *Roads: An anthropology of infrastructure and expertise*. Ithaca, NY: Cornell University Press.

Hill, A. 2018. *Media experiences*. London: Routledge.

Hjorth, L., and Richardson, I. 2017. Pokémon GO: Mobile media play, place-making, and the digital wayfarer. *Mobile Media & Communication*, 5 (1), 3–14.

Hopkins, J., and Turner, J. 2012. *Go mobile: Location-based marketing, apps, mobile optimized ad campaigns, 2D codes and other mobile strategies to grow your business*. Hoboken, NJ: John Wiley & Sons.

Humphry, J. 2019. "Second-class access": Homelessness and the digital materialization of class. In E. Polson, L. S. Clark and R. Gajjala (eds.) *The Routledge companion to media and class*, pp. 242–252. London: Routledge.

Humphreys, L., and Liao, T. 2013. Foursquare and the parochialization of public space. *First Monday*, 18 (1). https://firstmonday.org/ojs/index.php/fm/article/view/4966.

Ingold, T. 2011. *Being alive: Essays on movement, knowledge and description*. London: Taylor & Francis.

Ito, M., Okabe, D., and Anderson, K. 2009. Portable objects in three global cities: The personalization of urban places. In R. Ling and S. W. Campbell (eds.) *The reconstruction of space and time*, pp. 67–88. New Brunswick, NJ: Transaction Press.

Jansson, A. 2007. A sense of tourism: New media and the dialectic of encapsulation/decapsulation. *Tourist Studies*, 7 (1), 5–24.

Jansson, A. 2019. The mutual shaping of geomedia and gentrification: The case of alternative tourism apps. *Communication and the Public*, 4 (2), 166–181.

Larkin, B. 2013. The politics and poetics of infrastructure. *Annual Review of Anthropology*, 42, 327–343.

Lems, A. 2016. Placing displacement: Place-making in a world of movement. *Ethnos*, 81 (2), 315–337.

Leurs, E., and Hardy, A. 2019. Tinder tourism: Tourist experiences beyond the tourism industry realm. *Annals of Leisure Research*, 22 (3), 323–341.

Licoppe, C. 2017. From *Mogi* to *Pokémon GO*: Continuities and change in location-aware collection games. *Mobile Media & Communication*, 5 (1), 24–29.

Ling, R., and Campbell, S. W. 2009. Introduction: The reconstruction of space and time through mobile communication practices. In R. Ling and S. W. Campbell (eds.) *The reconstruction of space and time: Mobile communication practices*, pp. 1–16. New Brunswick, NJ: Transaction Publishers.

Massey, D. 2012. *For space*. Los Angeles: Sage Publications.

McQuire, S. 2016. *Geomedia, networked cities and the politics of public space*. Cambridge: Polity.

Miles, S. 2017. Sex in the digital city: Location-based dating apps and queer urban life. *Gender, Place & Culture*, 24 (11), 1595–1610.

Miller, D. 2013. *Tales from Facebook*. Cambridge: Polity.

Molz, J. G. 2012. *Travel connections: Tourism, technology and togetherness in a mobile world*. London: Routledge.

Molz, J. G., and Paris, C. M. 2015. The social affordances of flashpacking: Exploring the mobility nexus of travel and communication. *Mobilities*, 10 (2), 173–192.

Morley, D. 2017. *The migrant, the mobile phone, and the container box*. Oxford: Wiley Blackwell.

Murray, C., and Sixsmith, J. 1999. The corporeal body in virtual reality. *Ethos*, 27 (3), 315–343.

Norum, R. 2021. Time for representation: Mediating the moment in a mobile space. In A. Hill, M. Hartmann and M. Andersson (eds.) *Handbook to mobile socialities*. London: Routledge.

Norum, R., and Polson, E. 2021. Placemaking "experiences" during Covid-19. *Convergence: The International Journal of Research into New Media Technologies*, 27 (3), 609-624. https://doi.org/10.1177/13548565211004470.

Parks, L. 2001. Cultural geographies in practice: Plotting the personal: Global positioning satellites and interactive media. *Ecumene*, 8 (2), 209–222.

Pink, S., Lucena, R., Pinto, J., de Souza, A. P., Caminha, C., de Siqueira, G. M., de Oliveira, M. D., Gomes, A., and Zilse, R. 2019. Trust and knowing: Emerging technologies and mobility in the Global South. In R. Wilken, G. Goggin and H. A. Horst (eds.) *Location technologies in international context*. London: Routledge.

Plantin, J.-C., Lagoze, C., and Edwards, P. N. 2018. Infrastructure studies meet platform studies in the age of Google and Facebook. *New Media & Society*, 20 (1), 293–310.

Polson, E. 2016. *Privileged mobilities: Professional migration, geo-social media, and a new global middle class*. New York: Peter Lang.

Polson, E. 2018. "Doing" local: Place-based travel apps and the globally networked self. In Z. Papacharissi (ed.) *A networked self: Platforms, stories, connections*, pp. 159–174. London: Routledge.

Polson, E. 2019. Information superCalle: The social internet of Havana's WiFi streets. In R. Wilken, G. Goggin and H. A. Horst (eds.) *Location technologies in international context*. London: Routledge.

Rabih, J. 2020. Uber and the making of an Algopticon: Insights from the daily life of Montreal drivers. *Capital & Class*. https://doi.org/10.1177/0309816820904031.

Reig, A., and Norum, R. 2019. *Migrantes*. Barcelona: Ekaré.

Rubinstein, D., and Sluis, K. 2008. A life more photographic. *Photographies*, 1 (1), 9–28.

Salazar, N., Elliot, A., and Norum, R. 2017. Studying mobilities: Theoretical notes and methodological queries. In A. Elliot, R. Norum and N. B. Salazar (eds.) *Methodologies of mobility: Ethnography and experiment*, pp. 1–21. Oxford: Berghahn.

Scannell, P. 1996. *Radio, television and modern life*. London: Wiley.

Scholz, T. 2017. The people's disruption. In M. Graham and J. Shaw (eds.) *Towards a fairer gig economy*. London: Meatspace Press.

Schwartz, R., and Halegoua, G. R. 2015. The spatial self: Location-based identity performance on social media. *New Media & Society*, 17 (10), 1643–1660.

Shaw, L. A., Wünsche, B. C., Lutteroth, C., Marks, S., and Callies, R. 2015. Challenges in virtual reality exergame design. In *16th Australasian User Interface Conference (AUIC 2015) (CRPIT)*. Sydney: ACS.

Sheller, M. and Urry, J. 2006. The new mobilities paradigm. *Environment and Planning A*, 38 (2), 207–226.

Silverstone, R. 1999. *Why study the media?* London: Sage Publications.

Silverstone, R. 2005. The sociology of mediation and communication. In C. Calhoun, C. Rojek and B. S. Turner (eds.) *The Sage handbook of sociology*, pp. 188–207. London: Sage Publications.

Tsatsou, P. 2009. Reconceptualising "time" and "space" in the era of electronic media and communications. *PLATFORM: Journal of Media and Communication*, 1, 11–32.

Tsing, A. 2004. *Friction: An ethnography of global connection*. Princeton, NJ: Princeton University Press.

Urry, J. 2003. Social networks, travel and talk. *British Journal of Sociology*, 54 (2), 155–175.

Verhoeff, N. 2016. A tale of two times: Augmented reality as archival laboratory. In G. Fossati and A. van den Oever (eds.) *Exposing the film apparatus: The film archive as a research laboratory*, pp. 357–428. Amsterdam: Amsterdam University Press.

Waltorp, K. 2020. *Why Muslim women and smartphones*. London: Routledge.

Wellman, B. 2001. Physical place and cyber place: The rise of networked individualism. *International Journal of Urban and Regional Research*, 25 (2), 227–252.

Wilken, R. 2019. *Cultural economies of locative media*. Oxford: Oxford University Press.

Wilken, R., and Goggin, G. 2012. *Mobile technology and place*. New York: Routledge.

Williams, R. 1974. *Television*. London: Fontana.

Willson, M. A. 2006. *Technically together: Rethinking community within techno-society*. New York: Peter Lang.

Ziegler, F., and Schwanen, T. 2011. "I like to go out to be energised by different people": An exploratory analysis of mobility and wellbeing in later life. *Ageing and Society*, 31 (5), 758–781.

14

GEOGRAPHIES OF LOCATIVE APPS

Peta Mitchell, Marcus Foth and Irina Anastasiu

A large part of what makes a smartphone "smart" are its built-in location-based services: its ability to locate and be located. As researchers such as Frith (2015) have noted, the smartphone is a form of locative media, enabling the co-construction of "hybrid" physical–digital spaces and places through its geolocative infrastructures and mobile affordances. Researchers have also considered the smartphone to be an ever-more ubiquitous example of "geomedia" (McQuire 2016) or "spatial media" (Leszczynski 2015; Kitchin et al. 2017). With smartphone ownership reaching saturation-point in the Global North and growing rapidly, albeit unevenly, in advanced *and* emerging economies globally (Silver 2019), location-based apps and services, as well as the user-generated location data they elicit and collect, have become both big business and increasingly central to social governance.

In this chapter, we examine the geographies of locative apps—that is, their spatial infrastructures, affordances and emerging cultural economies rather than the mapping of global geographies of mobile app usage.[1] Our examination covers three themes: First, we trace the historical developments leading to the rise of mobile media and "hybrid" spatiality. Second, we discuss the spatial affordances and infrastructures of location-based apps and services and how these have contributed to an expanding location economy. Our final theme addresses the shape and nature of the emergent location industry and, in particular, its implications for privacy. We outline how locative apps and services have been central to the emergence of mobile geolocation as a tool of continuous surveillance, while offering in conclusion some prospective pathways to a more equitable (locative) data future.

The rise of mobile media and hybrid spatiality

In the mid-2000s, a body of research began to emerge bringing focus to the shifting spaces and spatialities engendered by networked and mobile media. Couldry and McCarthy's (2004) edited collection *MediaSpace* was a notable early work that highlighted the increasingly complex entanglements between media, space and place. Published in 2004, four years before the release of the second-generation iPhone 3G, which—with its app-rich environment and embedded GPS—would reconfigure the contours of what constitutes a "smart" phone, Couldry and McCarthy's book featured the mobile phone only marginally, as one among many media forms and artefacts (re)shaping physical, virtual and social space in the

DOI: 10.4324/9781003039068-17

early 2000s. Building on the concept of *mediaspace*, Humphreys's (2007) article on Dodgeball, an early location-based mobile check-in service launched in 2004 but defunct by 2009, gave critical insights into how location was emerging as a central driver within mobile ecosystems and mobile social networks in the mid–late 2000s. Unlike contemporary location-based mobile apps, which largely automate the capture of location information through a smartphone's built-in GPS or other location-capture infrastructure as we discuss below, Dodgeball required users to manually submit their location via text message, which would then be broadcast to the user's social network, again via text message (Humphreys 2007, 343). What Humphreys's year-long study of Dodgeball showed was that location-based mobile social networks had the potential to "strengthen, modify, and rearrange how urban public spaces and social connections are experienced" (2007, 344) thereby creating "third spaces" that sit between the virtual and the physical and that foster "habitual, dynamic, and technologically-enabled face-to-face interaction among loosely tied groups of friends" (2007, 355).

At around the same time, de Souza e Silva (2006) proposed the term "hybrid spaces" to describe how mobile technologies, like the mobile phone, were blurring the boundaries between virtual and physical spaces and places, and simultaneously reconfiguring both. By 2011, de Souza e Silva and Gordon had identified a new form of "location awareness" that they called "net locality," which they described as being "about what happens to individuals and societies when virtually everything is located or locatable," as well as "what individuals and societies can do with the affordances of this location awareness—from organizing impromptu political protests to finding nearby friends and resources" (p. 2). Here too, and like Couldry and Humphreys, Gordon and de Souza e Silva (2011) stress the hybrid spaces and places constructed in and through net locality: "location-aware mobile technologies," they maintain, "can change the way we experience both physical and digital spaces by configuring a new hybrid space, which is composed by a mix of digital information and physical localities" (p. 56).

Following on from this groundwork, much of the research on mobile media in the mid 2010s, like that of Frith (2015), Farman (2012) and Timeto (2015), stressed the collaborative, co-constitutive, and hybrid digital/physical form of placemaking occasioned by location-aware technologies and practices. Exemplifying the hybrid placemaking capabilities of socio-spatial mobile media that researchers were drawing attention to in the early–mid 2010s was the Livehoods project (livehoods.org), launched in 2012 (Cranshaw et al. 2012). Livehoods developed map-based visualizations of geotagged social media data drawn from Foursquare and Twitter, making visible the "geosocial overlay" of cities (Mitchell & Highfield 2017). While interest in mobile technology from design and human-computer interaction (HCI) predates this period (Jones & Marsden 2006), it was around this time that locative media applications became more common and sophisticated (Bilandzic & Foth 2012).

In parallel with the rise of hybrid socio-spatial affordances of mobile applications and tools, the mobile phone also enabled lay people to engage in geographic and planning based activities that were previously only accessible to experts with GIS training. This trend was called "neogeography" by Di-Ann Eisnor of Platial Inc., to describe a notion of "geography without geographers" (Foth et al. 2009, 103). The emergence of neogeography around 2005 saw the opening-up of geospatial technologies both to the general public and to academic disciplines beyond geography, particularly in the humanities, and was often couched in a rhetoric of democratization (Mitchell 2017). However, as Leszczynski (2014) has compellingly argued, neogeography was itself implicated in attempts to "monetize" spatial data and the geoweb, and its discourse of "newness" effectively worked to "depoliticize spatial media" (p. 70). While the term is not used as frequently these days, what remains current are the underlying tensions between the "democratization" and "monetization" of spatial technologies and data.

Alongside these changing socio-spatial mobile practices, critical to the emergence, widespread adoption and mainstreaming of this hybrid digital–physical engagement with location was the rise of the mass-market GPS-enabled smartphone following the 2008 release of the iPhone 3G. The second-generation iPhone was certainly not the first to have built-in GPS—the Benefon Esc! released nearly a decade earlier in 1999 was the first commercially available GPS-enabled phone (Mitchell & Highfield 2017). It did, nevertheless, establish GPS as a standard mass-market feature for smartphones and, with the launch of the App Store the same year, set the foundations for geolocation as an enabling platform for the app economy. As Frith (2015) notes, the three elements that have played key roles in defining and shaping location-based services are "location awareness, mobile internet, and app stores" (p. 42). Since then, location has not simply become embedded in our mobile phones, but also in the global digital data economy. Harvested through smartphone apps, often without the user's full awareness or informed consent, the personal location data of everyday users of mobile media have become highly monetizable and are today routinely sold and on-sold through third-party location monetization companies for various ends—from advertising to providing better public amenities to surveillance.

Since the early days of hybrid socio-spatiality, Wilken (2018; 2019) identifies that smartphone-enabled location-based services have moved through three "generations" or "iterations." Where the first generation required users to actively check-in, as in the case of Dodgeball, the second generation involved more "passive" or "ambient" disclosure of location on the part of the user, while the third, and current, generation is characterized by "ubiquitous geodata capture" (Wilken 2018, 25–26). In almost every case, Wilken (2018) notes, first- and second-generation services became defunct by the mid 2010s, replaced by the third-generation location-based services that now dominate the smartphone and app economy, and in which location has become fully integrated—from the interface to the "algorithmic processing, database population, monetization efforts and so on" (p. 26). In this process, geolocation has become, in effect, "a fully domesticated socio-technological assemblage working to connect and mediate bodies, places, platforms, and devices, and in doing so generating vast data stores of personal geographic information" (Mitchell 2020). In the following section, we outline the spatial affordances and infrastructures of locative apps and services and how these have contributed to a new spatial quality of interaction as well as setting the conditions for Wilken's "ubiquitous geodata capture" or what Kitchin (2015) has called "continuous geosurveillance."

The spatial affordances and spatial infrastructures of locative apps and services

In a seminal early work on the rise of "cyberplace" through networked communication, Wellman (2001) notes that the introduction of the mobile phone changed the way phone calls were likely to begin. Where a call to a fixed-line phone might occasion the question "Who is this?" a call to a mobile phone changed the fundamental question to "Where are you?"—a shift that, according to Wellman (2001), illustrates how "the context of place does matter" (p. 239). This relatively small change is also indicative of a new spatial quality of interaction with widespread repercussions. These repercussions have registered even more strongly with the advent of smartphones that incorporate a range of location-based services that can take advantage of new spatial affordances (Bilandzic & Foth 2009). Fröhlich et al. (2007) distinguish between four types of mobile spatial interaction: (1) applications that facilitate navigation and wayfinding; (2) mobile augmented reality applications; and (3) applications to create or (4) access information attached to physical places or objects.

Navigation and wayfinding are arguably among the most-used location-based services that smartphones offer. Two years after its launch in 2005, Google Maps debuted on the first iPhone (Gibbs 2015). Instant wayfinding capabilities at people's fingertips heralded a new era of never being lost again, although 20th century urban theorists such as Benjamin (1978) and Debord (1981) along with more recent researchers (Traunmueller et al. 2013; Foth 2016) have stressed the importance of becoming lost or disoriented to spatial experience. Since Google Maps went mobile, new features have been added, such as turn-by-turn GPS navigation, the visualization of traffic congestion, and restaurant and business reviews and their busy times. The ability to access such codified spatial knowledge—in many cases knowledge built on location data contributed by users—began changing how people navigate and negotiate the city, and contributing to what Graham (2005) has termed "software-sorted geographies."

Another popular but controversial feature added to Google Maps in 2007 is Street View. Offering users not only map-based navigation but also a series of interactive panorama photos taken from many streets across the world, Street View also raised privacy concerns. Street View—and more recently Live View—are also examples of mobile augmented reality (AR) applications (Aurigi & De Cindio 2008; Craig 2013), the second category of mobile spatial interaction. Mobile AR describes an interactive location-based feature that overlays images of the physical world with digital information rendered in real-time and in a spatially accurate 3D position. While Layar was an early mobile AR browser (Liao & Humphreys 2015), it was not until the 2016 location-based game Pokémon Go that mobile AR became widely popularized (Paavilainen et al. 2017), feeding into research on the spatial affordances and geographies of urban game play (Colley et al. 2017; Leorke 2019).

The third and fourth categories of mobile spatial interaction broadly cover locative apps for creating and accessing data attached to physical places or objects. One of the earliest examples was Urban Tapestries—a mobile location-based platform to connect people with the places they inhabit through their stories, experiences and observations (Silverstone & Sujon 2005). The spatial affordances of such locative apps have enabled a number of experimental implementations, ranging from location-based storytelling and location-based social interaction to crowd-sourced urban maintenance.[2]

These spatial affordances are enabled through the variety of spatial infrastructures for location capture that are embedded within the smartphone *and* our environment. Location capture today is most readily associated with ubiquitous GPS functionality on smartphones. This tendency to reduce geolocation or location capture to GPS alone, however, obfuscates less immediately apparent mechanisms for smartphone-enabled location tracking (including WiFi, mobile communication cell towers, and Bluetooth) that often work in concert, rather than in exclusivity. Far from halting location tracking, disabling a smartphone's GPS can uncover the plurality of hidden data capturing infrastructure through which an ever-more granular account of mobile phone users' outdoor *and indoor* movements is constructed.

Pioneering technologies, like satellite navigation systems/GPS or the Global System for Mobile Communications (GSM), paved the way for *trilateration* and *triangulation*—geometric calculations enabling receivers of satellite and cell tower signals to determine their own position based on signal frequency, strength, timing and/or angle from multiple emitters. GPS and GSM became commercially available in the 1980s and 1990s. In 2000, the US lifted the intentional accuracy reduction of GPS for civil and commercial use. These developments incentivized joint efforts to standardize location access to overcome the decade's high fragmentation across devices and smartphone operating systems (OSs), for example via the J2ME Location API first released in 2003.

Today, Apple and Google, which together control virtually the entire global smartphone OS market (Statcounter GlobalStats 2020), by default rely on a combination of GPS and cell tower triangulation, known as Assisted GPS. However, these methods become inaccurate or entirely unavailable inside buildings, tunnels, underground, or in areas with a low cell-tower density. GPS also quickly drains smartphone batteries, and tech-savvy or privacy-aware users disable it. Urban areas, home to the majority and most sought-after data subjects, require continuous and granular location data to maximize economic value for targeted advertising or the optimization of digital products and services.

A growing ecosystem of short(er)-range radio transmitters is exploited to this end, forming a fine-grained mesh of fixed, known locations against which to determine movement patterns using triangulation. This includes WiFi access points and low-cost, low-energy beacons based on Bluetooth that the smartphone identifies by quietly broadcasting its presence without initiating a formal connection that would alert the user. These installations proliferate in shopping centers, airports, office buildings and in public spaces—in town squares and on high streets. To make indoor location-tracking cheaper and more ubiquitous (and more covert), Google patented the use of existing power infrastructure inside buildings for indoor sub-room positioning (c.f. patent by Patel et al. 2013).

Efforts are also made to overcome inaccuracies or blank patches in this ecosystem, as well as the user's location privacy settings, by integrating standard motion-sensor data from the phone's gyroscope, accelerometer and magnetometer to accurately determine distance, direction, speed, or activity type in relation to a known geographical coordinate or indoor position (c.f. Google Patent by Norta et al. 2007). The enhanced location-tracking capabilities of these sensors tend to be underestimated. Narain et al. (2016) demonstrate how a driving route can be identified through "zero permission sensors" (p. 397)—data for which the OS does not require user permission before it can be collected—with 30–50% accuracy by comparing the generated route pattern to public road information. While this may seem insignificant, combining the data with other covert location-tracking mechanisms significantly increases this accuracy. Such approaches pose a significant threat to users' privacy (Narain et al. 2017).

Finally, advances in computer vision created new opportunities for *visual localization* methods, where algorithms allow for the estimation of a device's position by comparing the spatial features of a live photo or images captured via the phone's camera feed to those from a previously built database of geotagged images (c.f. Google patent by Steinbach et al. 2017). Any assumption that disabling GPS, WiFi or even Bluetooth on the smartphone will halt location tracking is misguided considering these ongoing efforts. In January 2018, Google phones were collecting location information inferred through nearby Bluetooth beacons even as Bluetooth was deactivated, as long as the location history feature was turned on (Yanofsky 2018). In August 2018, an Associated Press (2018) investigation found some Google services tracked location even *after* location history was turned off—testament to the economic value of location data and the lengths to which companies like Google will go to capture it. The plethora of indoor positioning startups, many targeting retailers (AngelList n.d.), points to a growing location economy, to which we now turn.

The location economy: Location intelligence, geoprivacy and surveillance capitalism

Alongside developments in location-aware technologies and location-based mobile services, a young but growing third-party "location intelligence" industry has emerged. Often also referred to as location analytics or location monetization, this industry reconfigures the geographies of locative media. In recent years, Foursquare—which was developed in 2008 by one of the founders of Dodgeball as a gamified check-in app and which is often considered to exemplify the idea of the "hybrid" spatiality of locative media (Frith 2013; Saker & Evans 2016)—has pivoted explicitly to become a location intelligence company (Martineau 2019), taglining itself "The Trusted Location Data & Intelligence Company." And yet, as Smith (2019) has noted, with its focus on "user-centric studies of audience geocoding" and large "mainstream location-based platforms such as Google Maps and Foursquare," much of the existing literature on locative media "neglects the important but often invisible role of third-party advertising servers and analytics industries that capture and analyse location data for a variety of political and commercial applications" (p. 1044). The availability of vast swathes of personal location data generated through smartphone apps and services has set the ground-work for a new kind of "geodemographics" that shift the focus from the generic postal code to the highly specific, and more personally intimate, "geocode" (Smith 2019, 1045). These "second-order" geodemographics enable new forms of algorithmic marketing and refined marketing metrics that can assess "audience 'lift'" achieved through intensively geo-targeted advertising (Smith 2019, 1045).

Gauging the size and contours of this industry and the market for personal location data is difficult since location data brokers and analytics firms do not always advertise themselves as such. Moreover, the app-based transactions between an individual's smartphone, the location-based app they are using, and a third-party location-monetization firm that the app may be sharing that individual's location with is hidden from the user. A 2018 report from Sudo Security Group (2018), a company developing a privacy enhancing app to track location tracking, found that many of the top free apps in Apple's App Store (including popular weather, parking and coupon apps boasting user bases of over 100 million) contained tracking code from one or more of 12 identified third-party location intelligence companies. The group also analyzed the pop-up justification that each app displayed to request access to a user's location data, finding that many apps did not disclose that granting location access meant the resulting data would be shared with or on-sold to third parties. Market reports for the location services and intelligence industries invariably indicate strong and continuing growth. The Geospatial Media and Communications (2019) *Location Intelligence Market Report* for 2019, for instance, states that the market for location data "has grown from nearly US$ 9 billion in 2014 to around US$ 22 billion in 2018 in terms of market size," and it is expected to double from 2018 to 2022. The report continues that the location intelligence industry "can be sub-divided into four major categories: a) hardware, b) software/platforms, c) 'location data and map content,' and d) 'solutions and services,'" with the latter two categories growing at the fastest rate and currently "account[ing] for nearly two-thirds of the market" (p. 3).

As our everyday mobile-mediated experiences of space and place are increasingly com-modified through the ubiquitous collection of personal geodata by locative apps and services, the meaning of location and our relationship to it also changes. Thatcher (2017) has high-lighted the intensifying commodification of everyday life brought about by the location-based app industry, to the extent that the term *location* now registers both the "everyday experience of being in a particular place at a particular time," and an emergent meaning of *location* as "digital commodity" (p. 2704). For developers of location-based apps, Thatcher continues, "location"—as a term and concept—"promises the ability to tie an individual's intentions and socio-economic information, as revealed through data mining, to a relatively precise spatio-temporal coordinate tuple of latitude, longitude, and timestamp" (p. 2704).

There is growing evidence to suggest that this distinction between location as (digitally mediated) experience of place and as digital commodity is beginning to collapse, or, rather, that there is increased public awareness of the variety of ends to which personal location data is being put to work within the data economy. In 2018, two major international news out-lets—the *Wall Street Journal* and *New York Times* (*NYT*)—published investigative reports into the smartphone app location data industry, with the discomforting headlines "Your location data is being sold—often without your knowledge" (Mims 2018) and "Your apps know where you were last night, and they're not keeping it secret" (Valentino-DeVries et al. 2018). The latter article reported that 17 of 20 apps tested by the *NYT* sent user location data to some 70 businesses. One app that the *NYT* article drew attention to was The Weather Channel app—this in turn spurred legal action, with the City Attorney of Los Angeles bringing a lawsuit against The Weather Channel for "deceptively" using its app to "amass its users' private, personal geolocation data" (Kelly 2019). Location is seemingly becoming a new legal frontier for governance and regulation, with a number of lawsuits and class actions being launched in recent years, including the Electronic Frontier Foundation (EFF) (2019) filing a class action against US telco AT&T for on-selling user location data to third parties such as "bounty hunters, car dealerships, landlords, and stalkers" without authorization. In 2021, the Federal Court of Australia ruled that Google had misled consumers about how the company collected and used personal location data, in a case brought by the Australian Competition and Consumer Commission (ACCC) (2021)..

As location becomes an increasingly dominant player in the platform and app economy, and as minor and major scandals relating to excessive tracking, leaks, hacks and inadvertent exposures of personal location data through smartphone apps and location-based services become more fre-quent,[3] the question of privacy has come to the fore in debates and discussion around locative media. These heightened threats to privacy have been well acknowledged and researched for at least two decades, with successive studies of mobile phone users' attitudes to privacy highlighting location as holding particular personal sensitivity (Barkhuus & Dey 2003; Zickuhr 2013; Martin & Nissenbaum 2019; Riedlinger et al. 2019). Taken as a whole, these studies not only reinforce the risks that location poses to personal privacy, but also that locative media require users to constantly negotiate tradeoffs between convenience and privacy. Granting a weather app access to a phone's GPS, for instance, can automate the provision of a weather forecast for the phone user's current location, making it more convenient than manual location input. In the current app economy, however, it is more likely than not that this exchange of location information will extend far beyond a transaction between app and user, and that the user will have no way of knowing the extent to which their location data has been traded and re-traded across the loca-tion economy.

Given the intensification of the location economy along with the changing practices, infrastructures and affordances of location in the era of smartphones, new attunements to the spatial aspects of privacy have come to the fore and, in recent years, there has been recog-nition of "geoprivacy" (Leszczynski 2017; Mitchell & Highfield 2017; Keßler & McKenzie 2018) as an emergent form of privacy of particular concern both for smartphone-using pub-lics and for public administrations. According to Leszczynski (2017), a "broadened concept" of geoprivacy can and must "account for the emergent complex of potential privacy harms and violations that may arise from a number of nascent realities of living in a (spatial) big data present," including the capture of individuals' personal location data and their spatio-tem-poral movements, the circulation of these data in and through the location economy, and the difficulties individuals face in controlling these "highly personal flows of spatial information about themselves in networked device and data ecologies" (p. 237).

Critically, over the same period, location has become central to what Shoshana Zuboff (2019) has termed "surveillance capitalism." Zuboff draws attention to the operational value of smartphone-derived location data within surveillance capitalism, as well as its heightened personal sensitivities, both of which lie in the inherent reidentifiability of supposedly "anonymized" or "anonymizable" location data. Zuboff quotes Princeton computer scientists Arvind Narayanan and Edward Felten as summing up location's simultaneously unique and uniquely sensitive selling-point: namely, that "there is no known effective method to anonymize location data, and no evidence that it's meaningfully achievable" (Zuboff 2019, p. 244). The current geosurveillant–geocapitalist moment we find ourselves in is not an unforeseen one. As early as 2003, Dobson and Fisher (2003) warned that advances in GIS and GPS technologies heralded the potential for a dystopian future based on real-time location control of humans—a form of bondage "extend[ing] far beyond privacy and surveillance" (p. 47). In their deliberately provocative paper, they called this looming form of location control "geoslavery," describing it as a "real, immediate, and global threat" (p. 47). Although the very worst excesses of Dobson and Fisher's vision of potential human bondage through location control may not (yet) have eventuated, as Obermeyer (2007) has suggested, the growing entanglements between geo-surveillance, consumer capitalism and governance have brought about a form of "volunteered (geo)slavery," requiring "new approaches to address spatial data privacy" and greater public awareness (p. 2).

Nowhere has this increased value of and heightened sensitivities around location data been more apparent in recent times than in the varying approaches countries took in developing contact-tracing apps in early 2020 during the COVID-19 pandemic (Morley et al. 2020). Where countries such as Poland, South Korea, India and Taiwan developed contact-tracing apps that tracked the location of smartphones, spurring global debates over privacy and reidentifiability, other apps, such as Australia's CovidSafe app and the contact-tracing framework developed by Apple and Google, stressed their *non-collection* of location data as a critical component of their privacy-protecting frameworks. One contact-tracing app—North Dakota's Care19 app—was even found to be sharing user location data with Foursquare, in violation of its own privacy policy (Morse 2020). In another contemporaneous example of the coextensivity of surveillance capitalism, the location intelligence industry and smartphone-derived personal location data, Mobilewalla—a "consumer intelligence" company that purchases app-generated location data from aggregators and claims "80–90% device coverage in the US"—published a report on the demographics and movements of Black Lives Matter protesters in cities across the United States (Doffman 2020). These very recent examples draw focus to a growing public awareness of how everyday digitally mediated locative practices are intimately tied to the location economy. They also signal the regulatory attention now turning to location intelligence and the location economy, particularly as they function as a supporting pillar within surveillance capitalism.

Conclusion

Locative apps and location-based services have quickly advanced to be taken for granted by many people in everyday life. The commercial frameworks many of these services are embedded within rely on location data being put to work within a vast (location) data economy. We stress that the "geographies" of locative apps, as we are defining them here, are the result of intertwined practices, affordances and infrastructures that shape and reshape our experience of place and space, our awareness and recognition of ourselves as data bodies and bodies of data, and our understandings of and attitudes towards (geo)privacy. Location-based services are a fast-paced field, and reactive policy strategies and regulatory instruments tend to lag behind the latest technology developments. However, transdisciplinary approaches recognizing the continued importance of hybrid space propose progressive and desirable pathways to avoid purely dystopian outlooks and outcomes. Leszczynski (2019) highlights geolocation's *affective* capacities that work to "perturb, animate, align, mobilize, organize, dis/assemble—other things, both human and other-than-human alike" (p. 208). Engaging with affect, Leszczynski continues, helps us to "grappl[e] with what geolocation 'does,' how it 'does' what it does, and to what ends" (p. 208).

Geolocation, as experienced through locative apps and services, can present—often simultaneously—an unwanted intrusion into personal privacy, a welcome perception of safety and security, and an opportunity to enact and engage with hybrid spatiality. Working through the affects, ambivalences and ambiguities of geolocation can help us find desirable pathways for a more caring and just spatially aware and spatially enabled society. Current and future research and development into promising concepts such as privacy-by-design (Langheinrich 2001), sousveillance (Mann et al. 2003; Foth et al. 2014), good data practices (Daly et al. 2019) and data sovereignty (Lynch 2020; Mann et al. 2020), as well as tactics and strategies for resisting geosurveillance (Swanlund & Schuurman 2018; Seidl et al. 2020), offer potential trajectories to ameliorate the negative consequences of the location economy.

Notes

1 Although outside the scope of this chapter, important research that maps the geographies of internet and mobile app use across the globe has been done by Lim et al. (2015) and Wu and Taneja (2016), among others.
2 See, for instance, *Mobile Narratives* (Wiesner et al. 2009), *TrainYarn* (Camacho et al. 2015) and *FixMyStreet* and *FixVegas* (Foth et al. 2011).
3 For reporting on recent location-data leaks and exposures on various apps and platforms see, for instance, Graham (2020) on Venmo, Hern (2018) on Strava, Whittaker (2019) on family tracking app Family Locator, and Martin (2019) on a range of dating app location-data leaks.

References

Associated Press. 2018. Google records your location even when you tell it not to. *The Guardian* (August 13). www.theguardian.com/technology/2018/aug/13/google-location-tracking-android-iphone-mobile.

Aurigi, A., and De Cindio, F. 2008. *Augmented urban spaces: Articulating the physical and electronic city*. Aldershot: Ashgate.

Australian Competition and Consumer Commission. 2021. *Google misled consumers about the collection and use of location data*. www.accc.gov.au/media-release/google-misled-consumers-about-the-collection-and-use-of-location-data.

Barkhuus, L., and Dey, A. K. 2003. Location-based services for mobile telephony: A study of users' privacy concerns. *Interact*, 3, 702–712.

Benjamin, W. 1978. A Berlin chronicle. In *Reflections: Essays, aphorisms, autobiographical writings*, pp. 3–60. New York: Schocken.

Bilandzic, M., and Foth, M. 2009. Mobile spatial interaction and mediated social navigation. In M. Khosrow-Pour (ed.) *Encyclopedia of information science and technology*, 2nd edn, pp. 2604–2608. Hershey, PA: IGI Global. https://doi.org/10.4018/978-1-60566-026-4.ch415.

Bilandzic, M., and Foth, M. 2012. A review of locative media, mobile and embodied spatial interaction. *International Journal of Human-Computer Studies*, 70 (1), 66–71. https://doi.org/10.4018/978-1-60566-026-4.ch415.

Camacho, T. D., Foth, M., Rittenbruch, M., and Rakotonirainy, A. 2015. TrainYarn: Probing perceptions of social space in urban commuter trains. In *Proceedings of the 27th Australian computer-human interaction conference (OzCHI 2015)*. New York: ACM. https://doi.org/10.1145/2838739.2838760.

Colley, A., Thebault-Spieker, J., Lin, A. Y., Degraen, D., Fischman, B., Häkkilä, J., Kuehl, K., Nisi, V., Nunes, N. J., Wenig, N., and Wenig, D. 2017. The geography of Pokémon GO: Beneficial and problematic effects on places and movement. In *Proceedings of the 2017 CHI conference on human factors in computing systems*, pp. 1179–1192. New York: ACM. https://doi.org/10.1145/3025453.3025495.

Couldry, N., and McCarthy, A. (eds.) 2004. *MediaSpace: Place, scale and culture in a media age*. London: Routledge.

Craig, A. B. 2013. Mobile augmented reality. In A. B. Craig (ed.) *Understanding augmented reality*, pp. 209–220. Boston, MA: Morgan Kaufmann. https://doi.org/10.1016/B978-0-240-82408-6.00007-2.

Cranshaw, J., Schwartz, R., Hong, J., and Sadeh, N. 2012. The Livehoods Project: Utilizing social media to understand the dynamics of a city. In *International AAAI Conference on Weblogs and Social Media 2012*. www.aaai.org/ocs/index.php/ICWSM/ICWSM12/paper/view/4682.

Daly, A., Devitt, K., and Mann, M. 2019. *Good data: Theory on demand*. Amsterdam: Institute of Network Cultures.

de Souza e Silva, A. 2006. From cyber to hybrid: Mobile technologies as interfaces of hybrid spaces. *Space and Culture*, 9 (3), 261–278. https://doi.org/10.1177/1206331206289022.

Debord, G. 1981. Theory of the Dérive. In K. Knabb (ed.) *Situationist international anthology*, pp. 50–54. Berkeley, CA: Bureau of Public Secrets.

Dobson, J. E., and Fisher, P. F. 2003. Geoslavery. *IEEE Technology and Society Magazine*, 22 (1), 47–52. https://doi.org/10.1109/MTAS.2003.1188276.

Doffman, Z. 2020. Black Lives Matter: US Protesters tracked by secretive phone location technology. *Forbes Magazine* (June). www.forbes.com/sites/zakdoffman/2020/06/26/secretive-phone-tracking-company-publishes-location-data-on-black-lives-matter-protesters/.

Electronic Frontier Foundation. 2019. *EFF Sues AT&T, data aggregators for giving bounty hunters and other third parties access to customers' real-time locations*. www.eff.org/press/releases/eff-sues-att-data-aggregators-giving-bounty-hunters-and-other-third-parties-access.

Farman, J. 2012. *Mobile interface theory: Embodied space and locative media*. New York: Routledge.

Foth, M. 2016. Why we should design smart cities for getting lost. In J. Watson (ed.) *The Conversation yearbook 2016: 50 standout articles from Australia's top thinkers*, pp. 109–113. Melbourne: Melbourne University Press.

Foth, M., Bajracharya, B., Brown, R., and Hearn, G. 2009. The second life of urban planning? Using neogeography tools for community engagement. *Journal of Location Based Services*, 3 (2), 97–117. https://doi.org/10.1080/17489720903150016.

Foth, M., Schroeter, R., and Anastasiu, I. 2011. Fixing the city one photo at a time: Mobile logging of maintenance requests. In *Proceedings of the 23rd Australian computer-human interaction conference (OzCHI 2011)*. New York: ACM. https://doi.org/10.1145/2071536.2071555.

Foth, M., Heikkinen, T., Ylipulli, J., Luusua, A., Satchell, C., and Ojala, T. 2014. UbiOpticon: Participatory sousveillance with urban screens and mobile phone cameras. In *Proceedings of the 3rd international symposium on pervasive displays*. New York: ACM. https://doi.org/10.1145/2611009.2611034.

Frith, J. 2013. Turning life into a game: Foursquare, gamification, and personal mobility. *Mobile Media & Communication*, 1 (2), 248–262. https://doi.org/10.1177/2050157912474811.

Frith, J. 2015. *Smartphones as locative media*. Cambridge: Polity.

Fröhlich, P., Simon, R., Baillie, L., Roberts, J., and Murray-Smith, R. 2007. Mobile spatial interaction. In *CHI '07 extended abstracts on human factors in computing systems*. New York: ACM. https://doi.org/10.1145/1240866.1241091.

Geospatial Media and Communications. 2019. *Location intelligence market report 2019*. https://geobuiz.com/geospatial-location-report/.

Gibbs, S. 2015. Google Maps: A decade of transforming the mapping landscape. *The Guardian* (February 8). www.theguardian.com/technology/2015/feb/08/google-maps-10-anniversary-iphone-android-street-view.

Gordon, E., and de Souza e Silva, A. 2011. *Net locality: Why location matters in a networked world*. Chichester: Wiley.

Graham, J. 2020. Venmo did what with my data? My location was shared when I paid with the app. *USA Today* (February 22). www.usatoday.com/story/tech/2020/02/22/venmo-noom-solitaire-shared-my-personal-info-data-firms/4795546002/.

Graham, S. D. N. 2005. Software-sorted geographies. *Progress in Human Geography*, 29 (5), 562–580. https://doi.org/10.1191/0309132505ph568oa.

Hern, A. 2018. Fitness tracking app strava gives away location of decret US Army bases. *The Guardian* (January 28). www.theguardian.com/world/2018/jan/28/fitness-tracking-app-gives-away-location-of-secret-us-army-bases.

Humphreys, L. 2007. Mobile social networks and social practice: A case study of Dodgeball. *Journal of Computer-Mediated Communication: JCMC*, 13 (1): 341–360. https://doi.org/10.1111/j.1083-6101.2007. 00399.x.

AngelList. n.d. *Indoor positioning startups.* https://angel.co/indoor-positioning.

Jones, M., and Marsden, G. 2006. *Mobile interaction design.* Chichester: Wiley.

Kelly, M. 2019. *The Weather Channel app unlawfully obtained user location data, says prosecutor.* www.theverge. com/2019/1/4/18168373/los-angeles-weather-channel-app-user-location-data.

Keßler, C., and McKenzie, G. 2018. A geoprivacy manifesto. *Transactions in GIS*, 22 (1), 3–19. https://doi. org/10.1111/tgis.12305.

Kitchin, R. 2015. *Continuous geosurveillance in the "smart city."* http://dismagazine.com/dystopia/73066/ rob-kitchin-spatial-big-data-and-geosurveillance/.

Kitchin, R., Lauriault, T. P., and Wilson, M. W. (eds.) 2017. *Understanding spatial media.* Los Angeles, CA: Sage.

Langheinrich, M. 2001. Privacy by design: Principles of privacy-aware ubiquitous systems. In *Ubicomp 2001: Ubiquitous Computing.* Berlin: Springer. https://doi.org/10.1007/3-540-45427-6_23.

Leorke, D. 2019. *Location-based gaming: Play in public space.* Singapore: Springer.

Leszczynski, A. 2014. On the neo in neogeography. *Annals of the Association of American Geographers* 104 (1): 60–79. https://doi.org/10.1080/00045608.2013.846159.

Leszczynski, A. 2015. Spatial media/tion. *Progress in Human Geography*, 39 (6), 729–751. https://doi.org/ 10.1177/0309132514558443.

Leszczynski, A. 2017. Geoprivacy. In R. Kitchin, T. P. Lauriault and M. W. Wilson (eds.) *Understanding spatial media*, pp. 235–244. Los Angeles, CA: Sage.

Leszczynski, A. 2019. Platform affects of geolocation. *Geoforum*, 107, 207–215. https://doi.org/10.1016/j. geoforum.2019.05.011.

Liao, T., and Humphreys, L. 2015. Layar-ed places: Using mobile augmented reality to tactically reengage, reproduce, and reappropriate public space. *New Media & Society*, 17 (9), 1418–1435. https://doi.org/10. 1177/1461444814527734.

Lim, S. L., Bentley, P. J., Kanakam, N., Ishikawa, F., and Honiden, S. 2015. Investigating country differences in mobile app user behavior and challenges for software engineering. *IEEE Transactions on Software Engineering*, 41 (1), 40–64. https://doi.org/10.1109/TSE.2014.2360674.

Lynch, C. R. 2020. Contesting digital futures: Urban politics, alternative economies, and the movement for technological sovereignty in Barcelona. *Antipode*, 52 (3), 660–680. https://doi.org/10.1111/anti.12522.

Mann, M., Mitchell, P., Foth, M., and Anastasiu, I. 2020. #BlockSidewalk to Barcelona: Technological sovereignty and the social licence to operate smart cities. *Journal of the Association for Information Science and Technology*, 71 (9), 1103–1115. https://doi.org/10.1002/ASI.24387.

Mann, S., Nolan, J., and Wellman, B. 2003. Sousveillance: Inventing and using wearable computing devices for data collection in surveillance environments. *Surveillance & Society*, 1 (3), 331–355. https:// doi.org/10.24908/ss.v1i3.3344.

Martin, K. E., and Nissenbaum, H. 2019. *What is it about location?* https://doi.org/10.2139/ssrn.3360409.

Martin, N. 2019. Another dating app has leaked users' data. *Forbes Magazine* (September). www.forbes. com/sites/nicolemartin1/2019/09/27/another-dating-app-has-leaked-users-data/.

Martineau, P. 2019. You may have forgotten Foursquare, but it didn't forget you. *Wired* (March 8). www. wired.com/story/you-may-have-forgotten-foursquare-it-didnt-forget-you/.

McQuire, S. 2016. *Geomedia: Networked cities and the future of public space.* Cambridge: Polity.

Mims, C. 2018. Your location data is being sold: Often without your knowledge. *Wall Street Journal* (March 4). www.wsj.com/articles/your-location-data-is-being-soldoften-without-your-knowledge-1520168400.

Mitchell, P. 2017. Literary geography and the digital: The emergence of neogeography. In R. Tally (ed.) *The Routledge handbook of literature and space*, pp. 85–94. New York: Routledge.

Mitchell, P. 2020. Geo-locations. In *Oxford research encyclopedia of literature.* Oxford: Oxford University Press. https://doi.org/10.1093/acrefore/9780190201098.013.979.

Mitchell, P., and Highfield, T. 2017. Mediated geographies of everyday life: Navigating the ambient, augmented and algorithmic geographies of geomedia. *Ctrl-Z: New Media Philosophy*, 7, 1–5. www. ctrl-z.net.au/journal/?slug=mitchell-highfield-mediated-geographies-of-everyday-life.

Morley, J., Cowls, J., Taddeo, M., and Floridi, L. 2020. Ethical guidelines for COVID-19 tracing apps. *Nature*, 582 (7810), 29–31. https://doi.org/10.1038/d41586-020-01578-0.

Morse, J. 2020. *Contact-tracing app caught sharing location data with Foursquare.* https://mashable.com/article/ care19-north-dakota-contact-tracing-app-sharing-location-data-foursquare/.

Narain, S., Vo-Huu, T. D., Block, K., and Noubir, G. 2016. Inferring user routes and locations using zero-permission mobile sensors. In *2016 IEEE Symposium on Security and Privacy (SP)*. Washington, DC: IEEE. https://doi.org/10.1109/SP.2016.31.

Narain, S., Vo-Huu, T. D., Block, K., and Noubir, G. 2017. The perils of user tracking using zero-permission mobile apps. *IEEE Security Privacy*, 15 (2), 32–41. https://doi.org/10.1109/MSP.2017.25.

Norta, H., Ashall, P., Oksanen, O., and Lammintaus, A. 2007. Mobile location devices and methods. USPTO 7266378. US Patent, filed December 12, 2002, and issued September 4. https://patentimages. storage.googleapis.com/53/9e/f7/130b5d0d41c39d/US7266378B2.pdf.

Obermeyer, N. 2007. *Thoughts on volunteered (geo)slavery*. www.ncgia.ucsb.edu/projects/vgi/docs/position/ Obermeyer_Paper.pdf.

Paavilainen, J., Korhonen, H., Alha, K., Stenros, J., Koskinen, E., and Mayra, F. 2017. The Pokémon GO experience: A location-based augmented reality mobile game goes mainstream. In *Proceedings of the 2017 CHI Conference on Human Factors in Computing Systems*. New York: ACM. https://doi.org/10.1145/ 3025453.3025871.

Patel, S. N., Truong, K. N., Abowd, G. D., Robertson, T., Reynolds, M. S., and Georgia Tech Research Corp. 2013. Sub-room-level indoor location system using power line positioning. USPTO 8392107. US Patent, filed June 28, 2007, and issued March 5, 2013. https://patentimages.storage.googleapis. com/c8/cd/92/5266afe06c9747/US8392107.pdf.

Riedlinger, M., Chapman, C., and Mitchell, P. 2019. *Location awareness and geodata sharing practices of Australian smartphone users*. https://eprints.qut.edu.au/132000/.

Saker, M., and Evans, L. 2016. Everyday life and locative play: An exploration of Foursquare and playful engagements with space and place. *Media, Culture & Society*, 38 (8), 1169–1183. https://doi.org/10. 1177/0163443716643149.

Seidl, D. E., Jankowski, P., Clarke, K. C., and Nara, A. 2020. Please enter your home location: Geoprivacy attitudes and personal location masking strategies of internet users. *Annals of the American Association of Geographers*, 110 (3), 586–605. https://doi.org/10.1080/24694452.2019.1654843.

Silver, L. 2019. *Smartphone ownership is growing rapidly around the world, but not always equally*. www.pewresea rch.org/global/2019/02/05/smartphone-ownership-is-growing-rapidly-around-the-world-but-not-a lways-equally/.

Silverstone, R., and Sujon, Z. 2005. *Urban tapestries: Experimental ethnography, technological identities and place*. www.lse.ac.uk/media-and-communications/assets/documents/research/working-paper-series/EWP07. pdf.

Smith, H. 2019. Metrics, locations, and lift: Mobile location analytics and the production of second-order geodemographics. *Information, Communication and Society*, 22 (8), 1044–1061. https://doi.org/10.1080/ 1369118X.2017.1397726.

Statcounter GlobalStats. 2020. *Mobile operating system market share worldwide*. https://gs.statcounter.com/ os-market-share/mobile/worldwide.

Steinbach, E., Schroth, G., Abu-Alqumsan, M., Huitl, R., Al-Nuaimi, A., and Schweiger, F. 2017. Visual localization method. USPTO 9641981. US Patent, filed May 16, 2011, and issued May 2, 2017. https://pa tentimages.storage.googleapis.com/4e/6c/fe/3dcc033c3c0663/US9641981.pdf.

Sudo Security Group. 2018. *Learn how apps track you*. https://guardianapp.com/research/ios-app-location- report-sep2018/.

Swanlund, D., and Schuurman, N. 2018. Resisting geosurveillance: A survey of tactics and strategies for spatial privacy. *Progress in Human Geography*, 43 (4), 596–610. https://doi.org/10.1177/0309132518772661.

Thatcher, J. 2017. You are where you go: The commodification of daily life through "location." *Environment & Planning A*, 49 (12), 2702–2717. https://doi.org/10.1177/0308518X17730580.

Timeto, F. 2015. Locating media, performing spatiality: A nonrepresentational approach to locative media. In R. Wilken and G. Goggin (eds.) *Locative media*, pp. 110–122. London: Routledge.

Traunmueller, M., Fatah gen Schieck, A., Schöning, J., and Brumby, D. P. 2013. The path is the reward: Considering social networks to contribute to the pleasure of urban strolling. In *CHI '13 extended abstracts on human factors in computing systems*. New York: ACM. https://doi.org/10.1145/2468356. 2468520.

Valentino-DeVries, J., Singer, N., Keller, M. H., and Krolik, A. 2018. Your apps know where you were last night, and they're not keeping it secret. *New York Times* (December 10). www.nytimes.com/intera ctive/2018/12/10/business/location-data-privacy-apps.html.

Wellman, B. 2001. Physical place and cyberplace: The rise of personalized networking. *International Journal of Urban and Regional Research*, 25 (2), 227–252. https://doi.org/10.1111/1468-2427.00309.

Whittaker, Z. 2019. *A family tracking app was leaking real-time location data.* https://social.techcrunch.com/2019/03/23/family-tracking-location-leak/.

Wiesner, K., Foth, M., and Bilandzic, M. 2009. Unleashing creative writers: Situated engagement with mobile narratives. In *OZCHI '09—proceedings of the 21st annual conference of the australian computer-human interaction special interest group: Design: Open 24/7.* New York: ACM. https://doi.org/10.1145/1738826.1738901.

Wilken, R. 2018. The necessity of geomedia: Understanding the significance of location-based services and data-driven platforms. In K. Fast, A. Jansson, J. Lindell, L. R. Bengtsson and M. Tesfahuney (eds.) *Geomedia studies: Spaces and mobilities in mediatized worlds,* pp. 20–40. New York: Routledge.

Wilken, R. 2019. *Cultural economies of locative media.* Oxford: Oxford University Press.

Wu, A. X., and Taneja, H. 2016. Reimagining internet geographies: A user-centric ethnological mapping of the World Wide Web. *Journal of Computer-Mediated Communication: JCMC,* 21 (3), 230–246. https://doi.org/10.1111/jcc4.12157.

Yanofsky, D. 2018. *Google can still use Bluetooth to track your Android phone when Bluetooth is turned off.* https://qz.com/1169760/phone-data/.

Zickuhr, K. 2013. *Location-based services.* www.pewresearch.org/internet/2013/09/12/location-based-services/.

Zuboff, S. 2019. *The age of surveillance capitalism: The fight for the future at the new frontier of power.* New York: Public Affairs.

15

DIGITAL SURVEILLANCE
AND PLACE

Ellen van Holstein

In the past few decades, surveillance and securitization have seen substantial changes with the introduction of a suite of digital technologies into the security arsenals of governments, corporations and citizens. These shifts have changed the governance mechanisms that direct control over spaces and people, and have profoundly impacted on people's experiences of safety, threat and privacy in place. In turn, technological and institutional responses to new forms of surveillance have reshaped the character of geographical work on surveillance. Importantly, geographers have moved away from research focused on dyadic relationships between governments and citizens to conceptualize more complex constellations of actors and technologies, and have adapted their methodological approaches to follow suit (e.g. Adey et al. 2013; Spiller 2016).

Surveillance means "to watch from above," and the concept has long been associated with an institution's control over an enclosed space such as a school, prison or factory (Galič et al. 2017). The interest of geographers in surveillance aligns with the discipline's longstanding interests in the role of the state, the production of territory and the circulation of power through governance technologies and practices. The insight into surveillance produced by geographers goes beyond the mere control over space that is implied in visual monitoring. Geographers also analyze how practices and technologies of surveillance shape the very spaces that are being monitored (Graham 1998; Koskela 2000; Kitchin & Dodge 2011; Pink & Fors 2017; Crampton et al. 2020). Put in other words, geographers understand that "[s]urveillance relates to, focuses on and projects itself into space, becomes inscribed there, and in the process contributes to the very production of the spaces concerned" (Klauser 2013, 275). Geographers study how, as digital technologies change the ways surveillance is practiced, the characteristics of place change too, including its power relationships, laws regulating property and privacy, opportunities for profit accumulation, and its affective atmospheres (Adey et al. 2013; Leszczynski 2016).

This chapter considers how geographical work changes as technologies digitize and surveillance practices change. The chapter demonstrates how geographers adapt their conceptualization of surveillance as digital technologies obscure the boundaries between the watcher and the watched. The chapter highlights the conceptual toolkits and methodological skillsets that allow media geographers to analyze how people are differently and unequally governed, empowered and curtailed as digital surveillance technologies are becoming an

DOI: 10.4324/9781003039068-18

increasingly constant and active presence in everyday spaces. The chapter considers aspects of surveillance ranging from commercial applications of digital surveillance to the collaboration of many ordinary people in surveillance. To start, the next section offers a discussion of the theoretical foundations of work on digital surveillance in geography.

From moulds to modulation

Geographers' engagement with the concept of surveillance predates the emergence of digital surveillance technologies. While surveillance practices have been prominent throughout history, modern forms of surveillance are distinct because of their emphasis on individuation, and because of the sheer volume and continuity of data collection that can be achieved (Poster 1990; Graham & Murakami Wood 2003; Lyon 2006). These changes have led to a quest, both in geographical work on surveillance and in the interdisciplinary field of surveillance studies, for suitable social theories to help understand digital surveillance, its subjects, technologies, spaces and effects. Both fields have been strongly shaped by Michel Foucault's *Discipline and Punish* (1991 [1979]) (for discussions see Dobson & Fisher 2007; Galič et al. 2017).

One of Foucault's central objectives was to analyze the workings of the modern state and its modes of government. In Foucault's view, a government's control over subjects is achieved by its ability to collect information. In line with this, Foucault based his thinking on power, discipline and surveillance in society at large on the practices and power relationships that govern institutions such as schools and hospitals. Foucault's thinking about surveillance was inspired by Jeremy Bentham's (1748–1832) architectural diagram of the panopticon. Bentham was a utilitarian philosopher and social reformer who developed the idea of the panopticon as a model for institutional buildings in which one overseer has constant visual access to all the subjects in its interior, while those subjects cannot see the overseer. In a panopticon, subjects are aware that they are under constant surveillance, but they do not know at which exact moment they are being watched. According to Foucault, this constant one-directional gaze combined with subjects' awareness of being judged against a set of norms and expectations makes subjects internalize those norms.

Foucault used Bentham's model to develop his theory of the disciplinary society, in which he argues that people increasingly govern themselves in modern society because of the presence, or threat, of constant surveillance and judgement (Foucault 1991). He focused on institutions such as schools and prisons and documented how institutions communicate expectations and rules, and he analyzed institutional practices such as record-keeping and reporting that create ways of knowing and measuring a population. Foucault argued that in performing these practices of surveillance, governments encourage individuals to fit into the normative mould. Foucault's thinking offered a radical break from scholarship that emphasized coercive qualities of surveillance.

Foucault's work became highly influential because of the technological developments that occurred shortly after the publication of his theory on the disciplinary society. Foucault developed his thinking on power and surveillance throughout the 1970s and 1980s, and this coincided with the introduction of computers and cameras into everyday spaces. The 1990s saw an exponential increase of CCTV cameras in public spaces in many countries including the US and the UK, and CCTV systems have many panoptic qualities (e.g. Fyfe & Bannister 1996; Koskela 2000). The bold presence of cameras in public spaces makes it obvious that one might be watched, while cameras do not disclose exactly when the watching occurs. Another panoptic quality of CCTV is its objective to punish people who present deviant behavior, or better still, to change people's behavior preventively so that no deviance occurs. These similarities between Foucault's theory and developments in surveillance technologies have led to a strong panoptic paradigm in studies of surveillance, and scholarship on CCTV can be considered the start of the contemporary geographical literature on surveillance (also see Graham & Murakami Wood 2003).

Since the introduction of CCTV into public spaces, surveillance has changed in multiple ways. Where the surveillant parties in Bentham and Foucault's panopticon were either a government institution or a corporation, contemporary surveillance is performed by complex partnerships consisting of governments, technologies and corporations. Furthermore, subjects under surveillance are increasingly encouraged to use digital technologies to collaborate in surveillance, for example, by monitoring themselves or their peers (Albrechtslund & Lauritsen 2013). With the collection of digital data, surveillance is "no longer limited to single buildings, and observations no longer limited to line of sight" (Gandy 1993, 23). This has inspired geographers and social theorists to critique Foucault's theory of the disciplinary society for being too preoccupied with visual forms of surveillance, and the theory is increasingly seen as unsuitable for the study of networked, algorithmic and multi-actor forms of surveillance (Poster 1990; Hardt & Negri 2000). Some adapt Foucauldian theories to new technologies and governance mechanisms, for example by putting forward the idea of a superpanopticon (Poster 1990), a sorting panopticon (Gandy 2003), or by combining various theoretical adaptations (Murakami Wood 2013). Others have abandoned Foucault's disciplinary society and its panoptic paradigm altogether to replace it with theories that emphasize the networked character of contemporary societies and their practices of surveillance.

Geographers are increasingly turning their attention to social theorists who conceptualize relationships between technology and society, especially Gilles Deleuze and Bruno Latour. While not explicitly focused on surveillance, the work of Deleuze and his collaborator Félix Guattari has been critically important for understanding the multiplicity and instability of digital forms of surveillance (Haggerty & Ericson 2000; Lyon 2006). Their concept of the assemblage captures systems of governance that lack clear boundaries and that consist of heterogeneous objects that come together to function as a whole in unforeseen ways (Patton 1994 cited in Haggerty & Ericson 2000). The assemblage has proven a fitting concept to analyze surveillance systems that exist at the intersections of various media that can be brought together and used for unforeseen purposes (Haggerty & Ericson 2000). Surveillance lurks in these systems as mere potential until an opportunity for surveillance presents itself and connections between systems are made.

Digital technologies have given rise to forms of surveillance that seem anonymous and benign because it does not target specific individuals. The collection of deidentified information is unlikely to spark resistance as privacy is predominantly understood as the protection of an individual against intrusion. Instead of honing in on individuals, data-driven surveillance involves collecting extensive information on individuals and dividing this information into segments. Ostensibly very personal experiences such as desires, fears and needs can be predicted based on combinations of people's age, education, gender, credit rating, recent purchases, marital status, postcode, etc. This information can be used to control people's collective behavior, for instance via targeted promotions and electoral advertisements. Deleuze (1992) introduced the concept on the "dividual" to describe the data doubles that emerge as people's individual information is sliced up. The concept marks the end of a period in the social sciences where the individual was deemed the smallest unit to which society can be reduced and explains the need for new forms of collective organizing and resistance in the face of this kind of divisive control.

Deleuze called the process by which large data sets are organized into categories and patterns to gain social control "modulation." Modulation allows surveillant parties to intervene in society's flows and rhythms unknowingly, because interventions are not presented as a response to an individuals' actions. Surveillance is carefully concealed so that people might experience the consequences of surveillance without knowing that surveillance occurred. The control of organizations and companies that collect data on users and customers is thus one-sided, and it forms a threat to democracy because it inhibits collective action and shared forms of control (Andrejevic 2007). In *Postscript on the Societies of Control* (1992) Deleuze stipulated that this kind of undetected surveillance is made possible by an illusion of freedom, and he suggested a break from Foucault's disciplinary society in which the subjects' awareness of surveillance is understood as an instrumental component of projects of surveillance.

A second philosopher who features prominently in geographical work on surveillance is Bruno Latour. Latour responded to Foucault's work by pointing out that surveillance is partial, fragile and prone to failure (see Albrechtslund & Lauritsen 2013). Where Foucault conceptualized all-seeing surveillant institutions, Latour argued that modern surveillance mechanisms do not see everything (pan), but very little (oligo) as digital sensors and devices function in oligopticons that only pick up particular kinds of information that are then isolated from their context (Latour & Hermant 1998). Latour's actor network theory (Latour 2005) conceptualizes the interactions between components in a network, and the theory is used in geographies of surveillance to understand how various people—such as staff, consumers and police—and material objects—such as digital sensors and smartphones—participate in surveillance (Adey 2004; Albrechtslund & Lauritsen 2013). Geographers use these theories of society and technology to analyze how digital surveillance shapes spaces, whether private (Kennedy & Strengers 2020), public (Minton 2018) or liminal (e.g. Crampton et al. 2020).

Ubiquitous computing and the de-territorialized subject

It is difficult to underestimate the reach of digital surveillance technologies and practices. Surveillant assemblages can include the data sets of multiple government and commercial parties, and surveillance systems thus transcend institutional boundaries. Collectively, the surveillant assemblage then consists of CCTV cameras, smart loyalty cards, number plate recognition technology, location and timestamped credit card information, public transport e-tickets, census data, social media profiles, our personal smartphones and the list goes on. This pervasiveness has brought on the "disappearance of disappearance" (Haggerty & Ericson 2000, 619), the disappearance of the possibility for anyone to go off the proverbial radar. Geographers study the accomplishment and the consequences of this omnipresence, its uneven spatial effects and its impact on subjective experiences of place (e.g. Vanolo 2014; Kitchin 2015; Sadowski & Pasquale 2015; Datta & Odendaal 2019).

The collection of large quantities of data, or Big Data, has motivated the collectors of that data to develop way to organize this data into meaningful and manageable units. When large quantities of data are involved, this sorting is commonly done by algorithms that are programmed to recognize patterns, that organize data by placing people into categories, or that sound an alarm when a combination of factors occurs (see Eubanks 2018). Geographers demonstrate how digitization has reworked the spaces and temporalities of surveillance. For example, when digital technologies allow authorities to collect excessive amounts of data to find people who break a certain rule, this information remains available to penalize people who deviate in other ways in the future (Swanlund & Schuurman 2019). This kind of design creep, where data is used for unintended purposes, makes surveillance mechanisms unpredictable. Data doubles and consumer profiles attach to a person unknowingly as they move through time and space, only to reveal their existence when access, for instance to credit, is denied (e.g. Graham & Murakami Wood 2003; Maalsen & Sadowksi 2019). Algorithmic recognition of patterns has led to forms of anticipatory governance where people are watched or confronted because they are expected to deviate from norms (Kitchin 2015). These reworkings of the sequencing of surveillance and discipline are possible because datafied subjects are never completely deterritorialized. Place continues to function as a key that can be used to re-assemble identities. For this reason, some have argued not to farewell the disciplinary society just yet and to focus on how surveillance strategies oscillate between discipline and control (Iveson & Maalsen 2019).

Shifts in the quantity, speed and continuity of data collection have opened up new objectives in surveillance (e.g. Graham 1998; Haggerty & Ericson 2000; Kitchin 2015). Surveillance is performed in the interest of security and law enforcement, but also for the creation of markets for new products and for the efficient and convenient delivery of a wide range of services. Scholars call this latter variety "surveillance capitalism" as it functions to extract data from consumers to add value to a business (Haggerty & Ericson 2000; Zuboff 2015). Geographers have built on the work of David Harvey to illustrate how this extraction is a form of capital accumulation by dispossession and how it alienates workers from the surplus value they create through their labor (Attoh et al. 2019). The parallel existence of surveillance for profit and discipline demonstrates that surveillance serves multiple interests simultaneously (Iveson & Maalsen 2019). The pervasiveness of surveillance and the mix of objectives it serves contribute to the acceptability of widespread intrusion. Leszczynski (2015) has put forward for instance that people do not resist surveillance and accept the necessity to disclose personal information to governments and businesses because it seems necessary for cost-effective service delivery. Geographers are interested in how such discursive constructs expand the spatial reach of surveillance and shift articulations of power. At the same time geographers highlight the subjective experience of surveillance by registering people's anxieties about being tracked and by analyzing spatial and emotional responses (Koskela 2002; Leszczynski 2015).

With the ubiquity of surveillance technologies, groups of people are being watched who were not monitored before. Haggerty and Ericson have argued that:

> [i]ndividuals with different financial practices, education and lifestyle will come into contact with different institutions and hence be subject to unique combinations of surveillance. The classifications and profiles that are entered into these disparate systems correspond with, and reinforce, differential levels of access, treatment and mobility.
>
> *(Haggerty & Ericson 2000, 618)*

Geographers take an interest in how surveillance interacts with existing axes of difference and inequality. While everyone is being watched in some form, the effects on different groups of people, such as welfare recipients, women and people for color, are starkly uneven (e.g. Koskela 2002; Leszczynski 2016). For instance, geographers have speculated and analyzed how surveillance is used to ban non-consumers or exclude people from spaces who are otherwise deemed undesirable (see Hatuka & Toch 2017). Geographers have also been persistently concerned with the possibilities of using surveillance technologies to increase the cost of insurance for those who can least afford it (e.g. Graham & Murakami Wood 2003; Maalsen & Sadowski 2019). The discipline's social justice concerns are going beyond issues of privacy and seek to understand how digital surveillance informs neoliberal governance regimes in which inequalities are actively reproduced while government responsibilities are devolved to the private sector.

Public-private surveillance assemblages

Social and cultural geographers have a rich tradition of researching citizenship and the changing role of the state as governments privatize public services and devolve responsibilities to local authorities and individual citizens. While securitization is deemed part of governments' responsibilities even in small government ideologies, surveillance is now commonly performed in collaborations between government and civil parties. For example, in their attempts to identify offenders, police increasingly request access to data that is collected by companies. Police departments have been known to use video footage of crowds posted to websites such as YouTube and Twitter to identify protestors or offenders and increasingly turn to digitally organized community groups on platforms such as Facebook and WhatsApp for assistance with surveillance (Kelly & Finlayson 2015; Van Holstein 2018). As a result, the boundaries between what counts as public and private have blurred and geographers analyze the shifts in power geometries this creates.

Surveillance digitization has led to increasingly complex relationships between governments and private companies as companies develop the technologies that gather the data that governments want to use. Concepts such as the "hybrid state" and "security assemblages" have been put forward to conceptualize partnerships between private parties and the state (e.g. Colona & Jaffe 2016) and help make sense of the incorporation of citizens, tech companies and a range of electronic devices into increasingly complex surveillance networks. A good recent example emerged when multiple police departments in the US were found to be collaborating with Amazon to access their customers' digital front door camera footage. Ring, Amazon's home surveillance company, was found to provide police with a portal where it can directly request access to footage from individual consumers. The company was also found to coach police officers in techniques to persuade consumers to grant access to this material, equipping police to sidestep the need for warrants and subpoenas. Because the police tap into an existing system, rather than create their own, they have effectively expanded the state's surveillance network without any of scrutiny or accountability that would otherwise be expected from a government institution (Haskins 2019; Perez 2019). The use of third-party digital information for surveillance is prompting legal scholars to reassess laws, such as the fourth amendment in the context of the US, that rely on three-dimensional conceptions of space and reasonable expectations of privacy to protect citizens' rights (e.g. Curry 1997). Geographers have important contributions to make to understanding the new relationships of power that digital technologies afford and how different conceptualizations of place shape those relationships.

The advent of digital technologies in surveillance coincided with the widespread privatization of public services and spaces (Graham & Murakami Wood 2003). The privatization of public services has changed how services are offered, for example by making services available first and most conveniently to premium paying customers. This process was first identified and conceptualized by geographers Graham and Marvin (2001) in their book *Splintering Urbanism*. The framework has provided insight into how data is used, for instance, to give certain customers priority in internet or telephone queues or to treat undesirable customers unfavorably. They point out that the widespread structural discrimination of entire categories of service users was made possible by the digitization of surveillance. The new digital surveillance assemblage coincides with a new political economy of consumer citizenship in which people's rights depend on their ability to pay for a service. Furthermore, for-profit services that are delivered based on consumer-generated data are most likely to suit people who are represented in that data set, and can overlook people who did not generate data, for example because they do not own a phone.

As digital technologies have blurred the boundaries between government and commercial surveillance practices, so too have they obscured the visibility of surveillance itself. As argued by Deleuze (1992), contemporary surveillance is carried out under the guise of freedom and empowerment. Tech companies and governments have an interest in making citizens feel like they have choice, for instance to contribute to safety in their neighborhood by volunteering their footage to police. Geographers have been very eager to point out how these changes effectively transfer state responsibilities to citizens. With the ostensible increase in freedom to do whatever one wants, the responsibility for social outcomes shifts away from public institutions and towards individual citizens. This creates new opportunities for surveillance and ways to legitimize it; Ring using the idea of the good citizen who takes responsibility for security is a case in point. Geographers analyze the processes through which services persuade consumers to self-monitor behaviors. Work has highlighted how smart energy meters give consumers insight into their energy usage in ways that create a sense of responsibility for an environmentally sustainable footprint (Levenda 2019), and geographers analyze how self-tracking devices change people's relationships to their bodies, health and surroundings (Pink & Fors 2017). Geographers analyze these shifting relationships between citizens and governments and have been consistently interested in citizens' seemingly voluntary contribution to the growth of surveillant assemblages.

Participatory surveillance

While any system of surveillance requires the participation of various actors and technologies, albeit only to internalize normative judgement (Albrechtslund & Lauritsen 2013), digital systems have made participation in surveillance widespread and acceptable. As people participate in surveillance by sharing personal information with various digital platforms and communities, they have been converted into suppliers of valuable data that can be commodified and otherwise capitalized on by corporate parties (Attoh et al. 2019). With the purchase and installation of each iWatch, FitBit, smart energy meter and home surveillance system, the internet of things expands (Maalsen & Sadowski 2019). Consumers are thus facilitating the strengthening of a leviathan surveillance network that for its corporate shareholder ownership is subject to very little scrutiny.

Contemporary surveillance networks and databases are not organized in a panoptic, top-down fashion. Individuals are not just disciplined; rather they actively participate in their own surveillance by contributing information to databases. These practices are captured in the concept "participatory surveillance" (Poster 1990). Participation raises important questions for geographers about scale and the circulation of power through surveillance. For example, Albrechtslund (2008) has argued that self-surveillance in the form of the sharing of information, activities and preferences can potentially empower and not only violate or exploit the user. Sharing information with wider networks of users can be an identity-shaping practice and a way to motivate oneself (Albrechtslund & Lauritsen 2013). In line with this, surveillance scholars interpret surveillance as a cultural practice that is normalized and rendered meaningful through cultural expressions such as reality television and consumer ratings (Staples 1997; Lyon 2018). Participation in surveillance thus becomes acceptable through wider digitally mediated social environments that comprise digital entertainment, convenience and gamification. It is not always obvious who are the powerful and the powerless ones in these networks (also see Molz 2006).

In addition to consumers generating and sharing data, participation in surveillance also takes more active forms. Some citizens are willing to use digital technologies to assist police departments in their work by becoming the proverbial eyes on the street (Kelly & Finlayson 2015). Citizens can turn to various apps to contribute to securitization, some of which were designed as communication platforms, such as Facebook and WhatsApp, and some of which were explicitly developed for securitization such as "Nextdoor" and Amazon Ring's "Neighbors." Geographers have pointed out that perceptions of risk and strategies for safety are shaped by a combination of personal experience, the shared knowledge of family, friends, neighbors, etc., and the impact of the media, and that security and fear are therefore dynamic, subjective and open to interpretation (England & Simon 2010). Given the subjective experience of security and risk and the role of media and technology therein, the digitization of surveillance practices shift perceptions of risk and responsibilities for security and surveillance (Van Holstein 2018). Algorithmically curated newsfeeds and online community spaces shape the information people are exposed to and this can strengthen existing prejudices and preferences. People's participation in surveillance, whether digitally or not, is thus always mediated by a digital social environment.

Communication technologies that facilitate and encourage people's participation in surveillance often feel secure and familiar because they already play a central role in people's everyday lives. Geographers are interested in people's relationships with technology and have highlighted how technologies are discursively rendered innocent and inherently beneficial (Kitchin & Dodge 2011). The discursive construction of technologies as benign, objective and solution-orientated has worked to stifle objections against their widespread use. People fearing digital surveillance are portrayed as "having something to hide," while surveillance mechanisms are increasingly a part of, or attached to, everyday practices such as shopping, travelling and talking to friends via digital platforms. This while digital surveillance technologies are consistently shown to fail to deliver the objectivity that they promise. For instance, CCTV cameras facilitate racial profiling, and the use of digital apps diverts people away from neighborhoods with ethnically diverse residents and business owners (Adey 2003; Leszczynski 2016). Combined with discursive constructions, the emotional position of technologies used for surveillance further de-politicizes surveillance while it has real effects on people's mobility and opportunities.

Recent and future geographies of digital surveillance

Technological innovation sees surveillance constantly move in new directions and into new spaces, and geographers follow this closely. For example, recent research on surveillance focuses on the penetration of surveillance networks into domestic spaces (Maalsen & Sadowski 2019; Strengers & Kennedy 2020). These explorations create important insights into how surveillance becomes a part of intimate spaces and relationships. This work also reinvigorates an interest in gendered relationships to technologies. Strengers and Kennedy (2020) for instance show how disparities in skill and confidence around technologies create unequal opportunities to engage in surveillance and argue that surveillance technologies become tools that widen gender disparity.

Another example of a novel direction in geographies of digital surveillance is work that analyzes advances in drone technologies used in surveillance. This collection explores concepts of verticality and cloud constellations used in digital surveillance technologies (Elden 2013; Garrett & Anderson 2018). As this work explores the possibilities and politics of drone use for objectives as varied as warfare and conservation, it explores how the spatial concept of volume and its visualizations become governance instruments (e.g. Monahan & Mokos 2013; Waghorn 2016). Additionally, a focus on surveillance for the purpose of environmental protection is amplifying existing calls in geography to move away from normative approaches to the study of surveillance as this work highlights that surveillance can equally erode and strengthen social justice.

As digital technologies continue to push the boundaries of what is possible in the field of surveillance, geographers continue to seek concepts that capture the ways in which surveillance reshapes social worlds. A series of 2019–2020 events, notably the protests in Hong Kong and Chile and the COVID-19 pandemic, will likely steer research on digital surveillance in new directions. The use of drones in these circumstances and the deployment of technologies such as facial recognition and temperature sensors raises important research questions about power dynamics and forms of resistance that emerge when adversaries use the same technologies (Leistert 2012).

Digitization of surveillance creates new opportunities and challenges for resistance. The concept of the individual plays a crucial role in debates about resistance to surveillance. Digital media facilitates individualization and oftentimes obscures the collective results of individual actions. For example, working with Uber drivers, Attoh and colleagues (2019) demonstrate how these workers create the company's data by driving around cities and how the same technology divides workers making it harder for them to collectively organize. During a strike one driver cannot know whether fellow drivers are striking. This brings to mind Deleuze's (1992) warning that unions can only remain relevant if they devise responses to forms of control that do not rely on an enclosed space. Resistance is not impossible, but it needs to adapt to the digital playing field. Geographers stress the possibility for sousveillance whereby citizens turn the gaze of surveillance up to those in power such as politicians and the police (Waghorn 2016). Other tactics and strategies for resistance discussed in this line of work are the minimization of exposure to surveillance, for example by using encryption, or by obfuscation where users deliberately add false information to databases to make them less accurate and therefore less valuable (Swanlund & Schuurman 2019). As the omnipresence of digital technologies creates feelings of powerlessness and inevitability in the face of surveillance, work on resistance is of critical importance moving forward.

To close, it is important that geographers reflect on their own role in surveillance. New digital methods available to geographers come with opportunities to experiment, and this opens geographers up to encountering unexpected information. For instance, geographers have asked participants to install apps that share the participants' locations with researchers (Hatuka & Toch 2017) and geographers who use social media platforms such as Facebook know that this makes it harder to draw boundaries around research practice both in terms of when research happens and where it is to take place (De Jong 2015). In this field of new possibilities, ethical slippage is likely to occur, and geographers have a heightened responsibility to reflect on their own research conduct in this light. After all, as geographers use existing technologies and networks in research, they too contribute to the expansion of a surveillant assemblage that is at best only partially under their control.

References

Adey, P. 2003. Secured and sorted mobilities: Examples from the airport. *Surveillance & Society*, 1 (4), 500–519.

Adey, P. 2004. Surveillance at the airport: Surveilling mobility/mobilising surveillance. *Environment and Planning A*, 36 (8), 1365–1380.

Adey, P., Brayer, L., Masson, D., Murphy, P., Simpson, P., and Tixier, N. 2013. "Pour votre tranquillité": Ambiance, atmosphere, and surveillance. *Geoforum*, 49, 299–309.

Albrechtslund, A. 2008. Online social networking as participatory surveillance. *First Monday* 13 (3). http s://doi.org/10.5210/fm.v13i3.2142.

Albrechtslund, A., and Lauritsen, P. 2013. Spaces of everyday surveillance: Unfolding an analytical concept of participation. *Geoforum*, 49, 310–316.

Andrejevic, M. 2007. *iSpy: Surveillance and power in the interactive era.* Lawrence, KS: University Press of Kansas.

Attoh, K., Wells, K., and Cullen, D. 2019. "We're building their data": Labor, alienation, and idiocy in the smart city. *Environment and Planning D: Society and Space*, 37 (6), 1007–1024.

Colona, F., and Jaffe, R. 2016. Hybrid governance arrangements. *European Journal of Development Research*, 28, 175–183.

Crampton, J. W., Hoover, K. C., Smith, H., Graham, S., and Berbesque, J. C. 2020. Smart festivals? Security and freedom for well-being in urban smart spaces. *Annals of the American Association of Geographers*, 110 (2), 360–370.

Curry, M. R. 1997. The digital individual and the private realm. *Annals of the Association of American Geographers*, 87 (4), 681–699.

Datta, A., and Odendaal, N. 2019. Smart cities and the banality of power. *Environment and Planning D: Society and Space*, 37 (3), 387–392.

De Jong, A. 2015. Using Facebook as a space for storytelling in geographical research. *Geographical Research*, 53 (2), 211–223.

Deleuze, G. 1992. Postscript on the societies of control. *October*, 59, 3–7.

Dobson, J., and Fisher, P. 2007. The panopticon's changing geography. *The Geographical Review*, 97 (3), 307–323.

Elden, S. 2013. Secure the volume: Vertical geopolitics and the depth of power. *Political Geography*, 34, 35–51.

England, M. R., and Simon, S. 2010. Scary cities: Urban geographies of fear, difference and belonging. *Social & Cultural Geography*, 11, 201–207.

Eubanks, V. 2017. *Automating inequality: How high-tech tools profile, police, and punish the poor.* New York: St Martin's Press.

Foucault, M. 1991 [1979]. *Discipline and punish: The birth of the prison.* London: Penguin.

Fyfe, N. R., and Bannister, J. 1996. City watching: Closed circuit television surveillance in public spaces. *Area*, 28, 37–46.

Galič, M., Timan, T., and Koops, B. J. 2017. Bentham, Deleuze and beyond: An overview of surveillance theories from the Panopticon to participation. *Philosophy and Technology*, 30 (1), 9–37.

Gandy, O. H. 1993. *The panoptic sort: A Political economy of personal information.* Boulder, CO: Westview.

Garrett, B., and Anderson, K. 2018. Drone methodologies: Taking flight in human and physical geography. *Transactions of the Institute of British Geographers*, 43 (3), 341–359.

Graham, S. 1998. Spaces of surveillant-simulation: New technologies, digital representations, and material geographies. *Environment and Planning D: Society and Space*, 16 (4), 483–504.

Graham, S., and Marvin, S. 2001. *Splintering urbanism: Networked infrastructures, technological mobilities and the urban condition.* London: Routledge.

Graham, S., and Murakami Wood, D. 2003. Digitizing surveillance: Categorization, space, inequality. *Critical Social Policy*, 23 (2), 227–248.

Haggerty, K. D., and Ericson, R. V. 2000. The surveillant assemblage. *British Journal of Sociology*, 51 (4), 605–622.

Hardt, M., and Negri, A. 2000. *Empire.* Cambridge, MA: Harvard University Press.

Haskins, C. 2019. *Amazon is coaching cops on how to obtain surveillance footage without a warrant.* www.vice. com/en_us/article/43kga3/amazon-is-coaching-cops-on-how-to-obtain-surveillance-footage-wi thout-a-warrant?xyz.

Hatuka, T., and Toch, E. 2017. Being visible in public space: The normalisation of asymmetrical visibility. *Urban Studies*, 54 (4), 984–998.

Iveson, K., and Maalsen, S. 2019. Social control in the networked city: Datafied dividuals, disciplined individuals and powers of assembly. *Environment and Planning D: Society and Space*, 37 (2), 331–349.

Kelly, A., and Finlayson, A. 2015. Can Facebook save Neighbourhood Watch? *The Police Journal*, 88, 65–77.

Kitchin, R. 2015. *Continuous geosurveillance in the "smart city."* http://dismagazine.com/dystopia/73066/ rob-kitchin-spatial-big-data-and-geosurveillance/.

Kitchin, R., and Dodge, M. 2011. *Code/space: Software and everyday life.* Cambridge, MA: MIT Press.

Klauser, F. R. 2013. Political geographies of surveillance. *Geoforum*, 49, 275–278.

Koskela, H. 2000. "The gaze without eyes": Video-surveillance and the changing nature of urban space. *Progress in Human Geography*, 24 (2), 243–265.

Koskela, H. 2002. Video surveillance, gender, and the safety of public urban space: "Peeping Tom" goes high tech? *Urban Geography*, 23 (3), 257–278.

Latour, B. 2005. *Reassembling the social: An introduction to actor network theory.* Oxford: Oxford University Press.

Latour, B., and Hermant, E. 1998. *Paris ville invisible.* Paris: Institut Synthélabo.

Leistert, O. 2012. Resistance against cyber-surveillance within social movements and how surveillance adapts. *Surveillance & Society*, 9 (4), 441–456.

Leszczynski, A. 2015. Spatial big data and anxieties of control. *Environment and Planning D: Society and Space*, 33 (6), 965–984.

Leszczynski, A. 2016. Speculative futures: Cities, data, and governance beyond smart urbanism. *Environment and Planning A*, 48, 1691–1708.

Levenda, A. M. 2019. Thinking critically about smart city experimentation: Entrepreneurialism and responsibilization in urban living labs. *Local Environment*, 24 (7), 565–579.

Lyon, D. 2006. *Theorizing surveillance: The panopticon and beyond.* Portland, OR: Willan Publishing.

Lyon, D. 2018. *The culture of surveillance: Watching as a way of life.* Hoboken, NJ: Wiley.

Maalsen, S., and Sadowski, J. 2019. The smart home on FIRE: Amplifying and accelerating domestic surveillance. *Surveillance and Society*, 17 (1–2), 118–124. https://doi.org/10.24908/ss.v17i1/2.12925.

Minton, A. 2018. The paradox of safety and fear: Security in public space. *Architectural Design*, 88 (3), 84–91.

Molz, J. G. 2006. "Watch us wander": Mobile surveillance and the surveillance of mobility. *Environment and Planning A*, 38 (2), 377–393.

Monahan, T., and Mokos, J. T. 2013. Crowdsourcing urban surveillance: The development of homeland security markets for environmental sensor networks. *Geoforum*, 49, 279–288.

Murakami Wood, D. 2013. What is global surveillance? Towards a relational political economy of the global surveillant assemblage. *Geoforum*, 49, 317–326.

Perez, S. 2019. *Over 30 civil rights groups demand an end to Amazon Ring's police partnerships.* https://techcrunch.com/2019/10/08/over-30-civil-rights-groups-demand-an-end-to-amazon-rings-police-partnerships.

Pink, S., and Fors, V. 2017. Being in a mediated world: Self-tracking and the mind–body–environment. *Cultural Geographies*, 24 (3), 375–388.

Poster, M. 1990. *The mode of information: Poststructuralism and social context.* Chicago, IL: University of Chicago Press.

Sadowski, J. 2020. *Too smart: How digital capitalism is extracting data, controlling our lives, and taking over the world.* Cambridge, MA: MIT Press.

Sadowski, J., and Pasquale, F. 2015. The spectrum of control: A social theory of the smart city. *First Monday*, 20 (7), 1–25.

Spiller, K. 2016. Experiences of accessing CCTV data: The urban topologies of subject access requests. *Urban Studies*, 53 (13), 2885–2900.

Staples, W. G. 1997. *The culture of surveillance: Discipline and social control in the United States.* New York: St Martin's Press.

Strengers, Y., and Kennedy, J. 2020. *The smart wife: Why Siri, Alexa, and other smart home devices need a feminist reboot.* Cambridge, MA: MIT Press.

Swanlund, D., and Schuurman, N. 2019. Resisting geosurveillance: A survey of tactics and strategies for spatial privacy. *Progress in Human Geography*, 43 (4), 596–610.

Van Holstein, E. 2018. Digital geographies of grassroots securitisation. *Social & Cultural Geography*, 19 (8), 1097–1105.

Vanolo, A. 2014. Smartmentality: The smart city as disciplinary strategy. *Urban Studies*, 51, 883–898.

Waghorn, N. J. 2016. Watching the watchmen: Resisting drones and the "protester panopticon." *Geographica Helvetica*, 71 (2), 99–108.

Zuboff, S. 2015. Big other: Surveillance capitalism and the prospects of an information civilization. *Journal of Information Technology*, 30, 75–89.

PART IV

Media and the politics of knowledge

16

RACE, ETHNICITY AND THE MEDIA

Absence, presence and socio-spatial reverberations

Douglas L. Allen and Derek H. Alderman

Race/ethnicity and media (geographies) are inseparably intertwined. Media, in various forms (e.g. music, news media, film, social media, literature), work to internalize, diffuse, and legitimize particular racialized visions of social and place difference. Media are sets of practices, technologies and places for communicating (and exchanging) narratives, images and even seemingly innocuous information—all of which socially construct people's identities in uneven and sometimes unjust ways (Craine 2007; Leszczynski 2015). The media does more than reflect racial and ethnic categories and hierarchies, but actively participates in their (re) production by selectively constructing and disseminating stories that essentialize the meaning of people, their lives and the places they inhabit, use and claim—often in ways that obscure the very social-spatial relations responsible for racial privilege and subordination.

The identity-constructing capacity of media is never realized outside an understanding of space. Prevailing conceptions of racial and ethnic difference are often tied to the different and often moralistic manner in which the media represents places associated with certain racial and ethnic groups. Indeed, media "functions as an act of communication... by and through which geographical information is gathered, geographical facts are ordered and our imaginative geographies are constructed" and made real (Craine 2007, 149). Media geographies, then, are implicated in the circulation of socio-spatial meanings about race/ethnicity and the symbolic and material production of place and landscape. This chapter draws attention to how this circulation of socio-spatial meaning through media is deeply imbricated with race, racism and anti-racism.

The relationship between race/ethnicity, media and geography is a complex one, encompassing at least four major dimensions. First, the media produces racial and ethnic exclusions (or inclusions) within its representations of society and space, and these images frame public thought, values and debate. Second, the landscape itself is a communicative tool, demonstrating that the media consists of geographic modes of circulating and institutionalizing ideas about racial/ethnic and place differences. Third, these media circulations are always embedded within wider social and spatial conditions and racial/ethnic power relations. This context shapes not just what we see, hear and read in the media but importantly the different ways we interpret it within our own racial/ethnic/place-specific worldview. Fourth, because

DOI: 10.4324/9781003039068-20

public interpretation and reaction is so important to the meanings constructed in the media, messages about race, ethnicity and place have socio-spatial reverberations or significant material consequences or effects for people, places and the wellbeing of both.

There is an iterative relationship between media and the production of race and ethnicity as well as racial/ethnic geographies. Media are both shaped by and in turn shape racial projects (Omi & Winant 1994) that contribute to the production of racialized geographies (Schein 2006). Media spatializes racial (mis)representations through the dissemination of stereotypes, helping to assign meaning and (lack of) value to places and people. From the perspective of critical race studies, media geographies participate in the dominant storytelling undergirding the construction and reification of white privilege, if not outright white supremacy (Solórzano & Yosso 2002). These master narratives perpetuate a "discursive violence" against marginalized populations (Jiwani 2009). Importantly, storytelling is not disconnected from lived realities and inequalities. Discursive violence produced through media is part of wider patterns of social, economic and physical trauma. The power and harm of these hegemonic narratives are readily evident, for example, in the use of various media to label particular neighborhoods as ghettos or slums and to label people within particular communities as dangerous or violent, ultimately opening them up to urban "renewal" and removal practices (Anderson 1987; Wilson & Mueller 2004; Nelson 2008; Hankins et al. 2012). This is exacerbated by how media are used, after-the-fact, to erase the harms and displacements resulting from media-aided gentrification (Zukin 2010). Because the media has participated in creating and concretizing racialized visions of society and place, it too often has been used to oppress and marginalize people of color (POC) by silencing their voices, erasing their presence and claims to space, and misrepresenting their lives and communities.

Media-created geographies assist in perpetuating racist social relations, but they also carry the potential for resisting exclusion (Schein 2006). Again, in the parlance of critical race studies, media can be deployed as a form of "counter-storytelling," making visible and heard the often-ignored identities and place-based experiences of those discriminated against (Solórzano & Yosso 2002). These counter-stories are embedded within wider traditions of political, social and economic activism. While much of the attention in media studies focuses on news media, resistant racial/ethnic geographies of media also arise through art, literature, film, music, internet blogs and social networking sites. These media shape our understandings of and engagements with the world and our sense of self, community and place. Thus, while media have been used to oppress racially marginalized communities, they also provide a means of expressing and asserting a sense of belonging and affirmation of being (differently) in the world (Woods 1998; Allen 2020; Allen & McCreary 2020; Brock Jr. 2020).

In this chapter we discuss the ways in which media, race and geographies are mutually constituted within the United States, focusing specifically on examples centering Black communities. We use a framework of absence and (affirmative) presence to show how media are implicated in both racial marginalization and racial liberation movements, as well as the production of places and landscapes that attend and shape those processes.

Media(ted) absences and presences: Media as a tool of erasure and misrepresentation

The media are often implicated in erasing the bodies, voices and knowledge claims of marginalized racial and ethnic groups, thus erasing the contributions of these communities to producing place and landscape and positioning them as outside of local, national and global imaginaries and normative codings of space. These absences in media and media spaces perpetuate a consequential amnesia, rendering POC as forgettable by (white) privileged communities. However, the media also works to give racially marginalized groups a negative, abjected-presence that reduces them to stereotypical depictions. The images of particular racialized bodies and communities are heavily mediated through a distorting, white spatial imaginary that leave the actual voices and lived complexities of the community absent even while they are visually present. In this section, we expand upon absence and abjected-presence within race and media, showing that socio-spatial media are implicated in this erasure and illustrating the socio-spatial reverberations of this erasure.

Absence: Erasure within media

Racial disparities in representation exist within various forms of media, particularly TV, film and news media. We are decades removed from the 1968 Kerner Report's call for greater representation of POC in media content and media production spaces, but these inequalities continue (Negrón-Muntaner et al. 2014). This lack of affirmative presence in media has socio-spatial consequences, "symbolically annihilating" the identities and experiences of those made invisible and producing a whitewashed vison of place (Alderman & Modlin 2008). The under-representation of Latinx and Black communities in media programming excludes these groups from the local and national imaginary of citizenship. Melissa Harris-Perry (2011) reminds us that the "struggle for recognition is the nexus of human identity and *national* identity" (p. 4, emphasis added).

This struggle over identity is evident in various forms of (spatialized) media. Carolyn Finney (2014) finds African Americans largely invisible within Great Outdoors narratives espoused by the US media. This Black invisibility is connected to the project of nation-building and the exclusionary way ideas of national cultural and natural heritage are narrated. In this way, the erasure of African Americans and their bodies and voices from our understandings of environmental histories and contemporary environmental movements is a form of environmental racism that marginalizes these communities' claims to national identity "and the rights and freedoms that come with that identification" (p. 42).

This absence is not just about visual representation but is also about obscuring the issues and concerns faced by POC. Patricia Hill Collins (2004) diagnoses this erasure through post-racial discourses, explaining that "new racism relies more heavily on the manipulation of ideas within mass media. The post-racial discourse presents hegemonic ideologies that claim that racism is over" (p. 54), limiting rhetorically what gets to "count" as racism. Black people (and one might include other POC as well) are thus put in danger because of the increased difficulty challenging racial projects that position Black people as "problems to their nation, to their local environments, to Black communities, and to themselves" (p. 54). This color-blind racism, as Eduardo Bonilla-Silva (2014) has defined it, seeks to shield those in power from accountability and justify disparities produced by structural racism. This has dire socio-spatial consequences for these racially marginalized communities that can become ignored during crises in which they disproportionately bear the burden. We see these consequences in the slow response of national media during the water crisis in Flint, Michigan (Jackson 2017). Or, how the delayed reporting of racial data on COVID-19 cases and deaths by local, state and national governments and late demands by the media for this information led to a long silence on the disproportionate impact of the pandemic on Black and Latinx communities (Kendi 2020). This absence was only exacerbated when the Trump Administration's Surgeon General, Dr. Jerome Adams, insinuated (not so subtly) that it was self-destructive Black behaviors that contributed to the disproportionate impacts while remaining largely silent on how racism in housing, employment, healthcare and transportation contributed to these impacts (Aleem 2020). These silences about the impacts of racism disproportionately play out spatially, positioning particular communities as expendable and justifying their continued exploitation.

Erasure also operates through spatial mediums such as landscapes. Landscapes are "discourses materialized" (Schein 1997, 663), spatial mediums that produce and circulate socio-cultural imaginaries and meanings through society (Schein 1997; Dwyer & Alderman 2008; Mitchell 2008; Allen et al. 2019). They hold the capacity to materialize values, histories and visions of place and society, and in privileging some imaginaries over others, they can have the effect of making absent the contributions of marginalized communities. Landscapes often reflect and reify the interests of the empowered, those with the material and social capacity to concretize their visions on the land (Mitchell 2003; Allen et al. 2019). The histories and visions of racially marginalized communities are often rendered mute through their absence and erasure in the landscape, replaced by a predominance of white narratives and white socio-spatial visions, particularly within Euro-American settler-colonial nations.

Abjected-presence: Misrepresentation within media

The portrayal of racially marginalized social actors and groups within the media often (re) produces inequality by constructing and perpetuating stereotypes. These misrepresentations obscure more than they reveal, reaffirming the privileged status of dominant classes and justifying the othering and exploitation of racially marginalized communities. Produced is an abjected-presence, where marginalized communities are visibly present but their voices and visions are occluded and the complexities of their lives are reduced to caricatures. These caricatures regrettably take on the power of social fact, gaining in authority as they circulate geographically and socially within and through public media (McElya 2007).

Contrast, for example, media coverage of the "War on Drugs" and the recent "Opioid Crisis." Media portrayals of Black communities and drugs position these communities as dangerous havens for criminals and legitimize discourses and tactics of war against these communities and their residents. The recent opioid crisis, however, which mostly affects white rural communities, has largely been depicted in the media as a health crisis that requires compassion for those suffering from an addiction not of their own doing (Nethernland & Hansen 2017; Shachar et al. 2020). While such representations of Black communities are, no doubt, partially a product of the absence of POC within media spaces as journalists, producers, writers and executives, the stigmatization of Black communities is also part of a broader national consciousness and discursive system, institutionalized and systemic to the point that increased diversity will most likely not be enough to remove the stigmatization of Black communities from media portrayals. Media are products of contextualized societies and geographies. While media certainly can and do challenge societal discourses, media more often reflect societal discourses, amplifying dominant, often marginalizing, narratives of race already circulating within society. These representations have social and material effects. Historical and continued depictions of Black people, particularly Black men, as dangerous criminals mediate how authorities view and treat them in space. Purse clutching in public, stigmatization for simply "shopping while Black" within commercial spaces, and police hyper-surveillance and violence in Black communities are products of an anti-Black national rhetoric circulated through the media, resulting in fear of Black men and perpetuating anti-Black violence (Oliver 2003; Day 2006; Smiley & Fakunle 2016; Brooms & Clark 2020).

This stigmatization of place accompanying media images often (re)produces harmful stereotypes about racially marginalized communities by situating these particular communities within landscapes of blight, criminality and despair. TV shows and films may increasingly make POC more visible but they nonetheless obscure and distort the socio-spatial conditions that racially marginalized communities are made to endure and the ones they create to resist. For example, according to George Lipsitz (2011), the edgy HBO show *The Wire* reifies the white spatial imaginary even as it has tried to depict the consequences of the "War on Drugs" within Baltimore, Maryland. Black people and real Black communities certainly have *a* presence in this show, but as Lipsitz argues, the misrepresentation of systemic conditions impacting the lives of Black Baltimoreans leaves *The Wire* "with the default positions inscribed in the white spatial imaginary: that people who *have* problems *are* problems" (p. 112). The absence of voices of the marginalized communities being portrayed and the absence of a Black spatial imaginary belie the presence of Black bodies and Black communities on screen. These abjected-presences subject racially marginalized communities to social and corporeal violence, enacted by individuals, society and the state, and show that this abjected-presence has spatial reverberations that are both social and material.

Media as a tool of resistance: Asserting (affirmative) presence

Existing alongside widespread examples of absence and abjected-presence is the appropriation of media as a tool for oppressed people to resist marginalization and assert presence, belonging and affirmation. While the media's role in erasure and misrepresentation must be addressed and challenged, we must also be mindful of how our own research is itself a circulation of knowledge, and thus avoid positioning people and communities as determined by their suffering. Geography, in particular, has a history of rendering invisible and mute racially marginalized voices within the discipline (McKittrick 2006; Hawthorne 2019). As a result, we must highlight and recognize how the (socio-spatial) visions, experiences and practices of racialized communities open up "possibilities for alternative, anticolonial, and liberatory forms of geographic knowledge and world-making" (Hawthorne 2019, 9).

In the following sections, we discuss some important examples of media being a racial liberation tool to assert an affirmative presence, a presence that affirms the humanity and dignity of racially marginalized people in the face of white supremacy, settler colonialism, and other forms of oppression. This affirmative presence not only challenges oppression and the processes of erasure and negative depiction traditionally found in the media, but also creates imaginative space for activists to envision more socially just and emancipatory articulations of society and space essential to equality and wellbeing. In particular, we focus on Black American activism in deploying music and social media to affirm Black belonging in the production of a more inclusive America.

Asserting presence through social media

Social media has become a powerful tool of social justice activism, amplifying activist messages and expanding the scale of local action. Hashtag activism demonstrates how marginalized communities transform social media platforms into resources for racial justice. Social media produce digital counterpublics, relational networks of people fused together around similar interests, issues or experiences that act as their own digital communities and digital spaces of exchange and cultural circulation. In this media space, the hashtag becomes a mechanism to amplify the counternarratives developed by/within the counterpublic, rendering them legible by the social media masses and magnifying the impact of this collective storytelling (Kuo 2018; Brock Jr., 2020; Jackson et al. 2020).

Social media has become vital to bringing attention to important racial justice issues and creating a dialogic space for Black people to demand action on issues harming their communities (Brock Jr. 2020; Jackson et al. 2020). Social media posts in the spring and summer of 2020 were filled, for example, with cries against racist violence, particularly state violence throughout the police and judicial system. As Jackson et al. (2020, 124) argue, social media activism is a tool of social movement organizing that can garner such attention that it becomes "an unavoidable issue for mainstream journalists and politicians" to which they must respond. It provides a platform for circulating news, images and videos through hashtags to shine (and maintain) a light on anti-Black violence and facilitates a network of community activists and concerned citizens that can be organized for both digital and traditional activism (Brock Jr. 2020; Jackson et al. 2020).

Cell phone videos posted to social media platforms, for example, have been used as a way of introducing a measure of accountability for the routine anti-Black violence experienced in encounters with police (Richardson 2017) and white antagonists that see "Black as nuisance" within public spaces (Henderson & Jefferson-Jones 2020). Cell phone videos, for example, have been prominently used to document attempts by white antagonists to police Black presence in public spaces such as in the cases of "BBQ Becky" and "Permit Patty" (Henderson & Jefferson-Jones 2020). While these incidents illustrate how Black people are made to appear not to belong in white controlled spaces, they show how Black digital communities can mobilize media to "bear witness" to these encounters and demand public accountability. These participant cell phone videos also seek to resist absence and abjection by turning the tools of surveillance back on those imbued with legal authority, "bearing witness" to police violence (Richardson 2017) such as the killing of Eric Garner in 2014, the shooting of Philando Castile in 2016 and the murder of George Floyd in 2020. These videos, shared on social media, mobilized Black witnesses and white allies to action in the form of digital and physical protests.

However, we must be careful of the uncritical expansion of the surveillance apparatus. Browne (2015) notes that "surveillance is nothing new to black folks" (p. 10) and the media have long been technologies of racial oppression and policing Black presence in public space. Despite the increase in videos of police violence, particularly via cell phones and police body cameras, there has been little accountability for the police officers involved. In addition, the selective leaking of body camera videos requires us to question if more surveillance is for the protection of communities of color or for protecting police departments and officers (Sacharoff & Lustbader 2017). Furthermore, participant videos, images and location data are increasingly used to identify protestors against police violence, illustrating how more surveillance can be turned against those fighting for racial justice (Leon 2020; Leopold & Cormier 2020; Ng 2020). It is not the absolute volume of images and video that will produce accountability, but where power lies in the making, storing and dissemination of these images. Who controls these images and what policies govern their use?

Social media can also project visions of geography. For example, the hashtag #FergusonIsEverywhere is reminiscent of Hunter and Robinson's (2018) remapping of the US as a series of "multiple Souths"; within the lived experiences of Black Americans there is not a region of the US "safe" from white supremacy and racist violence. Such re-envisioning of place reshapes relations of belonging, calling into question which regions/states (and thus populaces) are positioned as racist and which are given a pass. Furthermore, Brock Jr. (2020) notes how Black participation on social media not only challenges white supremacy and erasure of Black digital presence by "decentering whiteness as the default internet identity"

(p. 5), but also produces digital spaces to "extol the joys and pains of everyday life" in the face of anti-Black racism (p. 6).

Affirmative presence through musical performance

Sound, particularly music, plays an important role in producing and expressing cultural identity, social meaning and geographic imaginaries (Hudson 2006; Paiva 2018; Devadoss 2020). Music creates moments of rupture that can both suspend the status quo and forge a new sense of collective belonging among people and groups. These moments allow for transformative possibilities to the social and spatial order. Social justice movements have long used music to disrupt the socio–spatial norm and reshape power relations (Eyerman & Jamison 1998; Woods 1998; Fischlin & Heble 2003; Orejuela & Shonekan 2018).

Black communities have frequently deployed music as a form of protest and affirmation, resisting racial oppression and celebrating Black life from slavery to the present. The enslaved transformed music into a subversive mode of communication against planter power, using media to slow the pace of work, expressing solidarity with each other (particularly in the aftermath of planter violence), providing directions along the Underground Railroad, and even organizing slave revolts (Cruz 1999). The Blues extended this use of music as a form of resistance. Acting as an "almost exultant affirmation of life" (Richard Wright, quoted in Woods 1998, 19), the Blues was more than just entertainment or cultural expression. As Woods (1998) argues, it was an "evolving complex of social explanation and social action" (p. 29) that acted as an "ethic of survival, subsistence, resistance, and affirmation" (p. 27).

During the Civil Rights Movement, music not only galvanized activists and emotionally sustained protesters (Inwood & Alderman 2018), but it also shaped new "understandings of American democracy and American citizenship" (Rabaka 2016, 3). Artists, like Billie Holiday with the song *Strange Fruit* and Nina Simone in her musical thesis *Mississippi Goddam*, used music as a way of critiquing white supremacy, racist violence and the unfulfilled promises of American liberty. These songs are more than just interesting music; they are analyses and critiques of various spatialized oppressions. Wright (2018) argues that Holiday's song is a way of linking racial oppressions to spatial "systems of containment and exploitation" founded upon "the degradation of life and land" (p. 9). Similarly, Nina Simone's *Mississippi Goddam* is a forceful critique of the many geographies of oppression Black communities navigate, castigating particularly violent states like Alabama and Mississippi and noting that "this whole country is full of lies." These spatial critiques challenge existing racist visions of the country and posit a less generous, alternative vision of America, and in doing so demand places where Black lives are valued and celebrated.

Colin Kaepernick used the moment of the playing/singing of the US national anthem to critique police violence of unarmed Black people, challenging visions of America that erased its history of anti-Black racism. His actions sparked a wider protest movement of sitting and kneeling during the national anthem, revealing an already existing alternative sense of place experienced within racially marginalized communities across the US, particularly within Black communities. Kaepernick's silent protest (and society's reaction to it) has echoes of the silent, raised fist protest staged by Tommie Smith and John Carlos at the 1968 Olympics during the US national anthem. Aural media play a role in both of these protests. The American national anthem provides the site of protest (a time and space for the resistive act), and these athletes' refusal to engage with the anthem amplifies their critique of the US. Indeed, their silence enunciates the racial oppression in the US national project and asks: For whom does this anthem really play? The activism of athletes during the US national anthem reveals that musical performances can become sites of resistance to oppression, providing opportunities to challenge oppressive socio-spatial narratives and assert a Black sense of place often silenced or discredited.

More than simply a reflection (or critique) of society, music is a creative media that seeks to bring into being visions of society and space yet to be fully realized in practice. It serves as a way of asserting affirmative presence and refusing abjected-presence. Music festivals and parades, for example, have transformative potential and provide glimpses of more inclusive visions and enactments of community, society and space (Delgado 2016; Allen 2020; Allen & McCreary 2020). While music festivals can marginalize and solidify already existing exclusions, they also "can be sources of innovation and creativity... bringing rhythms that transform urban spaces" (Delgado 2016, 119). Analyzing the Lunar New Year Festival in San Francisco's Chinatown, Delgado notes how festivals shape community identity and produce broad relations of belonging that extend beyond local boundaries and that bring "recognition to this community's presence" (p. 163).

Allen (2020) finds similar dynamics in his research of Florida A&M University's (FAMU) homecoming parade. A historically Black College/University in Tallahassee, Florida, FAMU's homecoming remains one of the most significant events for FAMU and the broader Black community within Tallahassee. Allen shows how this parade, and particularly the Marching 100's participation in it, helps transform the vision of the city and for whom it matters. This parade has historically been a way of suturing together Black neighborhoods (Frenchtown and Southside) into a broader Black Tallahassee community, and FAMU's homecoming disrupts the dominant white spatial imaginary, allowing for a Black sense of place to emerge throughout the city. The parade affirms Black life and highlights the multitude of contributions of Tallahassee's Black community, contributions typically veiled and muffled in the narrative of Tallahassee. This socio-spatial transformation of the affective atmosphere of Tallahassee into an affirmation of FAMU and the Black community allows for different relations of belonging within the city that, for some FAMU students, allows for a mobility throughout the city they do not usually enjoy. The city feels safer, more inviting and more affirmative, reshaping the circulations of capital, people and discourses within the city (including within news media).

Johnson (2013) claims that "a right to visibility and mobility (in physical spaces) and demanding recognition and respect (in discursive spaces) addresses and redresses the injuries enacted by systemic spatial isolation and racism" (p. 168). Festivals and parades, like those highlighted by Delgado (2016) and Allen (2020), create ruptures in the status quo that assert a "right to visibility and mobility" (and one might also say audibility) for racially marginalized groups. They transform the socio-cultural vision and experience of the city and even alter material flows within the city. They constitute what McKittrick (2006, 137) has called a "participatory soundscape" through which a racially marginalized community "can say itself and its history." Festivals and parades, then, like the music utilized within them, can be viewed as socio-spatial media that provide opportunities for the "sayability" (or perhaps playability or audibility) of racially marginalized communities' affirmational counternarratives.

Conclusion

Though media, race and geography are indelibly imbricated and inextricable, we have relied on a large number of studies outside of the discipline of geography, and a socio-spatial analysis requires extending the central argument of these scholars. This is because, while many geographers studying media engage race at times and many scholars of race and media engage geography at times, there is little sustained engagement with race and media geographies as the central focus of the study. The literature feels largely sutured together, using an array of disparate research that, at times, requires a reinterpretation and conceptual extension of the original work to tease out the geographic and racial implications. It also demonstrates the importance politically as well as conceptually for geographers to engage scholars, particularly scholars of color, in race and ethnic studies. As a result, we call for a more sustained and deliberate focus on the socio-spatial implications of race and media. While many disciplines have engaged in research on race and various forms of media, geographers must take up this research and agenda to highlight the ways in which geography facilitates racial media projects through spatial circulations and the way racial media projects circulate socio-spatial meaning. We have presented one framework (absence/presence) through which to approach race and media geographies, but we recognize and wish to highlight the necessity for more theoretical and empirical engagement with the socio-spatial implications of race and media.

Future research trajectories should address: (1) the ways in which white spatial imaginaries are circulated through media resulting in the (neo)colonization and exploitation of local communities as well as nation-states; and (2) the ways in which racially marginalized communities resist these encroachments by asserting their own spatial imaginaries through various forms of media. Such an explicit study of race within media geography research will deepen our understanding of the circulations of racialized meanings and how these meanings are communicated through spatial mediums and attached to socio-spatial imaginaries and actual material landscapes. Furthermore, research that focuses on the ways racially marginalized communities use music, literature and other forms of media to assert and affirm alternative ways of being in the world offers scholars and activists insights into affirmative resistance practices as well as ways to adapt media for liberatory goals. Media geographies of race remain a compelling and under-researched, under-theorized field of geography, one that holds potential for producing insights for scholars and activists alike.

References

Alderman, D. H., and Modlin, E. A., Jr. 2008. (In)isibility of the enslaved within online plantation tourism marketing: A textual analysis of North Carolina websites. *Journal of Travel & Tourism Marketing*, 25 (3–4), 265–281. https://doi.org/10.1080/10548400802508333.

Aleem, Z. 2020. *The problem with the surgeon general's controversial coronavirus advice to Americans of color*. www.vox.com/2020/4/11/21217428/surgeon-general-jerome-adams-big-mama-coronavirus.

Allen, D. L. 2020. Asserting a Black vision of race and place: Florida A&M University's homecoming as an affirmative, transgressive claim of place. *Geoforum*, 111, 62–72. https://doi.org/10.1016/j.geoforum.2020.03.007.

Allen, D. L., and McCreary, T. 2020. Performing Black life: The FAMU Marching 100 and the Black aesthetic politics of disruption, presence and affirmation. *Cultural Geographies*, 1–15. https://doi.org/10.1177/1474474020931536.

Allen, D. L., Lawhon, M., and Pierce, J. 2019. Placing race: On the resonance of place with Black geographies. *Progress in Human Geography*, 43 (6), 1001–1019. https://doi.org/10.1177/0309132518803775.

Anderson, K. 1987. The idea of Chinatown: The power of place and institutional practice in the making of a racial category. *Annals of the Association of American Geographers*, 77 (4), 580–598. https://doi.org/10.1111/j.1467-8306.1987.tb00182.x.

Bonilla-Silva, E. 2014. *Racism without racists: Color-blind racism and the persistence of racial inequality in America*. New York: Rowman & Littlefield Publishers.

Brock, A., Jr. 2020. *Distributed Blackness: African American cybercultures*. New York: New York University Press.

Brooms, D. R., and Clark, J. S. 2020. Black misandry and the killing of black boys and men. *Sociological Focus*, 53 (2), 125–140. https://doi.org/10.1080/00380237.2020.1730279.

Collins, P. H. 2004. *Black sexual politics: African Americans, gender, and the new racism*. New York: Routledge.

Craine, J. 2007. The medium has a new message: Media and critical geography. *Acme*, 6 (2), 147–152.

Cruz, J. 1999. *Culture on the margins: The Black spiritual and the rise of American cultural interpretation*. Princeton, NJ: Princeton University Press.

Day, K. 2006. Being feared: Masculinity and race in public space. *Environment and Planning A*, 38, 569–586. https://doi.org/10.1068/a37221.

Delgado, M. 2016. *Celebrating urban community life: Fairs, festivals, parades, and community practice*. Toronto: University of Toronto Press.

Devadoss, C. 2020. Sounding "brown": Everyday aural discrimination and othering. *Political Geography*, 79, 1–10. https://doi.org/10.1016/j.polgeo.2020.102151.

Dwyer, O. J., and Alderman, D. H. 2008. Memorial landscapes: Analytic questions and metaphors. *GeoJournal*, 73 (3), 165–178. https://doi.org/10.1007/s10708-008-9201-5.

Eyerman, R., and Jamison, A. 1998. *Music and social movements: Mobilizing traditions in the twentieth century*. Cambridge: Cambridge University Press.

Fischlin, D., and Heble, A. (eds.) 2003. *Rebel musics: Human rights, resistant sounds, and the politics of music making*. Montreal: Black Rose Books.

Hankins, K. B., Cochran, R., and Derickson, K. D. 2012. Making space, making race: Reconstituting white privilege in Buckhead, Atlanta. *Social & Cultural Geography*, 13 (4), 379–397. https://doi.org/10.1080/14649365.2012.688851.

Harris-Perry, M. 2011. *Sister citizen: Shame, stereotypes, and Black women in America*. New Haven, CT: Yale University Press.

Hawthorne, C. 2019. Black matters are spatial matters: Black geographies for the twenty-first century. *Geography Compass*, 13 (11), 1–13. https://doi.org/10.1111/gec3.12468.

Henderson, T.-N. Y., and Jefferson-Jones, J. 2020. #LivingWhileBlack: Blackness as nuisance. *American University Law Review*, 69 (3), 863–914.

Hudson, R. 2006. Regions and place: Music, identity and place. *Progress in Human Geography*, 30 (5), 626–634. https://doi.org/10.1177/0309132506070177.

Hunter, M. A., and Robinson, Z. R. 2018. *Chocolate cities: The Black map of American life*. Oakland, CA: University of California Press.

Inwood, J. F., and Alderman, D. H. 2018. When the archive sings to you: SNCC and the atmospheric politics of race. *Cultural Geographies*, 25 (2), 361–368. https://doi.org/10.1177/1474474017739023.

Jackson, D. Z. 2017. *Environmental justice? Unjust coverage of the Flint water crisis*. https://shorensteincenter.org/environmental-justice-unjust-coverage-of-the-flint-water-crisis/.

Jackson, S., Bailey, M., and Welles, B. 2020. *#HashtagActivism: Networks of race and gender justice*. Cambridge, MA: MIT Press.

Jiwani, Y. 2009. Symbolic and discursive violence in media representations of Aboriginal missing and murdered women. In D. Weir and M. Guggisberg (eds.) *Violence in Hostile Contexts E-Book*. Oxford: Inter-Disciplinary Press. www.inter-disciplinary.net/publishing/id-press/ebooks/understanding-violence-contexts-and-portrayals/.

Johnson, G. T. 2013. *Spaces of conflict, sounds of solidarity: Music, race, and spatial entitlement in Los Angeles*. Los Angeles, CA: University of California Press.

Kendi, I. X. 2020. Why don't we know who the coronavirus victims are? *The Atlantic* (April 1). www.theatlantic.com/ideas/archive/2020/04/stop-looking-away-race-covid-19-victims/609250/.

Kuo, R. 2018. Racial justice activist hashtags: Counterpublics and discourse circulation. *New Media and Society*, 20 (2), 495–514. https://doi.org/10.1177/1461444816663485.

Leon, H. 2020. The 13 high-tech tools used by protestors and cops in their escalating battle. *Observer* (June 4). https://observer.com/2020/06/surveillance-technology-fueling-cops-vs-protestor-battle/.

Leopold, J., and Cormier, A. 2020. The DEA has been given permission to investigate people protesting George Floyd's death. *BuzzFeed News* (June 3). www.buzzfeednews.com/article/jasonleopold/george-floyd-police-brutality-protests-government.

Leszczynski, A. 2015. Spatial media/tion. *Progress in Human Geography*, 39 (6), 729–751. https://doi.org/10.1177/0309132514558443.

Lipsitz, G. 2011. *How racism takes place*. Philadelphia, PA: Temple University Press.

McElya, M. 2007. *Clinging to mammy: The faithful slave in twentieth-century America*. Cambridge, MA: Harvard University Press.

McKittrick, K. 2006. *Demonic grounds: Black women and the cartographies of struggle*. Minneapolis, MN: University of Minnesota Press.

Mitchell, D. 2003. Cultural landscapes: Just landscapes or landscapes of justice? *Progress in Human Geography*, 27 (6), 787–796. https://doi.org/10.1191/0309132503ph464pr.

Mitchell, D. 2008. New axioms for reading the landscape: Paying attention to political economy and social justice. In J. L.WescoatJr. and D. M. Johnston (eds.) *Political Economies of Landscape Change: Places of Integrative Power*. Dordrecht: Springer.

Negrón-Muntaner, F., Abbas, C., Figueroa, L., and Robson, S. 2014. *The Latino media gap: A report on the state of Latinos in US Media*. https://media-alliance.org/wp-content/uploads/2016/05/Latino_Media_Gap_Report.pdf.

Nelson, J. 2008. *Razing Africville: A geography of racism*. Toronto: University of Toronto Press.

Nethernland, J., and Hansen, H. 2017. White opiods: Pharmaceutical race and the war on drugs that wasn't. *Biosocieties*, 12 (2), 217–238. https://doi.org/10.1057/biosoc.2015.46.

Ng, A. 2020. *Geofence warrants: How police can use protesters' phones against them*. www.cnet.com/news/geofence-warrants-how-police-can-use-protesters-phones-against-them/.

Oliver, M. B. 2003. African American men as "criminals and dangerous": Implications of media portrayals of crime on the "criminalization" of African American men. *Journal of African American Studies*, 7 (2), 3–18. https://doi.org/10.1007/s12111-003-1006-5.

Omi, M., and Winant, H. 1994. *Racial formation in the United States: From the 1960s to the 1990s*. New York: Routledge.

Orejuela, F., and Shonekan, S. (eds.) 2018. *Black Lives Matter & music: Protest, intervention, reflection*. Bloomington, IN: Indiana University Press.

Paiva, D. 2018. Sonic geographies: Themes, concepts, and deaf spots. *Geography Compass*, 12 (7), 1–14. https://doi.org/10.1111/gec3.12375.

Rabaka, R. 2016. *Civil rights music: The soundtracks of the Civil Rights Movement*. Lanham, MD: Lexington Books.

Richardson, A. V. 2017. Bearing witness while Black: Theorizing African American mobile journalism after Ferguson. *Digital Journalism*, 5 (6), 673–698. https://doi.org/10.1080/21670811.2016.1193818.

Sacharoff, L. and Lustbader, S. 2017. Who should own police body camera videos? *Washington University Law Review*, 95 (2), 267–323.

Schein, R. H. 1997. The place of landscape: A conceptual framework for interpreting an American scene. *Annals of the Association of American Geographers*, 87 (4), 660–680. https://doi.org/10.1111/1467-8306.00072.

Schein, R. H. (ed.) 2006. *Landscape and race in the United States*. New York: Taylor & Francis.

Shachar, C., Wise, T., Katznelson, G., and Campbell, A. L. 2020. Criminal justice or public health: A comparison of the representation of the crack cocaine and opioid epidemics in the media. *Journal of Health Politics, Policy and Law*, 45 (2), 211–239. https://doi.org/10.1215/03616878-8004862.

Smiley, C., and Fakunle, D. 2016. From "brute" to "thug:" The demonization and criminalization of unarmed Black male victims in America. *Journal of Human Behavior in the Social Environment*, 26 (3–4), 350–366. https://doi.org/10.1080/10911359.2015.1129256.

Solórzano, D. G., and Yosso, T. J. 2002. Critical race methodology: Counter-storytelling as an analytical framework for education research. *Qualitative Inquiry*, 8 (1), 23–44. https://doi.org/10.1177/107780040200800103.

Wilson, D., and Mueller, T. 2004. Representing "neighborhood": Growth coalitions, newspaper reporting, and gentrification in St. Louis. *The Professional Geographer*, 56 (2), 282–294. https://doi.org/10.1111/j.0033-0124.2004.05602011.x.

Woods, C. 1998. *Development arrested: The blues and plantation power in the Mississippi Delta*. New York: Verso.

Wright, W. J. 2018. As above, so below: Anti-Black violence as environmental racism. *Antipode*, 1–19. https://doi.org/10.1111/anti.12425.

Zukin, S. 2010. *Naked city: The death and life of authentic urban places*. New York: Oxford University Press.

17

NATIONALISM, POPULAR CULTURE AND THE MEDIA

Daniel Bos

The media and popular culture play an integral role in how the idea of the nation has been developed, contested and contextualized over space and time. For media geographers there has long been an interest in how the media shapes both spatial imaginaries and identities (Burgess & Gold 1985; Adams 2009), and yet despite research on nationalism featuring in the subfield of political geography, the relationship between nationalism and the media has not warranted so much scholarly intrigue. While initial work on nationalism, defined here as "the territorial expression of identity: a sense of belonging to a group or community associated with a particular territory" (Mountz 2009, 287), has sought to pinpoint the emergence of the "nation," more recent theorizations have set out to "explore the geographies of nationhood manifest in everyday life" (Edensor & Sumartojo 2018, 553) and expressed through the media. On the other hand, communicative technologies and the media are transcending borders and expanding territorially bounded identities. Yet nationalism remains a potent force, evident in its resurgence and the rise of populism in a variety of geographic contexts (Ince 2019). Rather than dissipating, national identity and nationalism remain important and an enduring social identity for many people in the 21st century.

Competing theories on nationalism have emerged, including primordialism, perennialism and modernity, attempting to offer explanations as to how, why and when a nation is formed (see Ozkirimli 2017). Within modernist accounts, the media and communication—alongside and entangled with the emergence of the centralized bureaucratic state (Tilly 1992), standardized education, and intensifying industrial relations (Gellner 1983)—are noted as modern institutions and processes that have facilitated the continuing (re)production of the "nation" (Smith 1998). One of the most prominent scholars to consider the role of the media, Benedict Anderson (2016 [1983]), famously notes the role of "print capitalism" and the subsequent emergence of a standardized common language that created a sense of a political community, congruent with a territorial unit. In this sense, it was the arrival of the new technology of the printing press that enabled the circulation of national culture and language and nurtured a national consciousness. Anderson's work has been hugely influential within media studies, as it highlights the creation and distribution of a national imaginary via the popular press, which meant millions of people within a nation were able to read the same message at the same time. As such, Anderson (2006 [1983]) argues that nations are "imagined communities." This conceptualization of the nation and its manifestation in the cultural

DOI: 10.4324/9781003039068-21

practices of everyday life have proven foundational to much work tracing the relationship between nationalism and the media.

Advancing and complementing the cultural underpinnings of nationalism, Michael Billig's (1995) *Banal Nationalism* acknowledges the central importance of the "everyday" and the discursive and performative ways in which the nation is understood within prosaic contexts. Commenting on the media's role, Billig notes how the nation is "flagged" through the production, circulation and consumption of symbols, representations and routinized language—or deixis—located in print media, such as "we," "us," and "them," which provide a spatial reference point and promote the "self–other" dichotomy that establishes national identification. Despite a growing interest within geography in attending to the everyday manner in which nationalism is contested and negotiated in a variety of geographic contexts and across scales (Jones & Merriman 2009; Benwell & Dodds 2011; Koch & Passi 2016), there remains a limited systematic exploration of the relationship between media geographies and nationalism (but see Skey 2020 for a media studies perspective). This chapter aims to remedy this deficit and begins by reflecting on the everyday geographies in which the nation is (re)produced via the media and popular culture.

(Re)presenting the nation in the media and popular culture

The role of the media and popular culture in the study of nationalism took on new impetus in the 1990s with the emergence of the "cultural turn" in the social sciences. Central to this shift was an interest in the social construction of the nation—the everyday iterative processes and practices by which the commonsensical notion of a world divided into nations is propagated (Billig 1995; Edensor 2002)—and reproducing the "territorial trap" and the purported fixity and assumptions of nations as containers of societal relations (Agnew 1994).

As a result, questions of "when" and "what" is the nation have been superseded by nuanced critical reflections that seek to acknowledge *how* the "nation" comes to be (see Antonsich 2015). While nationalism studies had identified the processes of modernity, the role of media, cultural institutions, and modes of communication in creating and shaping a collective sense of national belonging, these studies often "privileged structure or form over content" (Skey2020). An emerging interest in the content of the media and popular culture identified how language, representation and performance (re)produce and communicate a sense of a distinct, bounded national territory. This new emphasis coincided with a burgeoning interest in taking popular culture and the media seriously within the broader discipline of geography.

While Tim Edensor's (2002) seminal work draws specific attention to the relationship between national identity and popular culture, a more sustained interest in the cultural practices and visual representations in which nationalism occurs has emerged more prominently through the interdisciplinary field of popular geopolitics. Here, scholars have noted the historical and contemporary influence of the media and popular culture as productive of popular understandings of nationalism, in which a world order of nation-states and the interactions between them are constituted with "everyday" representations, settings and practices (Sharp 2000; Saunders & Strukov 2018; Dittmer & Bos 2019).

It has been argued that the role of visual culture is central to this: existing in a variety of forms, operating in various geographic contexts and evoking affective relations that inform political orientations and identities. Saunders (2016, 13) argues the importance of the "national image," a "fluid, socially constructed view of the nation... which exists on both the domestic and foreign levels." Indeed, the field of popular geopolitics has stimulated a healthy

analytic focus on a wide range of popular culture and media texts, representations and per-
formances, including, but not limited to, print media (Falah et al. 2006), comic books
(Dittmer 2007), video games (Bos 2018a), photography (Foxall 2013), heritage (Waterton &
Dittmer 2016) and film (Carter & Dodds 2011).

The analysis of filmic representations of the nation has featured heavily in contributing to a
sense of national identity and diverse social constructions of the nation. Such work has drawn
attention to the visualization of place, people and quotidian landscapes that reaffirms concepts
of "self" and "other" (Edensor 2002); mobilizes gendered narratives of nationalism (An et al.
2016); and produces generic conventions evoking nationalistic tendencies (Carter & Dodds
2011). However, media geographers have critiqued the work of popular geopolitics through
its tendency to treat "texts as coherent, self-contained systems" and therefore "lack[ing] in
attention to the theoretical complexities therein" (Sharp & Lukinbeal 2016, 25). Recent
work, on the other hand, has attempted to go beyond representational accounts by attending
to the affective qualities and multimodal nature of film through which "[national] identity
can be received and negotiated" (Kirby 2019, 3). Attention has focused on the role of sounds
and music, emphasizing the multisensorial means by which national identities and geopolitical
sensibilities are evoked and communicated.

Whereas the predominant focus has enlisted textual deconstruction and the critical
interrogations of culturally mediated representations of the nation, less scholarly emphasis
has been placed on "going beyond the screen/text." This tendency to concentrate efforts
on a "finished" mediated text, or object, overlooks the wider array of political-economic
structures such as funding, distribution and marketing practices in which nationalism is
embedded and which draw upon "territorial appeals" (Coulter 2013; and see Ridanpää
2017; Bos 2020).

Building on earlier work emphasizing modern structures and institutes, such as the media,
as integral to the (re)production of nations (Anderson 1991; Gellner 1993), there has been
growing interest in and recognition of the power of cultural media institutions (Müller 2012;
Kuus 2020) such as television networks, video game developers, and film producers (Webber
2020), or more recently geospatial and mobile technologies (Lukinbeal et al. 2019), via
which cultures of nationalism emerge and are reconfigured.

Previous works have considered such institutions as homogenous entities and simply as
conduits for the dissemination of national and geopolitical narratives, but there has been
growing interest in exploring the internal orderings by which organizations are constituted
and "socio-material networks"; in other words, the arrangements of human and material
elements that work together toward a shared mission (Müller 2012). An illustrative example
is the consolidation and intensification of cooperation between the entertainment industries
and the military, termed the military-entertainment complex (Lenoir & Caldwell 2018).
Nationalistic media practices have a long history and are powerful weapons used explicitly
and implicitly during war for propaganda purposes. The "war on terror," for instance, has
seen the US Department of Defense providing military vehicles, equipment and personnel to
film producers in exchange for the ability to edit scripts and narratives (see Mirrlees 2017 for
examples). This demonstrates states' concerted propaganda efforts to promote national foreign
policy directives that glorify nationalistic perceptions and cultures of militarism to wider
publics. Such studies are vital in revealing the broader entanglements between the state and
entertainment industries, in which, at times, competing visions of national narratives and
images are sculpted and (re)written, but also silenced.

Next, I turn to what the media and mediated representations of the nation *do* and *where* by
acknowledging the everyday interactions, practices and experiences of mediated encounters.

Everyday nationalism and the media

Despite initial theoretical contributions exposing the "everyday" as a legitimate site for an understanding of how nationalism operates, such studies have been critiqued for prioritizing analysis of text and discourses over how people respond to them. This noticeable absence has troubled nationalism scholars, as there is a tendency to overlook *how* nationalism circulates and how the public receive these national narratives (Skey 2009; Benwell & Dodds 2011). Attending to matters of reception disrupts the notion that the nation is purely the result of macro-structural forces: it is "the practical accomplishment of ordinary people engaging in routine activities" (Fox & Miller-Idriss 2008, 537).

The processes of nationalism require detailed empirical investigation into how "ordinary" individuals are empowered as active producers and negotiators of national meanings (Antonsich 2016). In critiquing *Banal Nationalism* (Billig 1995), Skey (2009, 336) suggests the focus on the media overlooks "the complexity of the national audience," the socio-political contexts in which nationalism operates, and ultimately how it is received and negotiated by diverse internal and external national populations. A corrective to this has been encouragement of empirical studies, drawing upon audience and reception studies to explore processes of meaning-making as individuals and groups engage with popular (geo)political texts, objects and representations (Dodds 2006; Bos 2018b), and offering insights into the process of national identification vis-à-vis the media (Madianou 2005).

In the first instance, such an approach recognizes the heterogeneity of a "national audience" and how audience encounters are inflected through various subject markers, including class, ethnicity, gender and age, in which varying interpretations and understandings of the nation materialize, including among populations not classed as citizens of national territories (Koch 2016). A focus on the diversity of audiences acknowledges the agency of individuals and is essential to overcome theorizations extolling the power of the media without considering its actual efficacy, or the myriad ways by which national narratives become meaningful within the everyday socio-spatial contexts of people's lives. Moreover, as Dittmer and Larsen (2007) suggest, the lines between production and audience are blurred as processes of audience "feedback"—in this case, letters to editors of comic books—demonstrate a more reciprocal relationship in which popular nationalist imagery narratives get made.

This has led to an emphasis on the spaces and places in which mediated encounters with "the nation" operate and occur. These spatial contexts include national mass events (Rech 2015), but also spontaneously emergent, ceremonial and spectacular landscapes (Johnson 1995; Edensor & Sumartojo 2018). However, individuals also engage with popular culture and mediated texts and imagery in the mundane contexts of the home, and the scales and spatial contexts in which "nationalism is reproduced" exist at a localized level too (Jones 2008; and see Morley 2000). The burgeoning interest in domestic geopolitics aims to expose the socio-spatial context of the home in which a wide range of "agents, practices, objects, performativities and discourses… contribute to how geopolitics is rendered familiar, sanitised, embodied and enacted" (Woodyer & Carter 2020, 2). Focusing on the situated and domestic context of playing military-themed video games, Bos (2018b) considers the socio-material contexts and geopolitical encounters of play in which national identity is performed, and their sociality. Playing war online presents opportunities for individuals to express national allegiances verbally, as well as through the customization of avatars and the display of national flags. The home becomes an important site that shapes and informs mediated encounters, collapsing the public/private binary and revealing the multiscalar reproduction of the nation.

As a means of advancing interests in the everyday form in which the nation takes place, more recent work has begun to cover new terrain drawing attention to the affective qualities of nationalism (Militz & Schurr 2016; Antonsich et al. 2020). To this end, such work has presented opportunities to go beyond a focus on discursive and symbolic representations of the nation, to understand the affective intensity and forces of things, objects and the media. This draws particular attention to the *relational* and "the *processes* of emergence and intermittence, foregrounding and backgrounding, individualizing and collectivizing, presence and absence, through which national feelings, emotions and affects take hold (or not) in *and between* bodies of different kinds" (Merriman & Jones 2017, 600, emphasis in original).

The media arguably has become a key conduit in promoting affective resonances by which political sensibilities become mobilized and diffuse in and through wide segments of a "national audience" (Carter & McCormack 2006; Shaw & Wharf 2009). Closs Stephens (2016) outlines how the London 2012 Olympics became generative of "affective atmospheres" in which "national feelings" diffused and orientated individuals toward a national consciousness. National media institutes, such as the British Broadcasting Corporation (BBC), operated as "semiconductors," espousing the liveliness of the sporting event, and in doing so evoking a national togetherness through coalescing and circulating relations between national symbols, objects, narratives, emotions and bodies.

Such studies pose nuanced considerations, extending concerns with how mediated encounters of the nation unfold and are lived within everyday geographies. It is crucial to consider that, while such an emphasis seeks to challenge a skewed focus on critically attending to the cultural representations of the nation, it does not supplant the representational. Instead, "more than representational" accounts are essential for attending to the intermingling of practices, affects and things; the spatial and temporal contexts of these unfolding relations; their varying intensities; and how they work to form collective political identities and subjectivities (Müller 2015). Such thinking is allied to relational and posthuman ontologies that acknowledge the wider human and nonhuman assemblages in which meanings are communicated and constituted (Dittmer 2014; Weir 2018).

This calls forth a new paradigm for the study of the geographies of media and communication, by recognizing "the metaphysics of encounter" (Adams 2017), acknowledging the emergent relations between micro–macro, human–nonhuman and local–global in which mediated communication operates. As Adams (2017, 371) elaborates: "In the new metaphysics of encounter people engage with a wide range of different media and simultaneously encounter other people and things, near or far, still or mobile, perpetually redefining 'here' and 'there.'"

Such a turn raises pertinent questions around power: the capacities for different bodies to affect and be affected and how; the role of human agency and capacity within these relations; and the methodological challenges in empirically accounting for affective relations going beyond talk and text (Antonsich & Skey 2017). A focus on spatially mediated encounters presents an important avenue for considering how the "everydayness" of the nation comes into being.

Nationalism, globalization and the internet

Transnational mobility, mass transportation, supranational systems of governance and information and communication technologies (ICT) have heralded global interconnectivity in which the power and authority of nations and nationalism have been argued to be waning. However, the relationship between globalization and nationalism has also been claimed to be complementary rather than contradictory. The question, therefore, is not about the demise of the nation-state, but "how it is being reworked to remain salient among new socio-spatial formations; and how national identities are renegotiated and reconfigured in the age of globalisation (Biswas 2002)" (Antonsich 2015, 304). There has been growing interest in the role of the internet in questioning longstanding assumptions concerning a coherent and stable "nation" and national sense of identity (Eriksen 2007; Lu & Yu 2018). Indeed, the internet has allowed diasporic groups to engage in practices of long-distance nationalism, expanding social and cultural connections and sharing a sense of belonging across various geographic territories. Such studies show how the internet helps maintain relations and affiliations that are bolstering, rather than weakening, national identities.

Early work demonstrated how the structures and features of the internet, including language, domain names, hyperlinks and algorithms, reproduce and naturalize the division of the world into different nations (Eriksen 2007; Szulc 2017). Indeed, access to and regulation and governance of the internet are anchored in national territories. Internet-enabled streaming services such as Netflix still operate by and are aligned to national media policies, the viewing behavior of national audiences and matters of translation determining how they are shaped and perceived in different countries (Lobato 2019). In this sense, the internet is not abstract nor an aspatial domain, but interacts with territorialized national markets, audiences and regulations.

The emergence and popularity of (Western) social media platforms such as Facebook, Twitter, Instagram and YouTube contribute to newer politicized types of communication—nationalism 2.0—encouraging newer forms and distribution of user-generated content; the convergence and blurring of categories of production and consumption; and the governance and regulation by which nationalism is continually reworked (Fuchs 2019). Social media provides several affordances of interest for scholars concerned with media and nationalism. First is the ability to share user-generated content in which the nation can be symbolized in textual and visual forms and appear on numerous platforms. Second is the ability to facilitate transnational connections and networks with and between individuals across space and time, extending the notion of an "imagined community." The final affordance is enabling discussion and debate in which the nation is continually (re)made in both banal and remarkable ways (see Adams 2015, 396–397). In the following sections I will outline how the internet and social media are challenging and reconfiguring theorizations and practices of nationalism.[1]

The "national image" going viral

In an increasingly globalized and digitally mediated world, the national image has become more and more mobile, malleable and diffuse. This poses theoretical and methodological challenges. Within media and cultural geography, toolkits for the critical interpretation of visual communication and meaning are arguably ill-equipped to attend to the mass, mutable and multimedia nature of (national) digital imagery (Rose 2016a). The lower cultural barriers afforded by social media offer "bottom-up," creative and participatory opportunities for citizens in the visual and textual (re)articulation of nationalistic sensibilities (Pinkerton & Benwell 2014). Such digitally produced, displayed and performed images of nations can take myriad forms and include the production and sharing of images of national landscapes, architecture and events; the use of customizable overlays for Facebook profile pictures, including national celebrations and symbols such as flags; and the creation of new forms of digital imagery and expression, as in memes conveying nationalism (Ismangil 2019; Hodge &

Hallgrimsdottir 2019). This opens an important area of research in acknowledging the visual cultural practices and forms in which the nation is communicated.

Memes, for example, have become a staple aspect of internet media culture and "represent a truly banal, hidden form of nationalism" (Ismangil 2019, 243). Often taking visual form, memes usually involve humorous content transmitted via online interpersonal connections and can subsequently become a shared social phenomenon laden with particular nationalistic meaning and expression. Originally created as a character in an internet comic strip by Matt Furie in 2005, Pepe the Frog was strategically appropriated by the Alt-Right movement and featured heavily on sub-boards of the 4chan forum website. The cartoon's relatively simplistic aesthetics meant it was easily edited, shared and circulated. Offering a means of recruitment to the Alt-Right, Pepe the Frog served to propagate ethnic/white nationalism through the sharing of varying versions of the image via social media. Individuals who circulate Pepe the Frog do not necessarily share the Alt-Right's political-ideological values and meanings, but the character's affective properties have been weaponized in order to stoke agitation and tension within online and offline communities (Ash 2019).[2] Digital means of production and distribution and their speed have allowed the national image to go "viral," enabling competing visions of the nation and nationalism that have been utilized for a variety of political purposes.

Weaponizing affective nationalism

Central to understanding the relationship between social media and nationalism is acknowl-edging the manner in which it is being used, by whom and to what effect. The use of social media enables a variety of bottom-up processes by which the nation is discursively con-structed and performed. Indeed, social media and its affordances have been exploited via a range of strategic techniques, including sock puppets, trolls and bots, which have implications for the trustworthiness of the media, the increasing spread of misinformation, and further radicalization emerging from internet subcultures, all of which have profound implications for the mediated communication of nationalism. This has led to a growing interest in the power of social media, extending beyond its content to consider what social media *does* to feelings, experiences and performances of nationalism within and beyond digital cultures.

The affective modulation of social media has become part of a broader social-technical assemblage generating varying intensities and tonalities of nationalism, ranging from "luke-warm" (Mōri 2019) sentiments to the production of positively charged "happy affective atmospheres" (Closs Stephens 2016), which do much ideological work in foreclosing critical scrutiny and debates around nationalism. More attention is being paid to the weaponization of social media (Singer & Brooking 2018; Ganesh 2020), and how both state and nonstate actors are engaging with social media to manipulate and draw upon resurgent nationalist narratives for a range of political purposes.

Such practices have been instigated by nonstate actors outside of the national context. These include individuals in Macedonia producing clickbait for personal financial gain, but also more organized agencies such as "troll farms" or the Internet Research Agency in Russia, and their interference in and impact on the 2016 US presidential election (Bos & Dittmer 2019). The motives for such actions range from spreading ideologies to financial incentives, acquiring status and achieving a measure of control over powerful institutions. Such efforts have been seen as necessarily aligned with supporting a particular political candidate, but the spread of disin-formation has also been used to foster political divides in the US.

Such strategic manipulation has rested on affective appeals made possible by the affordances of social media, and the practices of individuals and groups who are using it to (re)imagine the nation. The Alt-Right, which has built a community through emotional and affective appeals, has become central to the promotion of ethnic and nativist narratives and ideologies. Alt-Right digital culture has involved the collective production and circulation of "white thymos,"[3] which describes "a complex of pride, rage, resentment, and anger that is created through informational and affective circuits that create the perception of a *loss* of white entitlement" (Ganesh 2020, 3, emphasis in original). Individual self-presentation via social media outlets has mobilized a collective imagined community premised on fear, and the purported subsequent need to protect white nativist identity politics. The Alt-Right has exploited the affordances of social media sites through the community production of visual memes and images propagating white victimization, and the practices of circulation and sharing that enable the flows of "white thymos" within a broader transnational network, connecting to the emotional dispositions and interpretative frameworks of a range of audiences. The affective and emotive appeal of social media is integral to the development of such online communities, but also to communicating and amplifying the inclusion and exclusion that are essential to the ideology of nationalism.

Transnationally mediated nationalism

The influence of migration and more recent modes of maintaining communication across space and time via social media have facilitated transnational connections that are redefining the sense of belonging to national territories. People see themselves as part of a group that shares common ideas, opinions, views and ideologies, with its reach extending beyond their national borders. More recently, studies have taken into consideration the role of the internet and social media in communicating, organizing and mobilizing, allowing the Alt-Right to become an international phenomenon, and positing an outlook that is transnational in focus (Hermansson et al. 2020). Indeed, the processes of transnationalism have been identified as being critical to the success of right-wing populism in Europe and beyond (Langenbacher & Schellenberg 2011). Paradoxically, such far right movements and groups oppose globalization and supranational systems of governance in favor of nativist concerns, and yet have increasingly engaged with politics at the transnational level—"globalized anti-globalists" (Grumke 2013).

The efficacy of such efforts to mobilize transnational connections has been contested. As Froio and Ganesh (2019, 531) argue, transnational exchanges between far right groups remain "moderate at best," with Twitter activity still predominantly aligned to, and within, national borders. The transnational scope and mobility of the far right movement are also still wedded to political parties and particular individuals—such as the president of the French far right populist party the National Rally, Marine Le Pen—in the mobilization of global far right discourse. How social media is being used to organize social groups and movements around nationalist ideologies, and the extent to which these appeals have transnational reception and mobilization, are important questions for media geography to take forward.

Conclusion

This chapter illuminates the continuing relevance of nationalism and the role of the media and popular culture in understanding the "nation." Research within media and political geography has explored the cultural politics of representing the nation and the manner and form in which a national image operates, and for what purpose. The interdisciplinary field of popular geopolitics has been noted to be key to the exploration of nationalism, although as others have argued, such work would benefit from a closer alignment and dialogue with media geography (Sharp & Lukinbeal 2015). Moreover, this chapter has argued that closer empirical attention needs to be paid to exploring how media encounters unfold within everyday geographies, and how the nation is understood, experienced, and internalized by an "ordinary"—yet heterogeneous and differentiated—public. Attending to media encounters enables opportunities to acknowledge the complex "more than human" networks and the variegated media ecology that are "perpetually redefining 'here' and 'there'" (Adams 2017, 371). The recent turn to acknowledging the affective, relational and material networks of media communication offers important contributions to recognizing, and critically attending to, questions of power, agency and national identity.

The role of globalization, and in particular new forms of mediated exchanges and the sociality presented by ICT, has been argued to present real challenges to national identification. Yet, as this chapter has shown, the transnational affordances of the internet and social media are not diminishing the role of nationalism, but actively complementing it. Indeed, social media has become a key arena in which nationalism is played out via the sharing and production of nationalistic content and debate, laden with affective and emotive power, and reinforcing shared community nationalistic interests and values extending beyond national borders.

In exploring this area further, important methodological and ethical questions remain. They include the speed and volume of data generation, and how both people and researchers react and respond to national events and discussions as they unfold and are discussed via social media. On Twitter around 500 million tweets are sent each day. The ability to engage effectively with the potential size of such data sets may require computational forms of recording and analysis (Sloan & Quan-Haase 2017). Issues of data accuracy, reliability and quality are important in social media research in the face of demographic characteristics and the presentation of "self" online. As Crosset et al. (2019) note when attempting to research and trace far right groups, these are often reliant upon amorphous networks and practices of anonymity. Current ethical guidelines when applied to social media data are still in their infancy and raise ethical issues, since participant anonymity cannot be secured when Twitter, for example, "will not allow tweets to be presented without usernames" (Sloan & Quan-Hasse 2017, 8). Nevertheless, social media research represents a fertile area of concern in which both political and media geographers are well placed to explore the pervasive and resurgent nature of mediated nationalism.

Notes

1 For the purpose of this chapter, I limit examples to the far right use of the internet, "referring [here] to extreme and radical right populist organizations as sharing three ideological cores: nativism, authoritarianism and populism". It is important to note that "within the far right organizational variants exist ranging from more to less established organizations" (Caterina & Ganesh 2019: 514).
2 Pepe the Frog has been reappropriated by activist groups as a symbol of resistance against state control, for instance in pro-democracy protests in Hong Kong (see Eliss 2019).
3 Thymos is defined "as the part of the soul that seeks recognition and redress of the injustices done to it" and "is the centre of rage, anger, indignation, and pride" (Ganesh 2020, 4).

References

Adams, P. C. 2009. *Geographies of media and communication*. New York: John Wiley & Sons.
Adams, P. C. 2017. Geographies of media and communication I: Metaphysics of encounter. *Progress in Human Geography*, 41 (3), 365–374. https://doi.org/10.1177/0309132516628254.

Agnew, J. 1994. The territorial trap: the geographical assumptions of international relations theory. *Review of International Political Economy*, 1 (1), 53–80. https://doi.org/10.1080/09692299408434268.

An, N., Liu, C., and Zhu, H. 2016. Popular geopolitics of Chinese Nanjing massacre films: A feminist approach. *Gender, Place & Culture*, 23 (6), 786–800. https://doi.org/10.1080/0966369X.2015.1058762.

Anderson, B. 2016 [1983]. *Imagined communities: Reflections on the origin and spread of nationalism*. Croydon: Verso.

Antonsich, M. 2015. Nations and nationalism. In J. Agnew, V. Mamadouh, A. J. Secor and J. Sharp (eds.) *The Wiley Blackwell companion to political geography*, pp. 297–310. Malden, MA: Wiley-Blackwell.

Antonsich, M. 2016. The "everyday" of banal nationalism: Ordinary people's views on Italy and Italian. *Political Geography*, 54, 32–42. https://doi.org/10.1016/j.polgeo.2015.07.006.

Antonsich, M., and Skey, M. 2017. Affective nationalism: Issues of power, agency and method. *Progress in Human Geography*, 41 (6), 843–845. https://doi.org/10.1177/0309132516665279.

Antonsich, M., Skey, M., Sumartojo, S., Merriman, P., Stephens, A. C., Tolia-Kelly, D., Wilson, H., and Anderson, B. 2020. The spaces and politics of affective nationalism. *Environment and Planning C: Politics and Space*. https://doi.org/10.1177/2399654420912445.

Ash, J. 2018. Media and popular culture. In J. Ash, R. Kitchin, and A. Leszczynski (eds.) *Digital geographies*, pp. 143–152. London: Sage.

Benwell, M. C., and Dodds, K. 2011. Argentine territorial nationalism revisited: The Malvinas/Falklands dispute and geographies of everyday nationalism. *Political Geography*, 30 (8), 441–449. https://doi.org/10.1016/j.polgeo.2011.09.006.

Billig, M. 1995. *Banal nationalism*. Thousand Oaks, CA: Sage.

Biswas, S. 2002. W(h)ither the nation-state? National and state identity in the face of fragmentation and globalisation. *Global Society*, 16 (2), 175–198. doi:10.1080/09537320220132910.

Bos, D. 2018a. Popular geopolitics and the landscapes of virtual war. In R. A. Saunders and V. Strukov (eds.) *Popular geopolitics: Plotting an evolving interdiscipline*, pp. 216–234. London: Routledge.

Bos, D. 2018b. Answering the Call of Duty: Everyday encounters with the popular geopolitics of military-themed videogames. *Political Geography*, 63, 54–64. https://doi.org/10.1016/j.polgeo.2018.01.001.

Bos, D. 2020. Popular geopolitics "beyond the screen": Bringing modern warfare to the city. *Environment and Planning C: Politics and Space*. https://doi.org/10.1177/2399654420939973.

Burgess, J., and Gold, J. R. 1985. *Geography, the media and popular culture*. London and Sydney: Croom Helm.

Carter, S., and Dodds, K. 2011. Hollywood and the "war on terror": Genre-geopolitics and "Jacksonianism" in The Kingdom. *Environment and Planning D: Society and Space*, 29 (1), 98–113. https://doi.org/10.1068/d7609.

Carter, S., and McCormack, D. P. 2006. Film, geopolitics and the affective logics of intervention. *Political Geography*, 25 (2), 228–245. https://doi.org/10.1016/j.polgeo.2005.11.004.

Carter, S., and Woodyer, T. 2020. Introduction: Domesticating geopolitics. *Geopolitics*. https://doi.org/10.1080/14650045.2020.1762575.

Closs Stephens, A. 2016. The affective atmospheres of nationalism. *Cultural Geographies*, 23 (2), 181–198. https://doi.org/10.1177/1474474015569994.

Coulter, K. 2013. Territorial appeals in post-wall German filmmaking: The case of Good Bye, Lenin! *Antipode*, 45 (3), 760–778. https://doi.org/10.1111/j.1467-8330.2012.01041.x.

Crosset, V., Tanner, S., and Campana, A. 2019. Researching far right groups on Twitter: Methodological challenges 2.0. *New Media & Society*, 21 (4), 939–961. https://doi.org/10.1177/1461444818817306.

Dittmer, J. 2007. The tyranny of the serial: Popular geopolitics, the nation, and comic book discourse. *Antipode*, 39 (2), 247–268. https://doi.org/10.1111/j.1467-8330.2007.00520.x.

Dittmer, J. 2014. Geopolitical assemblages and complexity. *Progress in Human Geography*, 38 (3), 385–401. https://doi.org/10.1177/0309132513501405.

Dittmer, J., and Bos, D. 2019. *Popular culture, geopolitics, and identity*, 2nd edn. Lanham, MD: Rowman & Littlefield.

Dittmer, J., and Larsen, S. 2007. Captain Canuck, audience response, and the project of Canadian nationalism. *Social & Cultural Geography*, 8 (5), 735–753. https://doi.org/10.1080/14649360701633311.

Dodds, K. 2006. Popular geopolitics and audience dispositions: James Bond and the internet movie database (IMDb). *Transactions of the Institute of British Geographers*, 31 (2), 116–130. https://doi.org/10.1111/j.1475-5661.2006.00199.x.

Dodds, K. 2008. "Have you seen any good films lately?" Geopolitics, international relations and film. *Geography Compass*, 2 (2), 476–494. https://doi.org/10.1111/j.1749-8198.2008.00092.x.

Edensor, T. 2002. *National identity, popular culture and everyday life*. Berg, Oxford: Bloomsbury Publishing.

Edensor, T., and Sumartojo, S. 2018. Geographies of everyday nationhood: Experiencing multiculturalism in Melbourne. *Nations and Nationalism*, 24 (3), 553–578. https://doi.org/10.1111/nana.12421.

Ellis, E. G. 2019. *Pepe the Frog means something different in Hong Kong—right?* www.wired.com/story/pepe-the-frog-meme-hong-kong/.

Eriksen, T. H. 2007. Nationalism and the internet. *Nations and Nationalism*, 13 (1), 1–17. https://doi.org/10.1111/j.1469-8129.2007.00273.x.

Falah, G.-W., Flint, C., and Mamadouh, V. 2006. Just war and extraterritoriality: The popular geopolitics of the United States' war on Iraq as reflected in newspapers of the Arab world. *Annals of the Association of American Geographers*, 96 (1), 142–164. https://doi.org/10.1111/j.1467-8306.2006.00503.x.

Fox, J. E., and Miller-Idriss, C. 2008. Everyday nationhood. *Ethnicities*, 8 (4), 536–563. https://doi.org/10.1177/1468796808088925.

Foxall, A. 2013. Photographing Vladimir Putin: Masculinity, nationalism and visuality in Russian political culture. *Geopolitics*, 18 (1), 132–156. https://doi.org/10.1080/14650045.2012.713245.

Froio, C., and Ganesh, B. 2019. The transnationalisation of far right discourse on Twitter: Issues and actors that cross borders in Western European democracies. *European Societies*, 21 (4), 513–539. https://doi.org/10.1080/14616696.2018.1494295.

Fuchs, C. 2019. *Nationalism on the internet: Critical theory and ideology in the age of social media and fake news*. London: Routledge.

Ganesh, B. 2020. Weaponizing white thymos: Flows of rage in the online audiences of the Alt-Right. *Cultural Studies*. https://doi.org/10.1080/09502386.2020.1714687.

Gellner, E. 1983. *Nations and nationalism*. Oxford: Blackwell.

Grumke, T. 2013. Globalized anti-globalists: The ideological basis of the internationalization of right-wing extremism. In S. von Mering and T. W. McCarty (eds.) *Right-wing radicalism today: Perspectives from Europe and the US*, pp. 13–22. London: Routledge.

Hermansson, P., Lawrence, D., Mulhall, J., and Murdoch, S. (eds.) 2020. *The international Alt-Right: Fascism for the 21st century?* Abingdon: Routledge.

Hodge, E., and Hallgrimsdottir, H. 2019. Networks of hate: The Alt-Right, troll culture, and the cultural geography of social movement spaces online. *Journal of Borderlands Studies*, 35 (4), 563–580. https://doi.org/10.1080/08865655.2019.1571935.

Ince, A. 2019. Fragments of an anti-fascist geography: Interrogating racism, nationalism, and state power. *Geography Compass*, 13 (3). https://doi.org/10.1111/gec3.12420.

Ismangil, M. 2019. Subversive nationalism through memes: A *Dota 2* Case Study. *Studies in Ethnicity and Nationalism*, 19 (2), 227–245. https://doi.org/10.1111/sena.12298.

Johnson, N. 1995. Cast in stone: Monuments, geography, and nationalism. *Environment and Planning D: Society and Space*, 13 (1), 51–65. https://doi.org/10.1068/d130051.

Jones, R., and Merriman, P. 2009. Hot, banal and everyday nationalism: Bilingual road signs in Wales. *Political Geography*, 28 (3), 164–173. https://doi.org/10.1016/j.polgeo.2009.03.002.

Kirby, P. 2019. Sound and fury? Film score and the geopolitics of instrumental music. *Political Geography*, 75. https://doi.org/10.1016/j.polgeo.2019.102054.

Koch, N. 2016. Is nationalism just for nationals? Civic nationalism for noncitizens and celebrating National Day in Qatar and the UAE. *Political Geography*, 54, 43–53. https://doi.org/10.1016/j.polgeo.2015.09.006.

Koch, N., and Paasi, A. 2016. Banal nationalism 20 years on: Re-thinking, re-formulating and re-contextualizing the concept. *Political Geography*, 54, 1–6. https://doi.org/10.1016/j.polgeo.2016.06.002.

Kuus, M. 2020. Political geography II: Institutions. *Progress in Human Geography*, 44 (1), 119–128. https://doi.org/10.1177/0309132518796026.=.

Langenbacher, N., and Schellenberg, B. 2011. *Is Europe on the right path? Right-wing extremism and right-wing populism in Europe*. Berlin: Friedrich Ebert Stiftung.

Lenoir, T., and Caldwell, L. 2018. *The military-entertainment complex*. Cambridge, MA: Harvard University Press.

Lobato, R. 2019. *Netflix nations: The geography of digital distribution*. New York: New York University Press.

Lu, J., and Yu, X. 2019. The internet as a context: Exploring its impacts on national identity in 36 countries. *Social Science Computer Review*, 37 (6), 705–722. https://doi.org/10.1177/0894439318797058.

Lukinbeal, C., Sharp, L., Sommerlad, E., and Escher, A. (eds.) 2019. *Media's mapping impulse*. Stuttgart: Franz Steiner Verlag.

Madianou, M. 2005. *Mediating the nation: News, audiences and the politics of identity*. London: UCL Press.

Merriman, P., and Jones, R. 2017. Nations, materialities and affects. *Progress in Human Geography*, 41 (5), 600–617. https://doi.org/10.1177/0309132516649453.

Militz, E., and Schurr, C. 2016. Affective nationalism: Banalities of belonging in Azerbaijan. *Political Geography*, 54, 54–63. https://doi.org/10.1016/j.polgeo.2015.11.002.

Mirrlees, T. 2017. Transforming transformers into militainment: Interrogating the DoD-Hollywood complex. *American Journal of Economics and Sociology*, 76 (2), 405–434. https://doi.org/10.1111/ajes. 12181.

Mōri, Y. 2019. Lukewarm nationalism: The 2020 Tokyo Olympics, social media and affective communities. *International Journal of Japanese Sociology*, 28 (1), 26–44. https://doi.org/10.1111/ijjs.12093.

Morley, D. 2000. *Home territories: Media, mobility and identity*. London: Routledge.

Mountz, A. 2009. Nationalism. In C. Gallaher, C. T. Dahlman, M. Gilmartin, A. Mountz and P. Shirlow (eds.) *Key concepts in political geography*, pp. 277–287. Los Angeles, CA: Sage.

Müller, M. 2012. Opening the black box of the organization: Socio-material practices of geopolitical ordering. *Political Geography*, 31 (6), 379–388. https://doi.org/10.1016/j.polgeo.2012.06.001.

Müller, M. 2015. More-than-representational political geographies. In J. Agnew, V. Mamadouh, A. J. Secor and J. Sharp (eds.) *The Wiley Blackwell Companion to Political Geography*, pp. 407–423. Malden, MA: Wiley-Blackwell.

Pinkerton, A., and Benwell, M. 2014. Rethinking popular geopolitics in the Falklands/Malvinas sovereignty dispute: Creative diplomacy and citizen statecraft. *Political Geography*, 38, 12–22. https://doi.org/ 10.1016/j.polgeo.2013.10.003.

Rech, M. F. 2015. A critical geopolitics of observant practice at British military airshows. *Transactions of the Institute of British Geographers*, 40 (4), 536–548. https://doi.org/10.1111/tran.12093.

Ridanpää, J. 2017. Culturological analysis of filmic border crossings: Popular geopolitics of accessing the Soviet Union from Finland. *Journal of Borderlands Studies*, 32 (2), 193–209. https://doi.org/10.1080/ 08865655.2016.1195699.

Rose, G. 2016. Rethinking the geographies of cultural "objects" through digital technologies: Interface, network and friction. *Progress in Human Geography*, 40 (3), 334–351. https://doi.org/10.1177/0309132515580493.

Saunders, R. A., and Strukov, V. (eds.) 2018. *Popular geopolitics: Plotting an evolving interdiscipline*. London: Routledge.

Sharp, J. 2000. *Condensing the Cold War: Reader's Digest and American identity*. Minneapolis, MN: University of Minnesota Press.

Sharp, L., and Lukinbeal, C. 2015. Film geography: A review and prospectus. In S. P. Mains, J. Cupples and C. Lukinbeal (eds.) *Mediated geographies and geographies of media*, pp. 21–35. New York: Springer.

Shaw, I. G. R., and Warf, B. 2009. Worlds of affect: Virtual geographies of video games. *Environment and Planning A*, 41 (6), 1332–1343. https://doi.org/10.1068/a41284.

Singer, P. W., and Brooking, E. T. 2018. *LikeWar: The weaponization of social media*. New York: Eamon Dolan Books.

Skey, M. 2009. The national in everyday life: A critical engagement with Michael Billig's thesis of Banal Nationalism. *The Sociological Review*, 57 (2), 331–346. https://doi.org/10.1111/j.1467-954X.2009. 01832.x.

Skey, M. 2020. *Nationalism and the media*. https://stateofnationalism.eu/article/nationalism-and-media/#a rticle.

Sloan, L., and Quan-Haase, A. (eds.) 2017. *The SAGE handbook of social media research methods*. London: Sage.

Smith, A. D. 1998. *Nationalism and modernism*. London: Routledge.

Szulc, L. 2017. Banal nationalism in the internet age: Rethinking the relationship between nations, nationalisms and the media. In M. Skey and M. Antonsich (eds.) *Everyday nationhood: Theorising culture, identity and belonging after banal nationalism*, pp. 53–74. London: Palgrave Macmillan.

Tilly, C. 1992. *Coercion, capital, and European states, AD 990–1992*. Oxford: Blackwell.

Waterton, E., and Dittmer, J. 2016. Transnational war memories in Australia's heritage field. *Media International Australia*, 158 (1), 58–68. https://doi.org/10.1177/1329878X15622079.

Webber, N. 2020. The Britishness of "British Video Games." *International Journal of Cultural Policy* 26 (2), 135–149. https://doi.org/10.1080/10286632.2018.1448804.

Weir, P. 2018. Networked assemblages and geopolitical media: Governance, infrastructure and sites in BBC Radio. *Geopolitics*, 25 (4), 1–31. https://doi.org/10.1080/14650045.2018.1465043.

Woodyer, T., and Carter, S. 2018. Domesticating the geopolitical: Rethinking popular geopolitics through play. *Geopolitics*, 25 (5). https://doi.org/10.1080/14650045.2018.1527769.

18
EUROCENTRISM/ORIENTALISM IN NEWS MEDIA

Virginie Mamadouh

News media by definition provide a window on the world beyond the direct reach of our physical body. They expand the scope of the world we are able to perceive, but paradoxically do this in a biased way: feeding our visual (and sometime auditory) systems at the expense of our other senses and producing and circulating biased representations of other places and peoples.

Eurocentrism and Orientalism are practices of ordering, bordering and othering (Van Houtum & Van Naerssen 2002) in knowledge production that shape our geographical imaginations and our geopolitical representations. Both are reinforced by uneven patterns in the media coverage of different places around the world. This chapter first introduces the notions of Eurocentricism, Orientalism and othering and how they pertain to (news) media. It then presents a number of examples from geographical studies. Finally it turns to the recent changes in information and communication technologies and media landscapes and how these impact the production, circulation and consumption of Eurocentric and Orientalist tropes.

Eurocentrism and Orientalism in popular geopolitics

Eurocentrism and Orientalism refer to two common biases in Western media. Eurocentrism points at the Western biases that underpin the selection and the framing of news, with greater attention and nuance for Western places than for places in other parts of the world, while Orientalism pertains to the negative stereotyping of non-Western places and people.

Eurocentrism presents European viewpoints as universal. Orientalism involves more specifically the othering of the Orient as the significant Other of Europe (or the West). Originally a current in 19th century European paintings focused on copying (Middle) Eastern styles and/or depicting (Middle) Eastern tableaus, it was also used to classify novels and travel writings dealing with those parts of the world. Since Edward Said published his seminal book *Orientalism: Western Conceptions of the Orient* in 1978 the term has been widely used in academic discourse to denote the Western gaze on the Orient and has now evolved into a broad movement towards postcolonial and decolonial studies (for geography see Jazeel 2014; 2015; Radcliffe 2017; Radcliffe & Radhuber 2020). Said analyzes how Western commentators (mostly authors of diaries and travel books in English literature) created, through lies,

DOI: 10.4324/9781003039068-22

confusions and/or sweeping generalizations, a stereotypical Orient and Oriental people. He argues that these Orientalist representations and the attitudes towards the Orient (Middle Eastern, Asian and/or North African societies) they convey are telling us much more about Western society than about the Orient they are supposed to describe. Orientalism implies a patronising attitude, portraying the Orient as an essentialized, static and undeveloped society, and legitimizing its subordination to Western imperial powers. For Said, this representation's main function is not to increase European knowledge and understanding of the Orient, but to strengthen the representation of European or Western society as developed, modern, rational, flexible and ultimately superior. As such it legitimates the subordination of the Orient to Europe, of the Orientals to the Europeans.

Orientalism is by definition Eurocentric, as it rests on a Eurocentric gaze on the relations between Europe and its Oriental Other; but Eurocentrism needs not to Orientalize, as it could also coexist with unawareness of the Other. However when translocal relations intensify, Eurocentrism can only be maintained through the active disregard of others' viewpoints. Orientalism is a powerful mechanism justifying and reproducing unequal power relations, especially when those who are Orientalized are exposed to those discourses and for lack of powerful alternative representations have largely interiorized the notion of their own inferiority (Fanon 1952).

In political geography and critical geopolitics, Eurocentrism and Orientalism have been widely acknowledged in the analysis of the geographical imaginations and geopolitical representations articulated and circulated by European/Western actors in general in their encounters with the rest of the world (Gregory 2004). In addition, Orientalism can be observed in othering processes at a smaller scale, between groups more similar to each other, when the Europeanness (or the Westernness) of certain groups and nations at the margins of Europe or the West is disputed (for example in Eastern Europe and Latin America respectively) and within Western societies (for example when descendants of certain groups of migrants are seen as foreign to the national community). There is a strong asymmetry in the binary, with a positive value being assigned to the Western and a negative value to the non-Western. This process fixes and essentializes differences, rejecting the potentialities of hybridity and change. Ordering devices become essentialized as fixed identities and reproduced through processes of social spatialization and spatial socialization (Paasi 1996; 2020); people tend to take many of these labels as natural and permanent. Orientalism also conceals differences among those seen as non-Western, reducing their being to the Western qualities they lack. But it also neglects differences within the West (a critique that can be addressed to Said's own account of Western Orientalism).

Othering processes are more diverse and might be compatible with less unequal relations than Orientalism (such as in the formation of national identities Dijkink 1996). These othering frames often invoke Orientalist tropes of various intensity, like in the nesting Orientalisms in the disintegrating Yugoslavia (Bakić-Hayden 1995). Likewise American Exceptionalism (othering Europe as its Other) could and should be distinguished from American Orientalism proper (about the non-Western Others) (Nayak & Malone 2009).

Finally othering processes such as Occidentalism and other forms of stereotypes based on race, religion or ethnicity could be mentioned here. While they also naturalize and essentialize differences, they differ however from Orientalism because they do not necessarily complement and justify highly unequal power relations. Quite the contrary. Occidentalism (in the Middle East or in East Asia) has frequently been rooted in resentment to the perceived domination by the West (not as a justification of exploitative relations). This does not, however, rule out that it may be used to justify violence in the short term and the intended submission and exploitation of the West in the long term. In any event, Occidentalism upholds similar processes of simplification, generalization and essentialization as Orientalism, and deserves critique and deconstruction by geographers too (Minca & Ong 2017).

In this chapter we focus on the media as vehicles of Eurocentrism and Orientalism in popular geopolitics and more specifically in news media. Geographers have discussed Orientalist discourses in popular culture drawing on a much wider plurality of sources, placing news media among many other sources such as cartoons, literature, movies, comic books, school books or video games. News media representations are nevertheless particularly important. They are often invoked to frame events in the world or to question those frames. For example attempts to foster class discussion of Said's discussion of Orientalism or of prejudices and stereotypes often engage media representations as familiar entry points (Ashutosh & Winters 2009; Hintermann et al. 2020).

It is nevertheless important to stress that Eurocentric and Orientalist perspectives affect academia at large, including the geographical scholarship we are dwelling on (for ideas about the critique of the Euro-Americanism of academic geography and aera studies and ideas to queer Eurocentric geographical knowledge production see Jazeel 2015). Indeed academic representations of world politics also often Orientalize people and places at the margins of the modern state system. Think of the "the gap" as opposed to "the integrated core" in *The Pentagon's Map* (Barnett 2003) or the non-Western civilizations in *The Clash of Civilizations* (Huntington 1996) (see Said 2001). Moreover most of the work of critical geopolitics is Eurocentric too, as it examines mostly Western news media, as if they were the only ones worthy of academic scrutiny. Exceptions will be discussed in the section on new media configurations.

Eurocentrism and Orientalism in news media

While Orientalism can easily be linked to racism (Banaji 2017), Eurocentrism is a more benign phenomenon rooted in a common-sensical expectation: news about nearby places/peoples is expected to matter more. To some extent we could argue that Eurocentrism in news media is an expression of the first law of geography as enunciated by Waldo Tobler (1970): "everything is related to everything else, but near things are more related than distant things." News media tend to focus on the places where their audiences dwell and to cover things more related to them. The persistence of Eurocentrism in media geographies is an artefact of their focus on geopolitical representations and media in particular places in the world, namely Western (national) media. Most transnational and global media are also largely Western based; they are often owned by Western firms and organized according to a market logic (or to a Western conception of journalism); they are serving primarily an audience in the West and are likely to select primarily news that is perceived as most relevant to that audience, especially when their main aim is to generate more income through the enlargement of their audience. Therefore they can be expected to cover news from Western places more frequently, more intensely and with more details, and to give voice to Western points of view.

Whether this "law" of proximity/distance is ethical or not in news coverage can be debated, but Eurocentrism is definitely problematic and detrimental when proximity/distance to the targeted audience(s) is obviously misrepresented. For example when places and peoples are invisible in the media despite strong interaction, for instance when activities in those places and the labor of these people is fundamental to the survival of the audience. Moreover media targeting audiences outside the West also often give disproportionate attention to events in the West.

This is true also of global media and news agencies monopolizing the exchange of international news (AFP, Associated Press, Reuters, United Press International). In the late 1970s and 1980s this biased political economy was at the center of a debate over media representations of the developing world. After decolonization in Asia and Africa, media concerns voiced by non-aligned nations in the 1970s and calls for a New World Information and Communication Order (NWICO) were followed by the installation of a commission by the United Nations Educational, Scientific and Cultural Organization (UNESCO). It was chaired by the Irish politician and winner of the 1974 Nobel Peace Prize Seán MacBride. In its report *Many Voices, One World* (published in 1980) the McBride Commission made an evaluation of structural imbalances and inequalities in the field of communications and presented 82 recommendations to create a NWICO (see Carlsson 2003). Apart from the agreement on the right to communicate adopted at the UNESCO General Conference of 1983, not much was realized. The report was highly controversial. Western countries led by the USA were hostile to NWICO and critical of what they saw as an infringement of the freedom of press and a bias against private ownership of media and communication. The US even withdrew its UNESCO membership in the 1980s. Despite follow-up actions (the International Programme for the Development of Communication (IPDC) launched in 1980, the World Summit on the Information Society (WSIS) in 2003 and 2005) as well as dramatic geopolitical and technological transformations, most of the structural imbalances identified in 1980 are still relevant. Four decades later the MacBride Report remains pertinent since new ICTs and social media are commodified while US transnational corporations are still hegemonic in global news flows and US corporations are still dominating the global information economy's market value, sales, profits and capital assets (Fuchs 2015), hence calls for a digital new world information and communication order (Thussu 2015). Although most of the literature on NWICO focuses on the political economy of communication, the underlying assumption is that these unequal power relations produce unequal geographies of news coverage.

On top of the differences in quantity, media coverage can be qualitatively different. Orientalism then also colors the news coverage and alters the qualities assigned to Western and/or to Orientalized (i.e. non-Western) actors and places, actively reproducing value hierarchies that justify unequal power relations between the West and the non-West. Evidently, such hierarchies also justify the disproportionate attention devoted to Western affairs in the media. In other words Orientalism normalizes Eurocentrism.

No doubt the ground-breaking study of Orientalism in critical geopolitics has been the seminal work of Scottish geographer Jo Sharp regarding the othering of Russia and the Soviet Union in the *Readers' Digest* (Sharp 1993; 1996; 2000). She scrutinizes how the popular general affairs magazine covered the Soviet Union from its inception to its implosion, and how these representations were producing American identity. During the first decades the attitude was ambivalent: some commonalities between Soviet and American values were underlined but American society was primarily portrayed as a classless society where a socialist revolution was redundant. After World War II differences were naturalized and essentialized, and during the Cold War the Soviet Union and communism were portrayed as evils threatening the American way of life: establishing them as America's significant Other.

Considering the post-Cold War period, Ó Tuathail (2002) examines the geopolitical reasoning in the case of the US response to the war in Bosnia in the summer of 1992, based on journalistic reports from leading US newspapers and transcripts of State Department press briefings. He distinguishes two contradictory storylines: "Balkan Vietnam" and "European genocide." Framing the war in Bosnia as a European genocide is underlining the need for an intervention of the international community including the USA, reactivating their commitment to prevent genocide. By contrast, framing it as a Balkan Vietnam is a warning against a military intervention, activating quagmire anxiety with the memory of the failed and tragic intervention of the USA in Vietnam in the 1960s and 1970s. Seeing Bosnia as a Balkan Vietnam can be said to Orientalize Bosnians (and other former Yugoslavs) much more than the other frame. It is othering Bosnia and Bosnians twice. First as a non-European place similar to Vietnam. Second Balkanism is already a widespread Orientalizing trope in European thought, essentializing interethnic hate and violence in South-Eastern Europe as a permanent and natural state of affairs in that part of the continent.

Robison (2004) finds similar processes at work in her study of the coverage of the Bosnian conflict in the British press, showing how Balkanism as an Orientalist discourse, and the framing of the war as "a humanitarian catastrophe requiring a humanitarian response," both impacted on policy options. She stresses the importance of geopolitical representations by concluding:

> despite a plethora of media coverage between 1992–1995 the international community failed to prevent the destruction of Bosnia as a country. There exists a real need to understand the extent to which representations of place in the media can impact on government responses to a crisis and for this reason any further work on how ideas of place are constructed can only be welcomed.
>
> *(Robison 2004, 397)*

There has been a sustained flow of publications analyzing geopolitical representations in popular geopolitics and in the media in particular, including visual representations. Maps and more specifically journalistic cartographies are also powerful carriers of Orientalist representations. In her longitudinal study of the cartography of Kurdistan through the analysis of 450 maps published in American quality newspapers and magazines from 1945 to 2002, Karen Culcasi (2006) shows how maps engage with geopolitics discourses and how the cartographic portrayal of the Kurds legitimates the dominant US geopolitical position on ongoing events, oscillating between violent rebel and backward victims, in equally Orientalizing discourses.

Othering can also target specific groups rather than specific places, regions or states. The representation of Muslim women in American media has been extensively researched: Fahmy (2004) studied Associated Press wire photographs of Afghan women during and after the Taliban regime, Falah (2005) studied the visual representation of Arab Women in US newspapers, and Rahman (2014) studied the portrayal of Pakistani women in the news magazine *Time* between 1998 and 2002 showing othering and Orientalizing alongside oversimplifying and decontextualizing. Remarkably, the othering and Orientalizing gaze can also be combined with positive assessment, for example in the news coverage about the Kurdish women fighters of the Women's Protection Unit, YPJ (Yekîceyên Parastine Jin) fighting Daesh (also known as Islamic State in the Levant) in the mid-2010s (Şimşek & Jongerden 2018). They problematize the portrayal of the Kurdish (feminist) struggle in US popular discourses as Orientalist because it silences the voices of women and decontextualizes their struggle and the politics of Rojava. Through Orientalism and gender stereotypes, the idea of the Middle East as backward and violent is reproduced.

Of the many examples of Orientalism that can be found in media, the portrayal of Africa as a dark, backward continent is probably one of the most enduring. The systematic study of the news coverage of the conflicts in Rwanda and in Bosnia from 1990 to 1994 in six major

American quality newspapers explored differences in the representations of two large-scale and violent ethnic conflicts and the response of the international community—one located in Europe, the other in Africa (Myers et al. 1996). They note a difference in share volume of coverage: more attention to Bosnia than to Rwanda. Moreover they compare the prevalence of the language of civil war, the language of savagery and the language of ethnicity and tribalism, observing that strategies and tactics are reserved for the description of Bosnian politics and savagery and tribalism for that of Rwandan politics. The Rwandan conflict is framed in more essentialist terms than in Bosnia, in terms of perpetual inter-tribal clashes in Sub-Saharan Africa, obscuring both the political dynamics of the ongoing violence and its geopolitical context. The authors also suggest strong connections between news framing and US geopolitics: residual frames from the Cold War (when covering Bosnia), the North-South divide (the Third World in need of help) and African marginality and irrelevance to US geopolitics after the end of the Cold War (when covering Rwanda). Although they acknowledge the othering processes at work in both cases, they conclude that othering is more profound when it comes to Rwanda than when it comes to Bosnia (and the Balkans). The language of tribalism is a key Orientalist trope that reproduces the notion of a savage and primitive Africa, stuck outside modernity.

A decade later David Campbell deals with the press coverage of the events in Darfur, Sudan and more specifically on visuals (Campbell 2007). He focuses on *The Guardian*, allegedly expecting to find there the most extensive and nuanced coverage of Africa in the British press. Nevertheless he also notices that the conflict is framed as placeless and timeless and that it impacts the international community and its inclination to intervene. Pictures have a tremendous influence: their aesthetic and emotional appeal ("a picture is worth a thousand words") as well as their claim to authenticity ("the camera doesn't lie") remain powerful, although often questioned and problematized—since controversies about pictures being staged has always been an issue and even more since technical tools to manipulate pictures have become extremely effective and widely affordable. Visual coverage has become a condition sine qua non to media coverage, up to the point that news media hardly cover events if no pictures or videos are available, and as a result these events are hardly acknowledged.

In his study, Campbell (2007) proposes the notion of visual economy to have a close look at the material circumstances under which photojournalists are working and the actual production of the newspapers. This entails a discussion of issues like access to the field, insurance and other restrictions, as well as whether they are able or not to sell their pictures to international agencies, and the technologies used for the transmission and publication of images. The (non-)availability of pictures constrains the options available at the editorial desk of the newspapers and for example might lead to the use as illustration of pictures taken far away from where the events discussed in the article took place (in this case, refugee camps in Chad rather than in Darfur). Campbell (2007) also stresses how the photographs produce a humanitarian visualization for Darfur, highlighting the victims over the perpetrators, reifying identities into fixed forms, framing the conflict into an eternal war between Arabs and Africans (rather than foregrounding political strategies and tactics) and naturalizing certain policy responses (humanitarian intervention) over others. In short, visuality is central to the production of a particular geographical imagination, not a mere illustration.

Beyond Eurocentrism and Orientalism

Next to these studies exposing Orientalism in Western media, some geographers have tried to address the other side of the story, either by studying media coverage trying to disrupt Orientalist discourses or by analyzing non-Western media.

In one of his many publications about the Bosnian war, Ó Tuathail (1996) also addresses Othering and Balkanism through an analysis of the work of the British journalist Maggie O'Kane reporting from Bosnia and more specifically from besieged Sarajevo between 1992 and 1996 in *The Guardian*. He shows how her work was different from mainstream reporting and how what he calls her anti-geopolitical eye disrupted the hegemonic geopolitical discourse about the Bosnian war, giving a voice to the victims rather than to politicians, and bringing the human consequences of the Bosnian conflict to the readers, shaping proximity and questioning responsibility.

Another issue is the way Orientalizing discourses circulate in media in Orientalized places. In Central Europe (Orientalized as less European than Western Europe) Kuus (2008) discusses how irony and self-deprecating humour were omnipresent in the coverage of the NATO invitation to join the military partnership in major Estonian newspapers in 2002. She stresses the parallel between Estonian attitudes towards joining NATO and Švejkian absurd obedience. This term referred to the 1920s Czech novels about *The Good Soldier Švejk* and his adventures in the Austrian Hungarian army. Švejk has become an emblematic figure of subversive resistance in Central Europe. Kuus' analysis stresses the ambiguous character of Švejk's behavior embodying both essentialist stereotyping of the Czechs (and other Central European nations) and the universal trope of clever individual resistance against the absurdity of bureaucracy.

Comparing the framing of the 2007–2008 global food crisis in Western newspapers and the English language Chinese newspaper *China Daily*, Gong and Le Billon (2014) contrast different blaming narratives. While the main Western newspapers in their sample (major world newspapers according to media database LexisNexis) framed the rising demand for food in China (and to a lesser extent in India) as a security threat, the Chinese newspaper articulates "a narrative built upon unequal power relations between the West and East and North and South" and blames the raise of biofuel.

To decenter the gaze of critical geopolitics, scholars have been reading the US War on Terror in the Arab World (Falah et al. 2006) and the Global South. For Tanzania, Sharp (2011) identifies key themes in the news coverage of the main English language newspaper *The African*: US dominance, nature of the US worldview, global perspectives on US attitudes, impacts on Africa/the Third World, and victimhood (about the victims of war and the importance of human security). She argues that we could speak of a subaltern geopolitics, reflecting back to the dominant geopolitics, decentring the hegemonic (American) geopolitical imagination of the US War on Terror.

For the Philippines, Woon (2014) examines the audience's interpretation of the representation of Mindanao (the southern major island of the Philippines) in national newspapers based in Manila in 2003, centered around the othering of Mindanao as an island plagued by chaos and conflicts and the associated othering of Muslims living mainly in that part of the country, a context marked nationally by heightened mobilizations for the independence of Mindanao and globally by the US War on Terror. Among his respondents he finds both corroborations of an Orientalist portrayal of Muslims and resistances to these frames, linked to readers' different experiences, positionalities and subjectivities. Audience research is indeed particularly important if we want to understand how Orientalist frames are influencing social relations (beyond foreign policy-making).

At the same time, Woon's study highlights that media also (re-)produce and circulate *domestic* geopolitical representations based on othering processes, where subregions and subgroups in the state are othered and even Orientalized vis-à-vis a local center. Eriksson (2008) analyzes the representation of Northern Sweden in Swedish national media, arguing that

Norrland is used as an abstract essentialized geographical category: it represents a backward and traditional rural space and is contrasted to equally essentialized urban areas representing Western modernity. Similarly Nwankwo (2020) reports essentializing discourses in the coverage of the farmer–pastoralist conflicts in Nigerian newspapers and Serrao (2020) draws attention to internal Orientalism in Brazilian social media regarding stereotypes and prejudice against *Nordestinos*, the inhabitants of Northeast Brazil. At the local scale, Qian, Qian and Zhu (2012) show in their analysis of the representation of the city of Guangzhou and of the Cantonese language the Othering of migrants in Guangzhou in local social media during the 2010 protests against a municipal plan to switch to Putonghua (Mandarin) on local television in order to promote a more cosmopolitan city image (the plan was withdrawn). Similar othering is at work in maps of Rio de Janeiro (both in Brazilian newspapers and on Google Maps) regarding the representations of favelas and their residents (Novaes 2014).

New media configurations: Globalization and digitalization

Originally media geographies of othering were looking primarily at conventional media and especially at newspapers. They were particularly important in the emergence of modern territorial states, national identities and standardized national languages. From 1990 onwards, however, globalization and digitalization changed the dynamics of news media through both technological innovations such as cable and satellite dishes, internet, mobile telephony and the rise of the new (social) media, and political processes such as the liberalization of broadcasting and publishing regulations.

Cable news changed the relation between political and media settings. *CNN International* became the emblematic channel for 24/7 news broadcasting in 1990 with its coverage of the Gulf War, creating the so-called "CNN effect." It has a temporal dimension as well: CNN and other 24/7 news outlets (in contrast to the evening news and broadsheet daily newspapers) force statespersons to react very quickly as news reaches the audience much quicker (almost in real-time). And it has a spatial dimension: CNN International aims at serving transnational audiences, undermining the naturalization of national foreign policy objectives.

Following CNN's success, other global news channels have been created: France 24, Al Jazeera (Qatar), TelesSur (Venezuela), Russia Today and Sputnik (Russia), China Global Television Network CGTN (formerly known as CCTV International) (China) or PressTV (Iran). Most are directly sponsored by a particular state, enjoying dissimilar degrees of press freedom. Some of these were particularly keen to deploy an alternative discourse to the one circulated by CNN as it was perceived as an American (read Eurocentric) sender: TeleSur's motto *Nuestro Norte es el Sur* (Our North is the South) clearly conveys the ambition to decenter the Eurocentrism of established global media.

Al Jazeera (based in Qatar) was a particularly important initiative, as acknowledged in the expression the "Al Jazeera effect." It again has two dimensions: a global and a local. First as opposed to CNN as an American (i.e. Western) voice, fighting its Eurocentrism and Orientalism, and giving a voice to non-Western actors. But it also serves as an independent news medium as opposed to national media accountable to autocratic rulers in the Arab World. Different language channels serve different audiences: Al Jazeera English serves the global audience (first effect), and the original Al Jazeera (in Arabic) serves the audiences in the Arab World (second effect). Youmans (2017) has documented the attempts and failures of Al Jazeera to establish itself as a mainstream news channel in the United States, stressing the paradox of such an ambition considering "Al Jazeera's derivative channels represent the Orient speaking back to its western authors," while serving an American audience would imply "de-Orientalizing, or over-Americanizing, its US-centered services" (Youmans 2017, 26).

That these channels offer different takes on the news can be experienced by those fortunate enough to access them and compare coverage, but whether they really differ in terms of Eurocentrism and Orientalism is far from established. In their comparison of four terrorist attacks on television news shows on four (inter)national channels, Gerhards and Schäfer (2014) found more similarities than expected. The few but notable differences were that CNN and Al Jazeera framed the attacks more as a geopolitical conflict ("War on Terror" frame) while the BBC (UK) and ARD (Germany) focused more on the victims ("crimes against humanity" frame), local reactions and rescue operations. Moreover although CNN and Al Jazeera both foregrounded the War on Terror between the USA/West and Islamist terrorists, Al Jazeera payed more attention to the motives of the perpetrators than CNN (Gerhards & Schäfer 2014). Comparing the representation of China and the USA in Africa in Al Jazeera English, the BBC and CNN Paterson and Nothia (2016) did not find significant differences regarding Orientalist tropes. China and the USA were framed differently but these representations were reproducing similarly disempowering stereotypes of Africans.

As part of the Chinese soft power deployment, Chinese television also went global targeting more specifically Latin America and Africa. Marsh (2016) puts to the test the ambition of the Chinese broadcaster CCTV to broadcast "African news from an African viewpoint" with a comparison between their program *Africa Live* and BBC World News' *Focus on Africa*. She observes the emergence of an alternative to the Western gaze reducing Africa to conflicts and famines, but signals a disproportionate attention to Chinese topics and Chinese interests: is Sinocentrism replacing Eurocentrism? In addition, Li (2017), comparing the coverage of CCTV Africa and Al Jazeera English of the 2014–2015 Ebola outbreak in West Africa, reports a different sensibility between the two non-Western global news providers, the former committed to positive news about Africa, the latter to being the voice of the Global South. She concludes that they "have augmented Africa's voices in the global arena but far from being Africa's own voices"; she is concerned that their predispositions, "be it source hierarchies or embedded solution-oriented rhetoric frame, can inflict on their representation of African affairs, and possibly feed into the production of a new kind of Africa's Otherness" (Li 2017, 129).

The impact of new information and communication technologies and new media on the circulation of Eurocentrism and Orientalism is even more difficult to assert than the plurality of voices on global news networks. The affordabilities of the new media (and social media in particular) are different from the conventional mass media, most fundamentally because they enable communication from many to many and blur the line between news producers and audiences, everyone being both.

New media might empower minorities subjected to Orientalist tropes to communicate in their own niche and to develop a common alternative take on ongoing events (for example young Dutch Moroccans on web forums after September 11, see Mamadouh 2001). Humor on Facebook posts can be a way to counter stereotyping and essentialization by exclusive majorities (for example Russian-speaking social media users in Latvia and Estonia in the light of the deterioration of the relations between the EU and Russia, see Juzefovics & Vihalemm 2020). More assertively, the mayor of Florence, Italy launched early in February 2020 the hashtag #Abbracciauncinese (Hug a Chinese) to counter the rise of Orientalizing representations of Chinese residents in mainstream media at the beginning of the COVID-19 pandemic, a campaign which sparked many posts on Twitter, Instagram or YouTube.

Social media can however also host niches for exclusionary discourses. In an online ethnography of four Czech language anti-immigrant Facebook pages, Doboš (2020) observes two Orientalist imaginative geographies that in his view frame migrants and their region of origin as Europe's Others: Islam and Muslims, Africa and African savages. He stresses that the fear of migrants' Otherness is complemented by the fear of Central Europeans of being the Other for Western Europeans. He focuses on the temporality of these imaginative geographies, framing difference as temporal difference (Muslims as medieval, African as prehistoric), reproducing the past into the future and ignoring the heterogeneity of temporality and the accidentality of a present event. His study also demonstrates that social media are allowing for bubbles in which Orientalizing discourses can be nurtured, fostered and naturalized, with little resistance, if any. Facebook pages can become places where members can seek validation and further substantiation of their Orientalizing prejudice.

Last but not least social media and conventional (mass) media constantly interact: newspapers and televisions channels have websites and Twitter feeds, but they also take cues from the new media, reporting on trending topics and rows on Twitter, and engaging sources on the ground through their mobile phones. With the diffusion of mobile internet citizens, journalism expanded (Pinkerton 2013) and social media have become important channels to cover grassroots protests or the effects of oppression and war (Iran in 2009, Tunisia, Egypt, Syria, Ukraine, Hong Kong…) although generally mediated by conventional media (picking up, selecting and framing these fragments). In detention on Manus from 2013 to 2019 for seeking asylum in Australia, the Iranian Kurd Nehrouz Boochani managed to get articles, poems and even a full length book (*No Friends but the Mountains* published in 2018) out of the country to testify about the horrendous living conditions in the camp in Papua New Guinea and to humanize their representation in Western media. While the representations of such stories in Western media run the risk of reproducing Orientalizing tropes, the authorities' success in supressing such voices altogether and in preventing victims from getting images and testimony out of the country (e.g. Yemen, but foremost Xinjiang) contributes to the invisibility of many tragedies.

On the other hand, websites and social media borrow stories from more conventional media and amplify their reach (Limonier 2018 for the diffusion of information from Russian news channels Sputnik and RT on news websites in West African countries), and narratives on social media forums and newspapers articles reinforce each other (Klinke 2016 on Russian cyber-bribes in the UK).

More complex and intertwined communicative arrangements

In conclusion, multilevel media analysis is much needed (Gilboa et al. 2016) to enhance our understanding of the circulation of media narratives, their multiplicity and their hybridity in general, and more specifically regarding ordering and othering frames such as Orientalism. Can social media users from the margin impact hegemonic representations produced in the center? Can social media successfully disrupt othering and Orientalizing discourses? Can local media contribute to national macro-regional and global media and successfully counterbalance their tendency to cover events in the political center more, and in a more nuanced way, than events in the peripheries? And foremost, does the plurality of media to which audiences are subjected reinforce or weaken their ability to resist Orientalizing tropes and to access more balanced news coverage? As yet, it seems that social media help in decentralizing Europe and demoting Eurocentric views by multiplying the broadcasted voices, but it is clear that they are used both to undermine and to advance Orientalist frames.

References

Ashutosh, I., and Winders, J. 2009. Teaching Orientalism in introductory human geography. *The Professional Geographer*, 61 (4), 547–560.

Bakić-Hayden, M. 1995. Nesting Orientalisms: The case of former Yugoslavia. *Slavic Review*, 54 (4), 917–931. doi:10.2307/2501399.

Banaji, S. 2017. Racism and orientalism: Role of the media. In P. Rössler, C. A. Hoffner and L. van Zoonen (eds.), *The international encyclopedia of media effects*. New York: John Wiley & Sons.

Barnett, T. P. M. 2003. *The Pentagon's new map: War and peace in the twenty-first century*. New York: Penguin Books.

Campbell, D. 2007. Geopolitics and visuality: Sighting the Darfur conflict. *Political Geography*, 26 (4), 357–382.

Carlsson, U. 2003. The rise and fall of NWICO: From a vision of international regulation to a relaity of multilevel governance. *Nordicom Review*, 24 (2), 31–67.

Culcasi, K. 2006. Cartographically constructing Kurdistan within geopolitical and orientalist discourses. *Political Geography*, 25 (6), 680–706,

Dijkink, G. 1996. *National identity and geopolitical visions, maps of pride and pain*. London: Routledge.

Doboš, P. 2020. Communicating temporalities: The Orientalist unconscious, the European migrant crisis, and the time of the Other. *Political Geography*, 80, 1–10.

Eriksson, M. 2008. (Re)producing a "peripheral" region: Northern Sweden in the news. *Geografiska Annaler: Series B, Human Geography*, 90 (4), 369–388.

Fahmy, S. 2004. Picturing Afghan women: A content analysis of AP wire photographs during the Taliban regime and after the fall of the Taliban regime. *International Journal for Communication Studies*, 66, 91–112.

Falah, G.-W. 2005. The visual representation of Muslim/Arab women in daily newspapers in the United States. In G.-W. Falah and C. Nagel (eds.) *Geographies of Muslim women: Gender, religion and space*, pp. 300–320. New York: Guildford Press.

Falah, G.-W., Flint, C. and Mamadouh, V. 2006. Just war and extraterritoriality: The popular geopolitics of the United States' war on Iraq as reflected in newspapers of the Arab World. *Annals of the Association of American Geographers*, 96 (1), 142–164.

Fanon, F. 1952. *Peau noire, masques blancs*. Paris: Éditions du Seuil.

Fuchs, C. 2015. The MacBride Report in twenty-first-century capitalism, the age of social media and the BRICS countries. *Javnost—The Public*, 22 (3), 226–239.

Gerhards, J., and Schäfer, M. S. 2013. International terrorism, domestic coverage? How terrorist attacks are presented in the news of CNN, Al Jazeera, the BBC, and ARD. *International Communication Gazette*, 76 (1), 3–26.

Gilboa, E., Jumbert, M. G., Miklian, J., and Robinson, P. 2016. Moving media and conflict studies beyond the CNN effect. *Review of International Studies*, 42 (4), 654–672.

Gong, Q., and Le Billon, P. 2014. Feeding (on) geopolitical anxieties: Asian appetites, news media framing and the 2007–2008 food crisis. *Geopolitics*, 19 (2), 291–321.

Gregory, D. 2004. *The colonial present: Afghanistan, Palestine, Iraq*. Malden, MA: Blackwell.

Hintermann, C., Bergmeister, F. M., and Kessel, V. A. 2020. Critical geographic media literacy in geography education: Findings from the MiDENTITY project in Austria. *Journal of Geography*, 119 (4), 115–126.

Huntington, S. 1996. *The clash of civilizations and the remaking of world order*. New York: Simon and Schuster.

Jazeel, T. 2014. Subaltern geographies: Geographical knowledge and postcolonial strategy. *Singapore Journal of Tropical Geography*, 35 (1), 88–103.

Jazeel, T. 2015. Between area and discipline: Progress, knowledge production and the geographies of geography. *Progress in Human Geography*, 40 (5), 649–667.

Juzefovičs, J., and Vihalemm, T. 2020. Digital humor against essentialization: Strategies of Baltic Russian-speaking social media users. *Political Geography*, 81, 1–11. https://doi.org/10.1016/j.polgeo.2020.102204.

Klinke, I. 2016. The Russian cyber-bride as geopolitical fantasy. *Tijdschrift voor Economische en Sociale Geografie*, 107 (2), 189–202.

Kuus, M. 2008. Švejkian geopolitics: Subversive obedience in Central Europe. *Geopolitics*, 13 (2), 257–277. doi:10.1080/14650040801991506.

Li, S. 2017. Covering Ebola: A comparative analysis of CCTV Africa's Talk Africa and Al Jazeera English's Inside Story. *Journal of African Cultural Studies*, 29 (1), 114–130.

Limonier, K. 2014. Russia in cyberspace: Issues and representations. *Hérodote*, 152–153 (1–2), 140–160.

MacBride, S., *et al.*1980. *Many voices, one world: Communication and society today and tomorrow.* Paris: UNESCO International Commission for the study of Communication Problems.

Mamadouh, V. 2001. Constructing a Dutch Moroccan identity through the World Wide Web. *The Arab World Geographer*, 4 (4), 258–274.

Marsh, V. 2016. Mixed messages, partial pictures? Discourses under construction in CCTV's Africa Live compared with the BBC. *Chinese Journal of Communication*, 9 (1), 56–70.

Minca, C., and Ong, C. E. 2017. Orientalism/Occidentalism. In D. Richardson, N. Castree, M. F. Goodchild, A. Kobayashi, W. Liu, and R. A. Marston (eds.) *International encyclopedia of geography: People, the earth, environment and technology.* New York: Wiley.

Myers, G., Klak, T., and Koehl, T. 1996. The inscription of difference: News coverage of the conflicts in Rwanda and Bosnia. *Political Geography*, 15 (1), 21–46.

Nayak, M. V., and Malone, C. 2009. American orientalism and American exceptionalism: A critical rethinking of US hegemony. *International Studies Review*, 11 (2), 253–276.

Novaes, A. R. 2014. Favelas and the divided city: Mapping silences and calculations in Rio de Janeiro's journalistic cartography. *Social & Cultural Geography*, 15 (2), 201–225.

Nwankwo, C. F. 2020. Essentialising critical geopolitics of the farmers–pastoralists conflicts in West Africa. *GeoJournal*, 85 (5), 1291–1308.

Ó Tuathail, G. 1996. An anti-geopolitical eye: Maggie O'Kane in Bosnia, 1992–93. *Gender, Place and Culture*, 3 (2), 171–185.

Ó Tuathail, G. 2002. Theorizing practical geopolitical reasoning: The case of the United States' response to the war in Bosnia. *Political Geography*, 21 (5), 601–628.

Paasi, A. 1996. *Territories, boundaries and consciousness: The changing geographies of the Finnish-Russian border.* Chichester: John Wiley.

Paasi, A. 2020. Problematizing "bordering, ordering, and othering" as manifestations of socio-spatial fetishism. *Tijdschrift voor Economische en Sociale Geografie*, 112 (1), 18–25.

Paterson, C., and Nothias, T. 2016. Representation of China and the United States in Africa in online global news. *Communication, Culture & Critique*, 9 (1), 107–125.

Pinkerton, A. 2013. Journalists. In K. Dodds, M. Kuus and J. Sharp (eds.) *The Ashgate research companion to critical geopolitics*, pp. 439–460. Farnham: Ashgate.

Qian, J., Qian, L., and Zhu, H. 2012. Representing the imagined city: Place and the politics of difference during Guangzhou's 2010 language conflict. *Geoforum*, 43 (5), 905–915.

Radcliffe, S. A. 2017. Decolonising geographical knowledges. *Transactions of the Institute of British Geographers*, 42 (3), 329–333.

Radcliffe, S. A., and Radhuber, I. M. 2020. The political geographies of D/decolonization: Variegation and decolonial challenges of/in geography. *Political Geography*, 78, 102–128.

Rahman, B. H. 2014. Pakistani women as objects of "fear" and "othering". *SAGE Open*, 4 (4), 1–13. https://doi.org/10.1177/2158244014556990.

Robison, B. 2004. Putting Bosnia in its place: Critical geopolitics and the representation of Bosnia in the British print media. *Geopolitics*, 9 (2), 378–401.

Said, E. 1978. *Orientalism: Western conceptions of the Orient.* London: Routledge and Kegan Paul.

Said, E. 2001. The clash of ignorance. *The Nation* (October). www.thenation.com/article/archive/clash-ignorance/.

Serrao, R. 2020. Racializing region: Internal orientalism, social media, and the perpetuation of stereotypes and prejudice against Brazilian Nordestinos. *Latin American Perspectives*. doi:10.1177/0094582X20943157.

Sharp, J. P. 1993. Publishing American identity: Popular geopolitics, myth and The Reader's Digest. *Political Geography*, 12 (6), 491–503.

Sharp, J. P. 1996. Hegemony, popular culture and geopolitics: The Reader's Digest and the construction of danger. *Political Geography*, 15 (6–7), 557–570.

Sharp, J. P. 2000. *Condensing the Cold War, Reader's Digest and American identity.* Minneapolis, MN: University of Minnesota Press.

Sharp, J. P. 2011. A subaltern critical geopolitics of the war on terror. *Geoforum*, 42 (3), 297–306. https://doi.org/10.1016/j.geoforum.2011.04.005.

Şimşek, B., and Jongerden, J. 2018. Gender revolution in Rojava: The voices beyond tabloid geopolitics. *Geopolitics*, doi:10.1080/14650045.2018.1531283.

Thussu, D. K. 2015. Reinventing "many voices": MacBride and a digital new world information and communication order. *Javnost—The Public*, 22 (3), 252–263.

Tobler, W. 1970. A computer movie simulating urban growth in the Detroit region. *Economic Geography*, 46 (S), 234–240.

van Houtum, H., and van Naerssen, T. 2002. Bordering, ordering and othering. *Tijdschrift voor Economische en Sociale Geografie*, 93 (2), 125–136.

Woon, C. Y. 2014. Popular geopolitics, audiences and identities: Reading the "War on Terror" in the Philippines. *Geopolitics*, 19 (3), 656–683.

Youmans, W. F. 2017. *An unlikely audience: Al Jazeera's struggle in America*. Oxford: Oxford University Press.

19

SEX, GENDER AND MEDIA

Marcia R. England

As representations of space, media reflect the sociospatial through the spaces and places portrayed and through mediated social and spatial interactions. Media help to understand individual and social experiences as well as how people understand geographies, whether lived or imagined. Media feed ideologies of sex and gender, and mediated understanding of gender and sex show up in a number of different ways.

Examinations of media representations uncover mediated and social constructions of interpersonal relationships and geographical imaginations. Rosalyn Deutsche (1991, 18) states: "representations are not objects at all, but social relations, themselves productive of meaning and subjectivity" (quoted in Cresswell & Dixon 2002, 4). This chapter discusses how media influence the constructed and intersecting identities of sex and gender. Within this chapter, I trace some of the important geographical ideas in the examination of sex, gender and media. Three questions I ask in this chapter are:

- How do media shape sex and gender?
- What are the mediated spaces of sex and gender?
- Where can feminist media geography go from here?

Qualitative methods (e.g. critical visual analysis) are typically employed to answer feminist geographical explorations of media. In critical visual analysis, Gillian Rose (2012, 20) asks geographers to contemplate the social in images including the production, the content and the consumption of an image: "the range of economic, social and political relations, institutions and practices that surround an image and through which it is used and seen." Feminist geographers seek to understand the social in visual images by examining not only the content of the image, but its wider context (Roberts 2016). The social relations depicted in filmic or "reel" representations often elucidate social relations in "real" life (Dixon et al. 2008; England 2018).

How do media shape sex and gender?

Media's power centers on their ubiquitous nature and their often subtle role in shaping society and individuals. The power of visual media stems from media normalizing gendered and sexualized structures and stereotypes. While subversive moments exist, media often

DOI: 10.4324/9781003039068-23

reproduce traditional notions of sex and gender. Mediated sociospatialities produce and reinforce these patriarchal structures. Douglas Kellner (1995, 5) states:

> Media stories provide the symbols, myths and resources through which we constitute a common culture and through the appropriation of which we insert ourselves into this culture.

Media are often not seen as constructs, but instead "how things are" (Hall 1995). Stuart Hall (1995, 18–20) argues:

> Media's main sphere of operations is the production and transformation of ideologies... Ideologies produce different forms of social consciousness, rather than being produced by them. They work most effectively when we are not aware that how we formulate and construct a statement about the world is underpinned by ideological premises; when our formations seem to be simply descriptive statements of how things are (i.e., must be), or of what we can "take-for-granted."... In modern societies, the different media are especially important sites for the production, reproduction and transformation of ideologies... But institutions like the media are peculiarly central to the matter since they are, by definition, part of the dominant means of ideological production.

Media portrayals of the everyday create a sense that it is a reflection instead of a construct and a commodified image and ideology. Phillip Green (1998, 16) states: "When ideological discourse 'works,' it does so by... seeming to be just a believable story about real people and their lives. Whatever social roles are eventually to receive us, visual culture is capable of presenting these roles as natural."

Media contribute to the formation of cultural and sociospatial identities by influencing how the viewers/readers/audiences see themselves, others and the spaces they inhabit. Representations of sex and gender roles demonstrate this—social constructions of sex and gender often demonstrate how patriarchy is upheld or subverted. Gender and sex often follow patriarchal scaffolds in many mediated images and spaces wherein manifestations of patriarchy upkeep traditional sex roles of the "active" male and "passive" female. Geography is key to understandings of how sex and gender are constructed in society. Media help to transmit these sociospatial constructions.

Representations of negative sex and gender stereotypes reinforce harmful sociospatialities of what, where, and who is to be valued. Affirming representations of sex and gender have the potential to destigmatize difference from the heteronormative and to legitimize identities typically not seen in mainstream media. Media can have liberating aspects, which destabilize patriarchal gender roles and understandings of sex. For Douglas Kellner (1987, 490), "Emancipatory popular culture subverts ideological codes and stereotypes, and shows the inadequacy of rigid conceptions that prevent insight into the complexities and changes of social life."

Feminist media scholars work to critique patriarchal ideologies. One such way is the Bechdel-Wallace Test, which is a cultural understanding of media. To pass the Bechdel-Wallace Test (created by Alison Bechdel and Liz Wallace), a piece of media has to have the following components:

1   it has to have at least two named women in it
2   who talk to each other
3   about something besides a man

The test first appeared in Bechdel's comic strip in 1985 and is inspired by Wallace's reading of the following Virginia Woolf passage from *A Room of One's Own* (2015 [1929], 74–75):

> But how interesting it would have been if the relationship between the two women had been more complicated. All these relationships between women, I thought, rapidly recalling the splendid gallery of fictitious women, are too simple. So much has been left out, unattempted. And I tried to remember any case in the course of my reading where two women are represented as friends... They are now and then mothers and daughters. But almost without exception they are shown in their relation to men. It was strange to think that all the great women of fiction were, until Jane Austen's day, not only seen by the other sex, but seen only in relation to the other sex. And how small a part of a woman's life is that... Suppose, for instance, that men were only represented in literature as the lovers of women, and were never the friends of men, soldiers, thinkers, dreamers; how few parts in the plays of Shakespeare could be allotted to them; how literature would suffer!

Kay Steiger (2008, 104) described the Bechdel-Wallace test as "the standard by which feminist critics judge television, movies, books, and other media." Neda Ulaby (2008, para. 7) argues that the test has saliency because "it articulates something often missing in popular culture: not the number of women we see on screen, but the depth of their stories, and the range of their concerns."

The Bechdel-Wallace test is a good litmus test, yet geographic research on media often sidesteps the influence media have on sex and gender, which then negates the power of media to influence those aspects. Two recent edited collections on media geographies and mediated spaces (Adams et al. 2016 and Mains et al. 2015) only touch upon sex, gender and media in one chapter between the two despite the growing field of feminist media geographies. In Mains et al.'s *Mediated Geographies and Geographies of Media* (2015), Ken Hillis and Michael Petit write on gay/queer men and mediated experiences (webcams, social media, etc.). The lack of discussion of sex and gender in media geography needs to be rectified as media feed into understandings of self and others, what is to be embraced and shunned, and where one should go or avoid.

What are the mediated spaces of sex and gender?

Media repeatedly strengthen gender and sex norms and behaviors in space. In addition, media portray and legitimize social relations. Mediated understandings of geography shape understandings of sex and gender because people and places are sexed and gendered and imbued with sociospatial codings. Identities and spaces are socially constructed and sociospatial norms create geographies of sex and gender. Constructions of identity and codings of space are continually in flux.

Media produce virtual geographies, which are read by viewers of visual media texts through the act of participation/viewing (Craine 2009). Virtual participation leads to sociospatial relationships where the virtual and the body interact:

> This engagement provides the biological self with the means to compensate for the loss of the corporeal in the virtual environment—the viewer/consumer can now simply transform that environment into a space more susceptible to human control. Thus the virtual digital environment becomes a fundamental part of human

experience—there is a literal projection of the human into virtual space thereby allowing the viewer/consumer to construct a spatial simulacrum of the previously invisible circulation of information through this simultaneous grounding and dislocating of the viewer/consumer's bodily experience.

(*Craine 2009, 236*)

Virtual spaces can be liberating as they provide for new interpretations of places, identities and experiences. Sexuality and gender can be altered in virtual spaces—one can be whoever and wherever one wants to be there.

The internet and other technologies produce and reproduce sociospatial power relations while simultaneously situating themselves as not a component of that power (Haraway 1988). This, like other media, creates an artifice where the interactions and images are seen as natural, instead of constructed. In a discussion of "new social spaces," Rob Kitchin (1998, 386) argues that the internet does not necessarily reflect the "formal qualities of geographic spaces." The internet and its cyberspaces are often constructed as a space free from the body because there are no "real" bodies in cyberspace. Bodies on the internet can only be images, but those images are often meant to reflect reality. There is, of course, fantasy prevalent on the internet. Fantastic spaces can be liberating. One is not bound by one's body. One can be anyone or anything.

Mediated bodies (and non-mediated bodies) are geographical. Bodies, while visceral and fleshy, are also socially constructed and their representations are key to social understandings of bodies and their "value." Steve Pile (1996, 186) argues: "the contours of the body are the contours of society" and therefore, necessary to examine. As I have argued elsewhere, "it is important to acknowledge just whose bodies are being portrayed and the ramifications of those representations" (England 2018, xviii).

The corporeality of mediated experiences demonstrates the intertwined relationship between viewer and media. Representations of the body portray a public exterior that contains an interior, private site, and as such have complicated sociocultural ramifications and are important to feminist geography as both public and private spaces. The body contains—all in the same bounded (and yet porous) package—multiple scales and social meanings. Depictions of bodies in visual media contribute to understandings of norms (e.g. what is appropriate in a public space versus a private space; who is deemed beautiful by society; and who is worthy of representation) in a variety of settings. Mediated bodies reflect social constructions and have spatialities. When the naturalization of the body is disrupted through understanding the social construction of the body, there are cascading effects. The Nature/Culture dualism feeds into the Emotional/Rational, the Public/Private and the Body/Mind binaries (Rose 1993; Duncan 1996). Deconstructing one creates a domino effect of deconstruction on the others.

Dualisms and their intersectionality with sex and gender are important to examine in feminist media geography. Geographies of sex show ideas of norm transgressions and deviances. Where and what kind of sex happens helps to define norms. Geographies of sex in media happen in a number of places—including print media, visual media (e.g. television, movies) and on the internet. Real experiences have a reciprocal relationship with mediated worlds. Representations of homosexuality and LGBTQ persons can be detrimental or uplifting. Images may resonate for the wrong reasons when stereotypes are hurtful and damaging. There can also be a sense of identity affirmation and normalizing when representation matches the individual watching. As such, viewers who identify with media images that contain homosexual, bisexual, pansexual, asexual and queer people see some of their experiences reflected (e.g. a season of MTV's *Are You the One?* that focused on queer dating). Media can also portray acceptance by family and friends or rejection by them.

Liminality is the state of between. Media are liminal spaces, which are spaces in, and thresholds between, dualisms. Liminal space occupies both sides of binaries at the same time or precariously perches between states, which can lead to transgression and deviance. As such, the liminality of media means that it can uphold or subvert patriarchy or occupy both sides of it at once.

Certain genres center on the liminal aspect of their narratives. For example, liminality is key to the "horrific" in horror texts. Characters and scenarios often both reify and decon-struct patriarchal constructions of sex and gender and their associated spatial manifestations. While horror films often portray violence against women, Carol Clover (2015) argues that these texts create a positive association with the female victim, or "Final Girl," who ulti-mately defeats her tormentor (e.g. *Halloween* (1978), *A Nightmare on Elm Street* (1984)). Yet, the Final Girl often subscribes to patriarchal constructions of virtue (e.g. virginity). Two films in particular that upended this tension were *Scream* (1996) and *The Cabin in the Woods* (2010). Although considered "meta-horror" due to the self-aware plot lines, both films broke the mold of the stereotypical aspect of purity to subvert patriarchy. Rhona Berenstein (1996, 5) notes that gender norms can be disturbed:

> It is as if the fiend's toying with, and mixture of, elements that usually remain separate, such as male and female gender traits, force or invite human characters to cross boundaries as well… it is also a generic space in which human characters, male and female, behave monstrously and transgress the social rules and roles that usually confine them.

The fragility of gender and sex social constructions is read by some critics as a celebration of overcoming patriarchy. Ken Gelder (2000, 3) argues, "Horror can sometimes find itself championed as a genre because the disturbance it willfully produces is in fact a disturbance of categories we may have taken for granted." Yet the punishment for transgression ends up actually reinforcing patriarchal norms in many cases. When patriarchal order is (re)established, the film concludes.

Stereotypes

Patriarchal stereotypes of masculinity and femininity as well as heteronormativity pervade traditional media and need to be uprooted by affirming representations. Harmful ideas of feminine and masculine standards of attractiveness are often represented in media. These images show a valuation of only certain types of bodies, of certain types of beauty. When these standards are challenged, the potential for the disruption of dualisms and traditional conceptions of gender increases. Some media contribute to conventional notions of gender and sex (e.g. sitcoms like *I Love Lucy, King of Queens, Everyone Loves Raymond*, etc.) while others take a more feminist stance on the portrayal of gendered bodies, behaviors and norms (*Roseanne, Modern Family, black-ish*, even *Full House*). Media which reinforce harmful stereo-types are slowly being uprooted by those that are more subversive, but patriarchal roots are deep.

Two stereotypes of patriarchal constructions of masculinity and femininity are the "tough guy" and the "damsel in distress." The tough guy is seen as a hypermasculine figure. For Bob Mondello (2014, para. 2), these stereotypes harken back to the beginning of cinema: "Admittedly, silent films used a kind of shorthand for American behavior—stereotypes, to allow directors to brush in characters quickly without dialogue: women were almost always domestic, delicate and passive, while men were outgoing, strong and active." The "tough guy" has been a common theme of movies. James Cagney and Humphrey Bogart were archetypes in movies such as *Angels with Dirty Faces* (1938) and *Casablanca* (1942) respectively. "Gangster" movies and film noir cemented the trope. The character of James Bond (starting as a movie character in *Dr. No* (1962) and continuing today) deserves a feminist media geography anthology all on his own for his harmful stereotypes. However, the tough guy eventually changed in a number of ways, most specifically his physical form. He is brawny—a manly man like the leading men of the 1980s (e.g. Arnold Schwarzenegger and Sylvester Stallone) or the savvy and (street) smart characters played by Harrison Ford (who stars in the *Indiana Jones* and *Star Wars* series), Vin Diesel (the *Fast and Furious, xXx*, and *Riddick* series) and Jason Statham (the *Transporter, Fast and Furious, Crank*, and *Mechanic* series) in the 2000s. Ben-Zeev et al. (2012, 54) put another spin on the "hypermasculine" males:

> Perhaps the most poignant prototype of masculinity as a negation of femininity is the hypermasculine man, depicted in a plethora of popular culture media. The hypermasculine male is characterized by the idealization of stereotypically masculine traits, such as virility and physicality, while concurrently rejecting traits seen as feminine and thus perceived as antithetical and even inferior to machismo, such as compassion or emotional expression.

Disruptions of traditional notions of masculinity and femininity are the "musculinized" female and the "sensitive" male. In these images, more masculine females are shown as more proficient and powerful in fighting and physical prowess (e.g. Sarah Connor from the *Terminator* series, the Slayers in television's *Buffy the Vampire Slayer*, Letty Ortiz from the *Fast and Furious* series). They are, to use Yvonne Tasker's (1993) term, "musculinized." While not necessarily feminist figures, musculinized females can destabilize gendered norms. They are not damsels in distress and not "just" the pretty girl to fill a role or cater to certain audiences. Musculinization can disrupt gender stereotypes by:

> restaging the relationship between women and violence as not only one of danger in which women are objects of violence but also a pleasurable one in which women retaliate to become the agents of violence and turn the table on their aggressors.
>
> *(Pinedo 1997, 6)*

While the action genre tends to have more musculinized women than other genres, the upending of gender stereotypes of powerful bodies has become popular in reality sport, such as the *Ninja Warrior* television series shown in several countries worldwide (e.g. fan admiration of contestant Jessie Graff in *American Ninja Warrior*).

While the sensitive male trope is often used to feminize a man, it can also be used to connote intelligence. Usually less physically imposing than the hypermasculine male, they are often seen as the brains to the brawn or an element of humor. The television series *M*A*S*H* was noted for its intelligent humor (as well as its misogyny) in the 1970s and 80s, but leading star Alan Alda received backlash for his portrayal of the increasingly sensitive protagonist, Hawkeye Pierce. Bruce Feirstein quipped in his satirical *Real Men Don't Eat Quiche* (1982): "We've become a nation of wimps. Pansies. Quiche eaters. Alan Alda types—who cook and clean and relate to their wives" (quoted in Ellsworth 2019, para. 4). While Feirstein's book was satire, there was salience to the idea that "sensitive" equaled "wimp." Fast forward to the popular 2000s sitcom *Big Bang Theory*, in which the stars are "wimpy" physicists. Decades later, sensitivity and intelligence are still often branded as weakness in a man and are often the source of humor in similar sitcoms.

Representations of LGBTQ experiences can help to "norm" them. They open up worlds to people who may not have direct experiences outside of heteronormativity. While sexuality stereotypes are common (often because they are easy for many to write/digest), the very presence of LGBTQ characters, actors and contestants in a variety of media provided room for change in the socio-political world. Changing "who" is/was present in media challenged many viewers but provided a non-confrontational space of learning about other people. Television sitcoms from *Soap* (Jodie Dallas) to *Will & Grace* (Will Truman and Jack McFarland) had gay characters that visited homes each week, slowly changing mindsets. In 2012, then US Vice President Joe Biden remarked: "I think *Will & Grace* probably did more to educate the American public than almost anybody's ever done so far. People fear that which is different. Now they're beginning to understand" (Brook 2019, para. 1).

When the first transgender male character was featured on a scripted television series, it was a pivotal moment in media history. The character's story resonated with viewers and was recognized by the GLAAD Media Awards, "which recognize and honor media for their fair, accurate and inclusive representations of the lesbian, gay, bisexual, transgender and queer (LGBTQ) community and the issues that affect their lives" (GLAAD 2020). GLAAD's Wilson Cruz stated, "When *Degrassi* introduced its large and loyal audience to Adam Torres, an authentic, multi-dimensional transgender character, the show not only made television history, but set a new industry standard for LGBT inclusion" (Adams 2013, para. 10).

Mental maps

Media inscribe mental maps by constructing norms in space. The previous sections detailed how media strengthen and instill ideologies of sex and gender. Certain, or sometimes many, public spaces are spaces of threat. Nancy Duncan (1996, 128) posits: "The public/private dichotomy (both the political and the spatial dimensions) is frequently employed to construct, control, discipline, confine, exclude, and suppress gender and sexual difference preserving traditional patriarchal and heterosexist power structures." Media fortify the dualism of Public/Private through images of the public as dangerous:

> Social critics, feminists, and academics all assert that the mass media contribute to the prevalence of fear of crime, and more specifically to female fear. They reason that the attention the media give crime and violence teaches women to fear, and continually reinforces those lessons through frequent portrayals of violence against women.
>
> *(Gordon & Riger 1991, 67)*

Texts that range from sitcoms to horror films to daily news outlets shape our constructions of space and place. Mediated geographies created mental maps of safety for women and LGBTQ communities. Mental maps are formed through everyday interactions with people and the media:

> Direct involvement with violence; the "but-nothing-happened" encounters; observation of other women's degradation; the impact of the media and cultural images of women; and shared knowledge of family, friends, peers, acquaintances, and co-workers all contribute to assessment of risk and strategies for safety.
>
> *(Stanko 1993, 159)*

News and images of violence against women and LGBTQ persons (e.g. "gay-bashing") create a sense of uncertainty about entering and participating in public spaces. Media, including news reports, movies and television, tell spatial mythologies through the images they broadcast or stream (Gordon & Riger 1991; Stanko 1993; Wekerle & Whizman 1994; Wilson et al. 1998). There has been pushback on these images. Reclamations of space and protests like Take Back the Night disrupt, challenge and protest sexual and domestic violence. But images in news feeds or in visual media combat those protests by shaping public space as a space to be avoided in the name of safety even though men are assaulted more than women in public spaces (Gerbner & Gross 1976; Gordon & Riger 1991; Wilson et al. 1998). Gerde Wekerle and Carolyn Whizman (1994, 4) point out: "Despite media focus on public violence and attacks by strangers, the most dangerous place, especially for women and children, is still the home."

Some media bank on geographies of fear. The aforementioned horror genre feeds into the discomfort created by media stories and images. Horror is criticized for objectifying women and for communicating misogynist messages (e.g. virginity saves lives and sex equals death). In many horror texts, sexual repression leads to the mutilation of women's bodies (foreshadowing the modern "incel" movement?). Yet William Schoell (1985) argues, "Scenes in which women whimper helplessly and do nothing to defend themselves are ridiculed by the audience, who find it hard to believe that anyone—male or female—would simply allow someone to kill them with nary a protest" (quoted in Clover 1994, 36).

Where can feminist media geography go from here?

Media (sometimes silently) shape our ideologies and norm our behavior—wherein their power lies. We are exposed to sexualized and gendered understandings of everyday space, which then influence our understandings of ourselves. As media become more entwined with daily life, the consequences of what they do to individuals and societies becomes paramount to examine. Social norms have changed as media increasingly permeate culture and society and change views of ourselves, others and the spaces we inhabit or imagine. Future research into media geography will need to incorporate the changes in how media are delivered to the viewer.

Mobile media inform conceptions of spaces, including geographical imaginations and mental mappings as we navigate the everyday (and the fantastic) through the lens of media. These new and familiar explorations create spaces that disrupt traditional sociospatial notions of sex and gender (e.g. David Morley's (1986) pioneering work on the sociospatial role of television in the home). Morley (1986) demonstrated that the television, especially the remote control, was key to the notion that "a man's home is his castle." The male head of the family "guarded" the remote and claimed the television as his domain. This is shifting with mobile media, in which media are mostly controlled (with some limitations perhaps placed on children) by their consumers. The mobility of media will only become more important as "rooted" media (such as cable) decline in popularity in favor of streaming services. The ability of media to be mobile—more so than ever before—shapes our interactions with spaces and each other. Media's literal place within our lives is changing. For many, the mobility of media has affected understandings of spaces that we inhabit as we consume media. It used to be that many homes had at least one television. With streaming services and apps, the television is no longer necessary to view television programming. DVRs have changed the temporality of media as well.

Reality television is another changing aspect of media. As scripted shows began to be seen as less profitable and "reality" gained popularity, more and more "unscripted" shows were produced. This is a phenomenon seen all over the world—"reel" shows that depict "real" life. Reality shows are starting to call their programs "unscripted" because of the nature of these programs. The moniker "reality" is often invalidated since many see a camera following participants around and filming their moves with producer influence as not a reflection of everyday life. Reality programming creates a reflection of the real world to some degree through its produced images in the reel world. Socioculturally, reality programming changes our individual and worldviews. But audiences are savvy. They constantly negotiate the distinction between the "reel" and the "real." That being said, entertainment is often more a factor than authenticity (e.g. none of the cast of Bravo's *Real Housewives of New York* are married).

While many reality shows have storylines that are scripted or heavily produced, there are still important aspects to examine. Many minorities (especially racial and sexuality) are produced as stereotypes in reality television (see Ragan Fox's experiences and analysis of his time on *Big Brother* (2018)). For example, *The Bachelor/Bachelorette* franchise is notoriously critiqued for its white heteronormativity. Yet LGBTQ visibility is at an all-time high in reality shows (e.g. *I am Cait, RuPaul's Drag Race*) (Lovelack 2019).

The examination of the "where" of reality television potentially further links geography to the genre. Location plays a key role in many reality television shows (*Big Brother* and *The Real World* are set in houses, *Love Island* is set in a tropical villa and *Paradise Hotel* in a tropical hotel). The "setting" for the show factors into the authenticity for the viewer. Tropical locations may detract more from "plotlines" than shows set more in houses. Yet fantasy sites are part of the allure and romance of reality television as it adds something beyond the mundane. Ironically, many reality television sets restrict use of outside media by its participants (*Big Brother, Love Island, America's Next Top Model*, MTV's *The Challenge*). Of note in future research may be the role of place in setting the tone for social interactions (e.g. does a home reproduce more familial or intimate relationships than does a hotel?).

Conclusion

Feminist geography should continue to focus on the effects media have on identities and spaces. Feminist media geography demonstrates how space forms sexed and gendered ideologies and behaviors. Representations in media can both concretize and destabilize understandings of sex and gender. Media have a liberatory potential in that they can undermine patriarchal relations and understandings or can undergird traditional, patriarchal ideologies and codings of space. By understanding how media produce sexed and gendered geographies, patriarchal ideologies and norms are scrutinized and possibly weakened. Media influence social encounters and accepted behaviors in those interactions. Visual media code spaces and inform sexed and gendered identities. Gendered codings, in turn, shape understandings of sex and sexuality and represent patriarchal constructions of masculinity and femininity. Yet, the potential exists for feminist media production and representation to destabilize harmful constructions and ideologies. There is still a lot of room for growth. Although media have brought to light sex and gender issues, they often contain remnants of harmful stereotypes and geographical imaginations. Charisse L'Pree (2013, para. 8) states:

> Media is not arbitrary, random, neutral or apolitical… It doesn't matter if the film was not meant to be "deep." Human lives are. Media characterizations have very real repercussions for real life. Even a film meant to be "just fun" can reject stereotypes and take on the difficult work of creating something that actually challenges what we think we know about people.

Understanding the role of media in gender formations and sexuality ultimately demands a need to understand how social power works both ideologically and spatially. It is key to examine how power infiltrates and informs sociospatial relations. When those roots of hegemony are discovered, a more inclusive society can flourish. Analysis of the role of media in creating, maintaining and disrupting patriarchy is more important than ever before in this time when calls for diversity, equity and inclusion are finally being heard.

References

Adam, N. 2013. *Sad turn for Adam on Degrassi in last night's episode*. www.glaad.org/blog/sad-turn-adam-degrassi-last-nights-episode-spoiler.

Adams, P. C., Craine, J., and Dittmer, J. (eds.) 2016. *The Routledge research companion to media geography*. New York: Routledge.

Anderson, B. 2006. *Imagined communities: Reflections on the origin and spread of nationalism*. New York: Verso.

Ben-Zeev, A., Scharnetzki, L., Chan, L. K., and Dennehy, T. C. 2012. Hypermasculinity in the media: When men "walk into the fog" to avoid affective communication. *Psychology of Popular Media Culture*, 1 (1), 53–61.

Berenstein, R. 1996. *Attack of the leading ladies: Gender, sexuality and spectatorship*. New York: Columbia University Press.

Binnie, J. 2001. The erotic possibilities of the city. In D. Bell, J. Binnie, R. Holliday, R. Longhurst and R. Peace (eds.) *Pleasure zones: Bodies, cities, spaces*, pp. 103–128. Syracuse, NY: Syracuse University Press.

Brook, T. 2014. *From Modern Family to Glee: How TV advanced gay rights*. www.bbc.com/culture/article/20140924-the-biggest-ally-of-gay-rights.

Clover, C. J. 2015. *Men, women, and chain waws: Gender in the modern horror film*, 2nd edn. Princeton, NJ: Princeton University Press.

Craine, J. 2009. Virtualizing Los Angeles: Pierre Levy, "The Shield," and http://theshieldrap.proboards45.com/. *GeoJournal*, 74 (3), 235–243.

Cresswell, T., and Dixon, D. 2002. Introduction: Engaging film. In T. Cresswell and D. Dixon (eds.) *Engaging film: Geographies of mobility and identity*, pp. 1–10. Lanham, MD: Rowman & Littlefield Publishers.

Dixon, D., Zonn, L., and Bascom, J. 2008. Posting the cinema: Reassessing analytical stances toward a geography of film. In C. Lukinbeal and S. Zimmerman (eds.) *The geography of cinema: A cinematic world*, pp. 25–47. Stuttgart: Franz Steiner Verlag.

Duncan, N. (ed.) 1996. *BodySpace: Destabilizing geographies of gender and sexuality*. London: Routledge.

Ellsworth, K. 2019. *Alan Alda, Feminist: M*A*S*H star was the sensitive '70s man*. https://groovyhistory.com/alan-alda-mash-feminist-sensitive.

England, M. 2018. *Public privates: Feminist geographies of mediated spaces*. Lincoln, NE: University of Nebraska Press.

Fiske, J. 1987. *Television culture*. London: Methuen.

Fox, R. 2018. *Inside reality TV: Producing race, gender, and sexuality on "Big Brother."* New York and London: Routledge.

Gelder, K. 2000. Introduction: The field of horror. In K. Gelder (ed.) *The horror reader*. New York and London: Routledge.

Gerbner, G., and Gross, L. 1976. Living with television: The violence profile. *Journal of Communication*, 26 (2), 172–199.

GLAAD. 2020. *31st annual GLAAD media awards*. www.glaad.org/mediaawards/31.

Gordon, M. T., and Riger, S. 1991. *The female fear*. Chicago, IL: University of Illinois Press.

Green, P. 1998. *Cracks in the pedestal: Ideology and gender in Hollywood*. Amherst, MA: University of Massachusetts Press.

Hall, S. 1995. The whites of their eyes: Racist ideologies and the media. In G. Dines and J. M. Humez (eds.) *Gender, race and class in media: A text-reader*, pp. 18–22. Thousand Oaks, CA: Sage Publications.

Haraway, D. 1988. Situated knowledges: The science question in feminism and the privilege of partial perspective. *Feminist Studies*, 14 (3), 575–600.

Hillis, K., and Petit, M. 2015. From webcams to Facebook: Gay/queer men and the performance of situatedness-in-displacement. In S. Mains, J. Cupples and C. Lukinbeal (eds.) *Mediated geographies and geographies of media*, pp. 261–272. Heidelberg: Springer.

Kellner, D. 1987. TV, ideology, and emancipatory popular culture. In H. Newcomb (eds.) *Television: The critical view*, pp. 471–503. Oxford: Oxford University Press.

Kitchin, R. M. 1998. Towards geographies of cyberspace. *Progress in Human Geography*, 22 (3), 385–406.

Lovelock, M. 2019. *Reality TV and queer identities: Sexuality, authenticity, celebrity*. New York: Springer.

L'Pree, C. 2013. *Race and gender in the Fast and Furious franchise.* https://charisselpree.com/2013/12/01/race-and-gender-in-the-fast-and-furious-franchise/.

Mains, S. P., Cupples, J., and Lukinbeal, C. (eds.) 2015. *Mediated geographies and geographies of media*. Heidelberg: Springer.

Massey, D. 2006. The geographical mind. In D. Balderson (eds.) *Secondary geography handbook*, pp. 46–61. Sheffield: Geographical Association.

Mondello, B. 2014. *Who's the man? Hollywood heroes defined masculinity for millions NPR.* www.npr.org/2014/07/30/336575116/whos-the-man-hollywood-heroes-defined-masculinity-for-millions.

Morley, D. 1986. *Family television: Cultural power and domestic leisure*. London and New York: Routledge.

Pile, S. 1996. *The body and the city: Psychoanalysis, space, and subjectivity*. London: Routledge.

Pinedo, I. C. 1997. *Recreational terror: Women and the pleasure of horror film viewing*. Albany, NY: State University of New York Press.

Roberts, L. 2016. Interpreting the visual. In N. Clifford (eds.) *Key methods in geography*. Thousand Oaks, CA: Sage.

Rose, G. 1993. *Feminism and geography: The limits of geographical knowledge*. Minneapolis, MN: University of Minnesota Press.

Rose, G. 2001. *Visual methodologies: An introduction to researching with visual materials*. Thousand Oaks, CA: Sage.

Stanko, E. 1993. Ordinary fear: Women, violence, and personal safety. In P. Bart and E. Moran (eds.) *Violence against women: The bloody footprints*, pp. 155–164. London: Sage Publishing.

Steiger, K. 2011. *No clean slate: Unshakable race and gender politics in The Walking Dead: Triumph of The Walking Dead—Robert Kirkman's Zombie Epic on Page and Screen*. Dallas, TX: Smart Pop.

Tasker, Y. 1993. *Spectacular bodies: Gender, genre and the action cinema*. New York and London: Routledge.

Ulaby, N. 2008. *The "Bechdel Rule," defining pop-culture character.* www.npr.org/templates/story/story.php?storyId=94202522.

Wekerle, G. R., and Whitzman, C. 1994. *Safe cities*. New York: Van Nostrand Reinhold.

Williams, L. (ed.) 2004. *Porn studies*. Durham, NC: Duke University Press.

Wilson, B. J., Donnerstein, E., Linz, D., Kunkel, D., Potter, J., Smith, S. L., Blumenthal, E., and Gray, T. 1998. Content analysis of entertainment television: The importance of context. In J. T. Hamilton (eds.) *Television violence and public policy*, pp. 13–53. Ann Arbor, MI: University of Michigan Press.

Woolf, V. 2015. *A room of one's own and three guineas*. Oxford: Oxford University Press.

20

MEDIA, BIOMES AND ENVIRONMENTAL ISSUES

Hunter Vaughan

Though debated due to its flexible and wide usage as a conceptual paradigm, notions of the "social construction of nature" (Castree & Braun 1998; Proctor 1998; Demerrit 2002) have become central to cultural studies and social science debates surrounding the environment and constitutive ways in which cultural practices position humanity within the natural world. Understandably because of its wide-reaching influence over cultural values and its multi-layered capacity for ideological messaging, popular screen culture has integrated this constructionist angle into the understanding and analysis of how films, television shows, streaming media and social media shape not only our environmental values and discourses, but the material environment itself. In these pages, and with aims of dismantling the nature/culture binary and pointing towards horizons of our digital media culture, I look at a range of environmental effects of primarily US screen media history and practice, from ways in which media practices impact specific biomes to how film and media industry history and textual tactics coincide with the perception of—and action surrounding—environmental issues.

Early ecocritical scholarship (Brereton 2004; Ingram 2004; Cubitt 2005) in film and media was heavily focused on issues of representation, aesthetics and narrativity. These studies surveyed ways in which popular film and television reaffirmed 20th century anthropocentric cultural values, laying out landscapes through aerial cinematography and piercing ocean depth through digital submersibles as the planet unfolded to the colonizing gaze of the visual consumer. Recent developments in ecocritical media studies and geography, which find overlapping interest in issues of technological infrastructure and social justice, have emerged according to a disciplinary terrain increasingly bridged between environmental humanities and social sciences. That the hierarchical order of empire—according to which Western, white, heteronormative men with capital stand atop a world order driven by exploitation, violence, slavery and biopower—was both reflected in and challenged by 21st century developments in globalization and analysis of transnational flows (Gustafsson & Kaapa 2013) has made clear the larger shared interests between environmental media studies and geography, as has a more material and geopolitical turn in environmental media studies (see Maxwell & Miller 2012; Bozak 2013; Cubitt 2016; Vaughan 2019).

Themes connecting climate change and environmental injustice to histories of global inequality have become not only apparent but central to the works of geographers such as Jennifer Gabrys and Kathryn Yusoff. Gabrys (2013) targets the layers of technological media

DOI: 10.4324/9781003039068-24

waste (or "digital rubbish") being generated by the era of electrical cultural practices, whereas Yousoff (2019) traces the reframing of geology as an object of study through many centuries of the violent displacement of Black and Brown peoples. Through very different lenses, both geographers ultimately relate back to a material notion of human, ecosystemic and resource rearrangement in ways that greatly benefit the horizon of environmental media studies. In this chapter, I offer a historical perspective on ways in which the social politics and global inequities of late colonialism and early globalization have been linked to the construction of a technological media infrastructure that has snaked the ocean floors with fiber optic cable, decorated landscapes with cell towers, and enshrouded the atmosphere in orbiting satellites. Today's digital culture results from mining the earth's organs for precious metals and penetrating its core to keep pace with rising energy dependency, and results in local and global distributions of toxic waste that profit the few while disproportionately endangering the many according to boundary lines based on race, gender and class. The potential global audience of this new and supposedly immaterial format for cultural distribution is not maintenanced via the beaming of mere signals on the ether or numbers in the cloud, but is carved from the disruption of geological layers and ecosystems, and deepened structures of human inequity.

Nonetheless, our burgeoning digital media apparatus has also provided for new horizons of spatial mapping, climate science and environmental interventions. LiDAR and other forms of remote sensing offer data gathering deep in glaciers, drones self-immolate in the rings of Saturn after beaming back images of a potential habitat for life elsewhere in the solar system, and submersible cameras peruse the ocean abyss. Meanwhile, our immediate surroundings and environments are increasingly controlled through machine-to-machine environmental sensing technologies (Gabrys 2015), and recent announcements by New York Governor Andrew Cuomo indicate that in the wake of the COVID-19 pandemic tech companies like Google will be entrusted to convert city recovery into the building of "smart cities" where data and surveillance allegedly make human existence more efficient. CGI-based data visualization and simulation has merged climate science and communication with animation, as infographics and short-form video replace prose and cinema as the predominant popular forms of communication. Meanwhile, traditional screen industry mechanisms like the star system adapt to shifting environmental values, with screen celebrities emerging as eco-warriors amidst a "post-political" era (Hammond 2017) where politicians emulate movie stars and the public sphere is replaced by the identity constructivism of social media.

In this chapter I map the connections between media practice, environmental issues and the geographical spaces of particular biomes, following a mostly chronological framework so as to situate visual culture and industry practices within the shifting forces of geopolitics and cultural norms and values. In doing so, I introduce a range of factors that shape screen industry practices, and offer a genealogy to the material impact of today's media apparatus.

Early

Derived in equal parts from entertainment and from science at the end of the 1800s, moving-image technologies were shaped by the industrial modernity of assembly-line consumer capitalism and the worldview of post-Enlightenment imperialism. Extending the 19th century's proliferation of image culture, through which colonial enterprise and racist social hierarchies had used mechanically reproduced images of advertising and tourism to package the world as something to be consumed through the eyes (Gunning 1989), cinema promised a more complete perceptual version of global tourism. Early ethnographic films such as

Robert Flaherty's *Nanook of the North* (1922) integrated Western cartography into direct filming of exoticized spaces and peoples (in this case the Inuit of Canada), bringing the global Other to the doorstep of urban American audiences. That Flaherty arrived in Canada prospecting for exploitable natural resources on behalf of his father's iron ore venture, and that the Inuit people had ceased to practice many of the cultural rituals he hoped to salvage on film, thereby prompting him to hire an Inuit actor to recreate them, were not part of the film's original narrative. This "narrative documentary," as it came to be the model for, was a noble savage construction placing the cultural Other as an index of a rough and uncivilized wilderness.

This two-pronged dichotomy (white Western filmmaker in control of image meaning, people of color framed through documentary codes of truth in the visual world as part of a wilder and separate nature), became the standard for early nature films such as *Simba* (1928) and the other films made by Martin and Osa Johnson in Africa (see Mitman & Cronon 1999, 5–35). As has been argued by cultural theorists (Said 1978; Hall 1997), image culture has been central to the "spectacle of the Other" (Hall 1997) crafted as a cultural corollary to the raced and gendered inequalities institutionalized through centuries of market capitalism, labor exploitation and military occupation. The connection between local and global social inequalities, founded in philosophical worldviews of the Scientific Revolution and instituted through the fossil fueling of armed expansionism, and the ongoing anthropogenic destruction of ecosystems and exploitation of natural resources, has increasingly been revealed and critiqued through feminist environmental histories (Merchant 1990) and decolonial cultural studies (Ghosh 2016). These connections must further be understood in relation to the geographical formations and industrial infrastructures that both shaped and were further shaped by the early cinema industry, which came into being amidst an urban boom facilitated by population shifts (in particular an influx of Eastern European immigration and the internal Great Migration of freed black southerners to northeastern industrial cities) and was interlinked infrastructurally and industrially with increased electrification and the rise of automobile culture, all of which depended on the burgeoning flow of fossil fuels.

During the decade of the Nickelodeon (roughly 1905–1915), while cinema technology was monopolized by Thomas Edison's Motion Picture Patent Company, film studios' design and architecture was shaped by the necessity for sunlight, steady electricity and other resource dependencies. Meanwhile, picture houses were mainly the purview of big cities, where screened shorts offered a paradoxical blend of forced spatial assimilation and cultural representation to diverse and otherwise marginalized populations (Musser 1991). As such, early cinema played a crucial cultural role in hand-holding the transition of the US and many other countries from agricultural to industrial nations, a problematic country/city tension covered by centuries of British literature (see Williams 1973) and central to many early feature films such as the overtly racist *Broken Blossoms* (dir. D.W. Griffith 1919) and F.W. Murnau's idyllic 1928 *Sunrise*. This tension would remain prominent through the twentieth and into the 21st century, only growing more focused on anthropogenic disruptions of biomes in recent non-narrative films such as *Baraka* (dir. Ron Fricke 1992) and *Manufactured Landscapes* (dir. Jennifer Baichwal, 2006). The early 1900s were also a time of great westward expansion, and while the move to California was partially fueled by a desire to escape the monopolistic practices of Edison's MPPC and the pervasive anti-Semitism of grounded northeastern institutions, Hollywood histories of late (McKim 2013; Jacobson 2015) have framed the move out west as one largely of environmental incentives. A new portrait emerges of geographical and geological asymmetry, with the skyscraping shadows of east-coast industrialization fading behind those drawn by southern California's allure of open spaces, long days with lots of sunlight and a diverse topographical concentration of desert, mountains and sea (McKim 2013, 51–52).

Despite a geographical shift, Hollywood built itself as both index and icon for capitalist modernity, perhaps best captured by the conveyor belt satire in Charlie Chaplin's wildly successful *Modern Times* (1936). With an assembly line manufacturing process, market-driven production, a sociopathic caste system promising unprecedented wealth to those at the top based on the exploitation of women and the suppression of people of color and other aspects of social difference, and an anthropocentric irresponsibility towards the raw materials and natural resources on which it depended, Hollywood was the ultimate manifestation of 20th century American capitalism. And, through specific genres such as the Western, mainstream screen culture managed to justify a violent past as a courageous myth of national origin (Bazin 2002 [1952]), to flatten out the problematic contradictions of an ideological system that would come to reign as neoliberalism (Wood 1989), and to foster popular perceptions of a value system (Schatz 1981) that included rampant environmental destruction. Meanwhile, far from the audience's eyes, an entire infrastructure of resource extraction and pollution was being laid across the nation to provide the material base for this culture of light and magic.

Classical

The Classical Hollywood era, extending roughly from 1927 (when sound was introduced) to 1949 (when the anti-trust Paramount Decree signaled the breakup of the studios' vertical integration of production, distribution and exhibition), laid the foundation for the most prominent ways in which contemporary screen culture overlaps with the environment. This includes normativized hierarchies, laid out through narrative content and formal aesthetics, that position humans as superior and meant for mastery of natural resources, biomes and wildlife, extending this to social hierarchies of patriarchy, white supremacy and heteronormativity; production norms and practices based on excessive travel, ecosystem disruption, low efficiency energy use and high waste production; an infrastructural life cycle that promotes environmental degradation at the service of mining and manufacturing, generates toxic wastes and necessitates distribution streams from highways to cable installation to satellite signaling; and the star system as a conveyance mechanism for social influence.

This industry's crystallization introduced new local, national and global pathways that impacted transportation flows, public health and ecosystem stability, and the geopolitics of wartime alliances and postwar spoils. Eastman Kodak, which established its monopoly on the patent technology and practices of film stock production in the early 1920s, was not only the nation's second-largest consumer of pure silver bullion (after the US Mint), but was also a cavernous abyss for water use and pollution. The Kodak Park Plant in Rochester, New York was propped strategically alongside Lake Ontario, from which it drew more than 12 million gallons of water daily for the annual production of 200,000 miles of film stock during the 1920s (Maxwell & Miller 2012, 73). By the end of the 20th century, when it was responsible for 80% of the world's film supply, Kodak Park was using 35 to 53 million gallons of fresh water per day. After being siphoned off of Lake Ontario, the water was run through the plant's elaborate chemical rinsing process and then dumped into the Genesee River, which extends through Rochester and another 157 miles down through New York and into Pennsylvania. Not until the Clean Water Act of 1972 were American factories forced to collect the majority of their wastewater in treatment plants—by that point, Kodak's dumping of post-production chemicals into the groundwater of New York made it the primary source of carcinogenic pathogens in the state, and Rochester was "ranked number one for overall releases of carcinogenic chemicals" from 1987 to 2000 (Maxwell & Miller 2012, 73).

In a more global geographical scale, the mainstream film industry proved crucial to the Good Neighbor Policy of the wartime era, an economic and cultural extension of banana republic-style imperialism focusing on trade relations with Latin America (which was seen as the US's hemispheric partner and, conveniently, the only geographical region—and therefore export market—not engulfed in war from the end of the 1930s to the mid-1940s). However, due to decades of negative stereotyping, Hollywood had badly hurt cultural relations with Mexico as well as Central and South American nations, an image problem they sought to rehabilitate through a number of films starring Latinx cast members (often Carmen Miranda) and featuring what were meant to be more positive stereotypes surrounding Latinx contributions to the American way of life: leisure travel, coffee and fruit (in particular bananas).

Consequently, trade and tourism flourished along this route and helped extend Hollywood's market while bolstering the long-rippling neocolonial economic imperialism of multinational corporations like the United Fruit Company. Across the Atlantic, a similar model was put into effect through the Marshall Plan, essentially trading postwar reconstruction support for opening European markets to American capitalism. This flow, powered by the technological wonders and financial profiteering of the military-industrial complex, saw a rise in output that was ideologically dubbed "prosperity" while being later understood scientifically as the "Great Acceleration" of climate change. Lastly, the expansion of the highway system and the flaring out of the suburbs in the 1950s gave rise to widescale pesticide use (the environmental and health consequences of which were documented and critiqued by Rachel Carson in her seminal 1962 *Silent Spring*) and the multiplex culture of New Hollywood; as "white flight" (or the relocation of higher income urban white populations to the suburbs) built a geography of racial segregation, movies became bigger, louder and costlier, with exhibition gimmicks like 3D and IMAX growing out of the seemingly infinite construction and manufacturing meant to meet (and to inspire) consumer growth.

Meanwhile, in conjunction with the rise of photojournalistic weeklies like *Life Magazine* and the introduction of television, the postwar advertising industry reshaped consumer capitalism, promoting excess and conformity in the name of progress. Despite Brown v. the Board of Education desegregating American schools in 1954, and strong antiwar and women's movements pushing progressive social values, conventions and clichés of social inequality were maintained through advertising codes, which coupled major consumer products like automobiles and fashion with the exoticization of people of color, the objectification of women, the celebration of heteronormativity and—in combination with these— the human domination of landscapes and resources (see Sturgeon 2008). With the expansion of the suburbs, the rise of television as a dominant screen format and the crystallization of hegemonic power through the filters of advertising, the postwar era laid the urban infrastructural groundwork, wired to a central communications industry via televisual media and guided by the ethical void of the free market, that would come to reign over the post-Cold War era of a global digitized neoliberalism.

New

In 1968, astronaut William Anders took a photograph of a distant and half-darkened earth over the horizon of the lunar surface; dubbed "Earthrise," this photograph had profound ramifications on popular perceptions of our place in the universe, and lent itself to many philosophical and narrative musings of the emerging environmental movement. The first

Earth Day, in 1970, was tied to an image campaign, spread across *Life Magazine* and other major photo weeklies, largely focused on the symbolism of gas masks in order to render visible the hidden hazards of environmental destruction. Despite rising public concern and increasing visibility of massive biome-impacting industrial disasters like the Santa Barbara oil spill and the Love Canal hazardous waste scandal, it fell to the electronic production and dissemination of images to shape scientific, political and cultural concerns around environmental issues. It wasn't until satellite photographs provided visual evidence of ozone depletion that NASA scientists fully understood the scale of the crisis (Dunaway 2015, 199), but by this point the wheels of neoliberalism were too perfectly oiled: screen media's most concerted effort to express a collective environmentalist stance, the 1990 *Earth Day* television special, reduced environmental action to green consumerist individual choices such as recycling.

Finis Dunaway's 2015 *Seeing Green*, aptly subtitled "The Use and Abuse of American Environmental Images," offers perhaps the most comprehensive and poignant study of visuality in the 20th century environmental movement. However, that it concludes with a discussion of Al Gore and Davis Guggenheim's startling success with *An Inconvenient Truth* (2005) only demonstrates how quickly media technologies and practices change. Moreover, Dunaway's focus on the meaning of images reveals how enshrouded within the folds of media and communication history is the material base of visual culture. The images that revealed the ozone hole were among the earliest signs of what our electronic environmental media industry would become: a network of satellites, fiberoptic cables, signal routers and televisual screens through which the computing industry would place the capacity for complex imaging production and access in the pockets of everyday consumers. Not without its costs, though, and very much in a continued industrial transformation of our planet's geography and atmosphere.

The end of the Cold War brought new master narratives of globalization, accompanied by digital technologies that could condense space, hasten time and collapse borders like never before. A mounting genre of films around the turn of the millennium—experimental ethnographic documentaries like *Baraka*, feature Hollywood films such *Traffic* (dir. Steven Soderberg 2000), *Babel* (dir. Alejandro González Iñárritu 2006) and other Anglo-centric transnational multi-narratives, and direct confrontations with Western imperialism like *In a Better World* (dir. Susanne Bier 2011)—attempted to capture this through transnational flows of cast and crew and a nomadic movement between locales. Though much scholarship of this era tried to keep up with the consolidation and fragmentation of national and regional cinemas during this time, a critical turn also looked at opportunities offered by new media for the emergence of haptic and sensory modes of expression that brought out of the margins the voices of diasporic and Indigenous peoples (Marks 1999). Such non-places and silenced spaces have also become spotlighted in the wake of recent natural and environmental justice disasters, for example the social geographical implications of Hurricane Katrina through such disparate formats as the television documentary series (see Spike Lee's *When the Levees Broke*, 2006) and feature-length narrative magical realism (*Beasts of the Southern Wild* (dir. Benh Zeitlin 2012)).

In addition to this oft-concealed impact of globalization on those at its margins, we must add the burdens and transformations placed on the natural resources in the earth's core, the biomes of its surface and the atmosphere maintaining its life. I argue that the more poignant geographical impacts of media globalization were occurring through the carving of multinational economic agreements, resource mining and tech manufacturing, and the laying of a global infrastructure for digital culture.

Media, biomes and social justice

Any look at the landscape of the new millennium cannot escape the global grid of mediatized and manufactured technology that links our modes of communication, energy extraction and use, and everyday behavior. From the subterranean mapping systems and video displays of deepwater oil drilling and shale fracking apparatuses, to the in-dash visualization of engine-operation and miles-per-gallon calculation in automobiles, to remote thermostat controls like Nest and other smart home-control systems, digital technology is part and parcel of the current shifts in ecological exploration, extraction, use, conservation, preservation and sustainability. Moreover, we have entered a radically new era of "mining": digital technologies' exponential need for precious metals produced a 21st century Gold Rush in the Coltan mines of central Africa, while data mining through these devices has transformed methods of marketing research and consumer (as well as private citizen) surveillance—a dependency on digital interface and trust in the Big Brother of tech firms that has only increased under the COVID-19 pandemic.

As Jussi Parrika (2015, 50) writes, the "materiality of information technology starts from the soil and the underground." This largely metallic materiality begins in mines of countries like the Democratic Republic of Congo, and thus the material origin of the New Digital World Order of exploitation, imperialism and environmental injustice. In 2001 the UN Security Council issued a statement condemning surrounding nations, including Rwanda, Burundi and Uganda, for using the instability of the Congolese civil war to pilfer Congo's natural resources, including its Coltan reserves, which they sold to tech manufacturers in order to help fund ongoing military exercises; over a decade later this was still continuing, prompting the Hague Center for Strategic Studies, under the project supervision of Marjorlein de Ridder, to publish an 85-page report in May 2013 on "Coltan, Congo, and conflict." Unregulated mining practices have caused profound environmental distress, while the exploitation of labor and resources in an unstable political atmosphere has positioned smart technology as one of the main culprits in an ongoing civil war in one of the least stable parts of Africa. These raw materials then enter the global shipping routes to labor campuses (or "sweatshops"), outsourced manufacturing hubs that attract mostly female workers from impoverished rural spaces, through which multinational companies can exploit favorable labor regulations and import/export tariffs at the cost of workers' rights and welfare and the environmental degradation caused by constant shipping and poor waste management.

Precious metals and human labor are not the only resource integral to our digital culture, a set of norms and behaviors increasingly defined by paradigms of immediacy and permanence and guided by an industrial logic of obsolescence. In terms of our information and entertainment content, we expect immediate and limitless access, while at the same time expecting the expanding network of online archiving services such as Dropbox and iCloud to guard our personal data in perpetuity. The apparent invisibility of where our content and data come from (orbital satellites and undersea cables) and where it is stored (data farms) permits us to enjoy a sense of virtuality that is, paradoxically, quite real and overwhelmingly material. Moreover, maintaining our seemingly boundless universe of digital ether requires vast amounts energy from what are still mostly coal-powered grids, cooling chemicals and other peripheral resources, cycling constantly through infrastructures made of plastics, cements and precious metals. These "invisible" places, from Congolese mines to Taiwanese labor campuses to Silicon Valley superfund sites to Icelandic data farms, are the undergirding geographical and environmental justice imprints of today's global digital media production and practice.

Energy concerns (which are vast, longstanding and urgent and, from the displaced families of flailing coal towns to the splatter-paint proliferation of desert solar farms, have profound geographical consequences) aside, the geographical incisions of this "immaterial" cultural form's expansive materiality run deep and wide, and include devastating environmental justice repercussions of digital byproduct pollution and waste disposal. The incredible potential for storage and the supposed democratization of production and distribution empowered by our new digital tools not only necessitates the generation of an increasingly constant energy supply, but also plays a key role in the proliferation of packaging and technological waste that is determining the tail end of the digital imperialism chain.

Firstly, the last 25 years have overflowed with social justice violations due to semiconductor and microprocessor manufacturing, a problem well documented in Jan Mazurek's seminal *Making Microchips* and the edited collection *Challenging the Chip*, a series of essays documenting the proliferation of Super Fund sites that sprang up on the underbelly of Silicon Valley. Secondly, the end of this life cycle, which is a cradle of its own, is in the digital dumping grounds of under-developed nations. Freighters of American e-waste deposit their contents in villages of nations such as Kenya and China, where mostly child and woman workers pick through it, risking exposure to toxic chemicals and gases that increases greatly as they begin to melt the materials down in order to salvage precious metals. In these villages, less than two decades of this practice has led to heightened levels of cancer and other disease, and chemicals have seeped into the soil and polluted water sources so as to contaminate agriculture and drinking water, leading to birth deformities and other health problems. Like the geopolitical and social justice connotations of Coltan mining and electronics contract manufacturing, the waste disposal and recycling of digital media technology reeks of imperialism in the age of globalization, in which wealthy societies can outsource their consumer dirty work to poorer nations, who then are further crippled through the heightened psychological and physical health detriment of unregulated salvage practices.

This is only one facet of the larger infrastructural, geographic and environmental burdens our digital media culture imposes on natural resources, biological welfare and ecosystem stability. Nicole Starosielski (2015; Starosielski & Walker 2016) has documented the diverse ways in which the fiber optic cables through which 80% of the global economy flows have altered our landscapes and seascapes, and shaped our communities. This global industry, which pushes conformed consumer habits and cultural values across the globe, also helps to shape community agricultural practices and forest and mountain landscapes in upstate New York through the expansion of internet service cables (Starosielski & Walker 2016), while trans-Pacific cable networks shape local marine ecosystems and behaviors in Guam, and their landing sites and connection points shape the architectures of highways in O'ahu and towns in the Philippines (Starosielski 2015, 141–169).

Such infrastructural issues are the flipside of the Janus-like coin of access and agency that make digital media seemingly democratizing in its empowerment of environmental science, communication and justice advocacy. Citizen science apps and online sharing platforms are helping to deconstruct the institutionalized hierarchy of technical expertise and epistemology, and can empower modes of grassroots resistance—for example by helping Gulf coast residents to bypass the media wall set up by BP following the Deepwater Horizon catastrophe. Similar strides have been taken through the expanded use of drone cinematography; the problematic infrastructural base of new media technologies has not only forged new mechanisms for climate science (see uses of LiDAR in studying the deep time of Arctic glaciers), but has also enabled wide-ranging potential for the use of mobile cameras to reveal the very infrastructures and environmental damages, from mining practices to species disruptions, that big tech has tried to keep hidden.

Similarly, social media has proved central to iconic acts of organized environmental justice such as the #NoDAPPL protests at Standing Rock (Johnson 2017), serving to amplify specific environmental issues and to facilitate on-the-ground action planning. The social media phenomenon surrounding Greta Thunberg demonstrated the paradoxes of this, as her online celebrity revealed both social media's potential for the amplification of environmentalist causes as well as ways in which it stands as a battleground to reinforce deeply entrenched values and inequalities (see, for example, the photo-cropping exclusion of Ugandan climate activist Vanessa Nakate, by the Associated Press, of a photo taken of climate youth activists at the World Economic Forum in January 2020, leaving only the inclusion of Thunberg and three other white women from the Global North). In doing so, the imperial supremacy developed through centuries of exploitation and carefully crafted politics of representation is inflected onto the geographical framing of a global activism movement meant to be founded on principles of inclusion and equality.

Thunberg's digital celebrity comes on the coattails of a new version of environmental celebrity embodied by politicians whose environmental stances have brought them attention (Al Gore, Alexandra Ocasio-Cortez), conservation icons easily turned into screen stars and legends (Steve Irwin, Diane Fossey), and Hollywood celebrities using their clout to espouse environmental positions (Meister 2015, 283–285). The latter category exemplifies how mainstays of the classical moving-image industry have become updated for the digital era, where star personas have rebranded themselves (and reinvigorated their relevance and, consequently, their fan base) in different ways. While reality-star-turned-president Donald Trump disbands the EPA and lightens regulations on fossil fuel companies, Leo DiCaprio flies the planet in a private jet to make films about his own concern for climate change; Shailene Woodley brings attention to environmental injustice by Facebook live-streaming her arrest at Standing Rock; Matt Damon has embraced short-form comedic video to support his Water. org project; and Angelina Jolie has converted her screen celebrity into international UN collaborations on human rights and environmental preservation and conservationists; and climate deniers take turns churning the iconography of Thunberg, whose boat-and-train tour of the Global North became a rallying cry for local voices caught in a global problem, through the messaging mills of Instagram and Twitter.

The degree to which these figures in fact generate shifts in cultural values and mobilize change is debatable, as their fame still depends largely upon conventional neoliberal market dynamics and environmentally destructive screen production norms formulated a century ago. Moreover, while their messaging may engage with environmental issues, there is no proof that they—like directors such as James Cameron who have crafted their auteur image around progressive environmental stances, while relying on technofetishistic brands of innovation that have profound environmental footprints—are willing to invite substantial systematic change to the industry and cultural values that buttress their popularity, privilege and profit. While environmental topics and stances may be splashed across big and little screens across the world, this amounts to little but greenwashing, image rebranding and issue trending as long as it remains business as usual on the manufacturing and production side of screen media.

The water wars that made Los Angeles possible and decimated Inyo Valley and the surrounding agricultural ecosystems and communities were only the beginning: Hollywood and other major national industries would colonize rural areas and reshape urban spaces with the building of studios and rerouting of resources. During the postwar expansion of global markets the studios would give way to a more free flow of independent productions, being sold off to banks and then subsumed by multinational corporations, with television shows and news, feature fiction and documentary films, and streaming series and specials filmed in diverse locales where not only day passes for night but Toronto passes for New York and New Zealand for Middle Earth. During the 1990s, the decade following the end of the Cold War and trumpeting the acceleration of neoliberal global capitalism, tax rebates and other fiscal plans in US states and nations and municipalities around the world lured major productions through financial incentives, prompting migrations of media production communities in some cases and, in many more, the travel and temporary installation of dozens and even hundreds for runaway productions that often badly traumatize local ecosystems.

During the decade that was launched by the *Earth Day* television special, DiCaprio—arguably the celebrity face and voice most affiliated with the climate debate—starred in *Titanic* (dir. James Cameron 1997) and *The Beach* (dir. Danny Boyle 2000). The former, a testament to NAFTA's model of globalization, set up a "100 Days Studio" on the Pacific coast of the northern Mexican town of Popotla; the production's chlorine treatment of the water on set led to the pollution of surrounding seawater, decimated the local sea urchin industry, and reduced overall fish levels by a third (Maxwell & Miller 2012, 70). *The Beach*, in order to derive its pristine shots of an untouched exotic paradise, displaced local flora and flattened a natural barrier dune in pursuit of the image of paradise that would subsequently bring so much tourism to Phi Phi Leh beach that in 2019 a two-year tourism moratorium had to be enacted to try to resuscitate the local coral reef system.

These are but two exemplary anecdotes of media production's ongoing contribution to anthropogenic destruction of the environment, a problem that has not eased with the transition to digital practices. As introduced above, the increased digital dependency of recent media has meant only a different dynamic of strain on natural resources and production of toxic waste and greenhouse gases. *Avatar* (dir. James Cameron 2009), marketed as the first fully digital film, embodies the hypocrisy of this shift: a surface narrative critiquing the destructive pillaging of natural resources and implementing a myth of eternal life on an alternate planet once we have squandered this one, the film that billed itself according to the immateriality of the digital relied heavily on analog materials and production methods, necessitated global travel for large groups of cast and crew, depended on a transnational flow and maintenance of massive amounts of digital data 24 hours a day every day, and gave rise to the construction of alternate exhibition platforms and amusement park experiences.

While film studios and media corporations create sustainability executive positions and stage green initiatives across their websites, the reality is that the 21st century has seen only fragmentary uphill movement, largely on behalf of European governmental funding bodies and independent production consultants, towards the cultural shift that might be enacted were media industries to think radically about environmentally conscientious practice. In the meantime, we are merely distracted by the glamor of celebrity, misdirected by corporate greenwashing and seduced by the catharsis of textual therapy offered by our screen spectacles.

There is no myth that more convincingly allows us to take a rain check on accountability than the myth of eternal life—which in *Avatar*, in 2009, we can see as a gift to the anxious viewers of the era of accelerated climate change, a virtual escape from the consequences of the present. In a moment of severe environmental discord and rising anxiety, our willing merger with the digital belies an inherently languorous assumption built into our use of the term "Anthropocene": that we have squandered the planet we inherited, and that any salvation lies in a technosolutionist embrace of the virtual lives offered via digital media. We may indulge in fantasies of space colonization and white male salvation in the face of

Indigenous oppression, and may send drones to seek life on other planets as we shift our social interactions to the digital ether, but we only get one Earth. We are living out our only role, many millennia into this film that we will not get to see the end of. We do not get to stand up and walk out, having cathartically exercised our anxieties and fears without consequence; there is no sequel. While the social politics and geographical effects of digital culture must certainly be reckoned with, it is time we also turn our eyes to the hidden media infrastructures that have radically altered the compositions and livelihoods of our planet.

References

Bazin, A. 2002 [1952]. *Qu'est-ce que le cinéma?* Paris: Les Editions du Cerf.

Bozak, N. 2013. *The cinematic footprint.* New Brunswick, NJ: Rutgers University Press.

Brereton, P. 2004. *Hollywood utopia: Ecology in contemporary American cinema.* Bristol: Intellect Books.

Castree, N., and Braun, B. 1998. The construction of nature and the nature of construction: Analytical and political tools for building survivable futures. In B. Braun and N. Castree (eds.) *Remaking reality: Nature at the millennium*, pp. 3–42. New York: Routledge.

Cubitt, S. 2005. *EcoMedia.* Amsterdam: Rodopi.

Cubitt, S. 2016. *Finite media.* Durham, NC: Duke University Press.

De Ridder, M., Usanov, A., Auping, W., Lingemann, S., Espinoza, L. T., Ericcson, M., Farooki, M. 2013. Coltan, congo, and conflict. *The Hague Center for Strategic Studies*, 20 (3). https://doi.org/10.13140/RG.2.2.35662.05441.

Demerrit, D. 2002. What is the "social construction of nature"? A typology and sympathetic critique. *Progress in Human Geography*, 26 (6), 766–789. https://doi.org/10.1191/0309132502ph402oa.

Dunaway, F. 2015. *Seeing green: The use and abuse of American environmental images.* Chicago, IL: University of Chicago Press.

Gabrys, J. 2013. *Digital rubbish.* Ann Arbor, MI: University of Michigan Press.

Gabrys, J. 2015. *Program earth.* Minneapolis, MN: University of Minnesota Press.

Ghosh, A. 2016. *The great derangement: Climate change and the unthinkable.* Chicago, IL: University of Chicago Press.

Gunning, T. 1989. An aesthetic of astonishment: Early film and the (in)credulous spectator. *Art and Text*, 34 (Spring), 31–45.

Gustafsson, T., and Kääpä, P. (eds.) 2013. *Transnational ecocinema: Film culture in an era of ecological transformation.* Bristol: Intellect Books.

Hall, S. (ed.) 1997. *Representation: Cultural representations and signifying practices.* London: Sage Publications and Open University.

Hammond, P. 2017. *Climate change and post-political communication.* London: Routledge.

Ingram, D. 2004. *Green screen: Environmentalism and Hollywood cinema.* Exeter: University of Exeter Press.

Jacobson, B. 2015. *Studios before the system.* New York: Columbia University Press.

Johnson, H. 2017. #NoDAPL: Social media, empowerment, and civic participation at Standing Rock. *Library Trends*, 66 (2), 155–175.

Maxwell, R., and Miller, T. 2012. *Greening the media.* Oxford: Oxford University Press.

Mazurek, J. 1998. *Making microchips: Policy, globalization, and economic restructuring in the semiconductor industry.* Cambridge, MA: MIT Press.

McKim, K. 2013. *Cinema as weather.* London: Routledge.

Meister, M. 2015. Celebrity culture and environment. In A. Hansen and R. Cox (eds.) *Routledge handbook of environment and communication*, pp. 281–289. New York: Routledge.

Merchant, C. 1990. *The death of nature.* San Francisco, CA: Harper.

Mitman, G., and Cronon, W. 2009. *Reel nature: America's romance with wildlife on film.* Seattle, WA: University of Washington Press.

Musser, C. 1991. Ethnicity, role-playing, and American film comedy: From *Chinese Laundry Scene* to *Whoopee* (1894–1930). In L. Friedman (ed.) *Unspeakable images: Ethnicity and the American cinema*, pp. 39–81. Urbana, IL and Chicago, IL: University of Illinois Press.

Parrika, J. 2015. *A geology of media.* Minneapolis, MN: University of Minnesota Press.

Proctor, J. D. 1998. The social construction of nature: Relativist accusations, pragmatist and critical realist responses. *Annals of the Association of American Geographers*, 88, 352–376. https://doi.org/10.1111/0004-5608.00105.

Said, E. 1978. *Orientalism*. New York: Pantheon Books.

Schatz, T. 1981. *Hollywood genres*. New York: McGraw-Hill.

Smith, T., Sonnefeld, D. A., Pellow, D. N., and Hightower, J. (eds.) 2006. *Challenging the chip: Labor rights and environmental justice in the global electronics industry*. Philadelphia, PA: Temple University Press.

Starosielski, N. 2015. *The undersea network*. Durham, NC: Duke University Press.

Starosielski, N., and Walker, J. (eds.) 2016. *Sustainable media*. New York: Routledge.

Sturgeon, N. 2008. *Environmentalism in popular culture: Gender, race, sexuality, and the politics of the natural*. Phoenix, AZ: University of Arizona Press.

UN Security Council. 2001. *Report of the Panel of Experts on the Illegal Exploitation of Natural Resources and Other Forms of Wealth of the Democratic Republic of the Congo*. https://reliefweb.int/report/democratic-republic-congo/report-panel-experts-illegal-exploitation-natural-resources-and.

Vaughan, H. 2019. *Hollywood's dirtiest secret: The hidden environmental costs of the movies*. New York: Columbia University Press.

Williams, R. 1973. *The country and the city*. New York: Oxford University Press.

Wood, R. 1989. Ideology, genre, auteur: Shadow of a doubt. In R. Wood (ed.) *Hitchcock's films revisited*, pp. 288–302. New York: Columbia University Press.

Yusoff, K. 2019. *A billion Black Anthropocenes or none*. Minneapolis, MN: University of Minnesota Press.

INDEX

actor-network theory 1, 199
advertising 5, 10, 62, 102, 107, 109, 110, 111, 112, 132, 142, 152, 176, 185, 188, 257, 260
affect 6, 50, 51, 56, 57, 134, 135, 136, 137, 138, 139, 140, 142, 191, 216, 221, 222, 224, 226, 227, 252
African Americans 38, 211
AIDS 140, 155
Airbnb 172
Al Jazeera 12, 239, 240
algorithms 1, 55, 56, 114, 152, 153, 154, 156, 185, 188, 198, 199, 203
Alibaba 150
alt-Right 226, 227,
Amazon 201
Anthropocene 265
Apple 172, 176, 188
Apple Watch 176
AR (see augmented reality)
assemblage theory 1, 11, 185, 198, 199, 201, 202, 204, 224
audience 10, 12, 37, 97, 101, 102, 107, 140, 149, 188, 223, 238, 239, 251, 257
augmented reality 82, 177, 185

Baidu 24, 150
BBC 224, 240
bearing witness 214
Bentham, Jeremy 197, 198
big data 85, 199
biomes 256, 258
BIPOC 165
Bluetooth 80, 82, 167, 169,
body cameras 214
bordering 232
bots 226
Bytedance 67

cartography 5, 10, 81, 123, 124, 258
CCTV cameras 12, 168, 197, 198, 199, 203, 239, 240
celebrities 146, 150, 257, 264
cewebrities 66, 67
cell phone (see mobile phone)
cell phone videos 12, 214 238
censorship 8, 9, 19–26, 39, 45, 63, 64, 66, 70, 71, 151
China 9, 20, 21, 22, 24, 26, 37, 60–71, 147, 149, 150, 238, 239, 240
Chinese media 60–71
clickbait 106, 107, 226
CNN 106, 111, 239, 240
co-presence 97, 161
corporeality (see embodiment)
counter-stories 210
COVID-19 7, 8, 29, 37, 38, 46, 47, 70, 172, 173, 190, 211, 240, 257, 262
critical race studies 142, 210
cybercafe 25
cyberplace 185
cyberspace 2, 7, 19, 20, 26, 63, 66, 71, 109, 248

Dailymotion 149–150
data economy 185, 190
Deleuze, Gilles 134, 136, 198, 199, 202, 204
DigiPlace 164
Digital divides 2, 8, 9, 19, 29–47, 74, 78
Digital geographies 7
digital media 2, 7, 8, 10, 11, 29, 38, 46, 47, 74, 78, 84, 102, 114, 173, 175, 178, 204, 256, 257, 262, 263
digitalization 12, 109, 110, 121, 239
disciplinary society 197, 198, 199
disconnection 161
discourse analysis 98

discrimination 201
discursive violence 210
driverless vehicles 172
drones 204, 257, 266

education 4, 11, 29, 31, 32, 34, 38, 39, 40, 45,
46, 47, 83, 125, 149, 152, 153, 154, 155, 156,
200, 220
embodiment 6, 9, 12, 75
emotion 6, 80, 113, 120, 140, 200, 203, 215,
224, 227, 237, 250
emplacement 173, 179
energy use 259
equity (see inequity)
essentializing 12, 209, 233, 235, 236, 239, 240
ethnicity 4, 5, 11, 38, 209, 210, 233, 237
Eurocentrism 232–241
everyday life 83, 118, 122, 125, 162, 167, 173,
176, 177, 178, 215, 220, 221, 253
exploitation 212, 215, 234, 258, 259, 262, 264

Facebook 8, 12, 24, 40, 44, 96, 109, 114, 176,
201, 203, 225, 240, 241, 264
face-to-face 2, 7, 53, 54, 155, 184
fake news 10, 106–115, 151
fantasy 98, 107, 248, 253
feminist/feminism 5, 10, 140, 245, 246, 247, 248,
249, 250, 252, 253, 258
festivals 216
fiber optic cables 4, 29, 263
film 2, 5, 7, 10, 63, 118–127, 138, 140, 142,
147, 209, 210, 213, 222, 245, 249, 251, 253,
256–261, 264, 265, 266
filtering 20, 45, 63, 70, 71, 149
first law of geography 234
FitBit 176, 202
Foucault, Michel 5, 9, 64, 197–199
Foursquare 163, 176, 184, 188, 190

games 1, 2, 29, 49, 146, 155, 165, 166, 177, 222,
223, 234
gender 5, 11, 12, 38, 47, 97, 100, 122, 138, 140,
166, 198, 203, 222, 236, 245–254, 257
geocoding 188
geolocation 11, 185, 189, 191
geopolitics 6, 7, 97–100, 124, 221–222, 223, 227,
232–238, 240, 256, 259
geoprivacy 187–189
geospatial technologies 184
globalization 5, 12, 109, 110, 113, 114, 140, 224,
227, 239, 257, 261, 265
Google 22, 24, 79, 110, 145, 149, 150, 151, 152,
153, 163, 165, 168, 186, 187, 188, 189, 190, 239
Google Maps 163, 165, 169, 186, 188
GPS 1, 74, 80, 82, 83, 84, 163, 166
Grindr 175
GSM 186

hacking 8, 9, 20, 49–57
hashtag activism 213
hegemonic narratives 210
heteronormativity 249, 251, 253, 259, 260
Hollywood 121, 258–260, 261, 264
home 2, 4, 5, 8, 20, 83, 132, 133, 135, 138,
139, 147, 162, 167, 178, 201, 202, 223, 251,
252, 253
humor 226, 240, 250, 251
hybrid reality 177
hybrid spaces 11, 164–167, 184, 191

identity 2, 11, 37, 49, 53, 56, 57, 97, 113, 122,
135, 136, 140, 162, 164, 202, 209, 211, 215,
216, 220, 221, 222, 223, 225, 227, 228, 235,
248, 257
ideology 1, 64, 66, 70, 246
imagined community 6, 100, 225, 227
imperialism 257, 260, 262, 263
inequality 5, 7, 37, 164, 200, 256, 260
infrastructure (see media infrastructure)
information technology 29, 85, 262
Instagram 147, 166, 172, 225, 240, 264
internet 2, 4, 7, 8, 9, 19–26, 29–47, 55, 56,
60–71, 74, 109, 110, 113, 114, 138, 141,
146, 147, 149, 150, 151, 163, 175, 185,
201, 210, 214, 224–225, 226, 227, 239, 241,
248, 263
internet of things 85, 161, 167, 202
internet policy 45–46
Internet Research Agency 114, 226
intersectionality 11, 248
iWatch (see Apple watch)

journalism 6, 10, 95, 99, 101, 102, 106–115, 151,
234, 241

Kaepernick, Colin 215
Kodak 259

landscapes 5, 9, 10, 12, 40, 49, 50, 51, 57, 106,
118, 120, 121, 122, 132, 133, 136, 137, 138,
141, 209, 210, 212, 213, 217, 222, 223, 225,
232, 256, 257, 258, 260, 262, 263
Latinx 211, 260
Latour, Bruno 198, 199
learning management systems 155
LGBTQ persons 100, 140, 251, 252, 253
liminality 249
location data 173, 175, 183, 185, 187, 188–190,
214
location economy 183, 187, 189, 190, 191
location intelligence 187–190
location-aware devices 11, 176, 184, 187
location-based services 163, 164, 185, 189
locative apps 11, 183–191
Lyft 175

MacBride Report 235
maps (see cartography)
media
activism 214
frames 96, 99, 233, 234, 237, 238, 241
genres 122, 147, 149, 155, 249, 250, 252, 253,
   259, 261
infrastructure 257, 266
production 141, 211, 253, 262, 265
regulation 65
stereotypes 4, 12, 97, 210, 212, 213, 233, 234,
   236, 239, 240, 245, 246, 248, 249–251,
   253, 260
types 33, 37
mediated mobility 172–176
memes 109, 146, 225, 227
mental maps 82, 138, 251–252
microblogs 9, 45, 63, 67, 68
military 40, 154, 222, 223, 236, 238, 258,
   260, 262
mobile media 11, 163, 172, 173, 178, 179, 183,
   184, 185, 252
mobile phone 24, 32, 61, 77, 161–163, 165, 166,
   183, 184, 185, 186, 189, 241
mobilities, mediated 173–175, 176, 178, 179
mobility turn 173
modernity 3, 108, 221, 239, 259
more-than-representational theory 6
movement (see mobility)
MTurk 175
music 2, 29, 37, 38, 81, 96, 142, 147, 149, 152,
   153, 209, 213, 215–216, 217, 222

Nanook of the North 258
national image 221, 225–226, 227
nationalism 12, 66, 71, 96, 220–228
neogeography 184
neoliberalism 140, 259, 260, 261
Netflix 152, 155, 156, 225
networked infrastructures 166–169
news media 2, 6, 29, 100, 102, 107, 108, 114,
   141, 209, 210, 211, 232–241
newsfeeds 203
New York Times 24, 106, 111, 112, 189
NicoNico 149, 156
NWICO 235

Orientalism 232–241
Othering 12, 97, 98, 212, 232–241

panopticon 11, 20, 197, 198
participant videos 214
participatory mapping 165
patriarchy 246, 249, 254, 259
people of color 210, 258, 259, 260
Pepe the Frog 226
photography 3, 5, 118, 222

placemaking 11, 172, 178–179, 184
Pokemon Go 165, 177, 186
pollution 71, 259, 263, 265
popular culture 12, 139, 161, 220–223, 227, 234,
   246, 247, 250
pornography 20, 25, 151
postcolonial/decolonial theory 122, 142, 232
power geometries 4, 201
power relations 5, 12, 96, 125, 136, 141, 168,
   196, 197, 209, 215, 233, 235, 238, 248
privacy 11, 20, 45, 49, 52, 53, 54, 67, 101,
   152, 169, 179, 183, 186, 187–190, 196, 198,
   200, 201
protest 25, 50, 97, 190, 201, 204, 214, 215, 239,
   241, 252, 264
public space 75, 76, 147, 162–163, 167, 184, 187,
   197, 214, 248, 251, 252

race 4, 5, 11, 12, 38, 97, 122, 142, 165, 166,
   209–217, 233, 257
racial justice (see social justice)
racial segregation 260
racism 96, 209, 211–212, 215, 216, 234
reality TV 139, 202, 252–253
resistance 9, 11, 21, 26, 112, 113, 137, 198, 204,
   215, 216, 217, 238, 241, 263
RFID tags 80, 82, 161, 167, 169, 175
right-wing extremism 12, 114

satellites 1, 4, 29, 79, 165, 167, 172, 186, 239,
   257, 259, 261, 262
securitization 60, 196, 201, 203
self-expression 51, 145
selfies 147
sense of place 6, 12, 118, 215, 216
sexuality 142, 248, 251, 253, 254
smart cities 257
smart homes 83
smartphones 9, 30, 36, 40,46, 60, 78, 84, 149,
   161–163, 167–169, 175, 185, 186, 189,
   190, 199
SMS 67, 162
social justice 12, 200, 204, 213, 215, 256,
   262, 263
social media (see also specific services)
social norms 8, 12, 162, 252
social power relations (see power relations)
sousveillance 191, 204
spatial media 183, 184, 188, 211, 216
storytelling 178, 186, 210, 214,
streaming 10, 29, 31, 37, 38, 39, 46, 133, 135,
   141, 145–156, 225, 252, 256, 265
stunts, online 10, 147, 155
surveillance 9, 11, 19, 20, 24, 25, 52, 53, 54,
   55, 56, 66, 96, 101, 102, 168, 169, 175, 178,
   183, 185, 187, 189, 190, 196–204, 212, 214,
   257, 262

Taskrabbit 175
Telegraph 3, 4, 108
Television 1, 2, 4, 5, 7, 10, 60, 63, 65, 96, 122,
    132–143, 150, 161, 202, 222, 239, 240, 241,
    247, 248, 250, 251, 252, 253, 256, 260, 261, 265
temporality 242, 252
Tencent 9, 63, 67, 150
text messages 184
textual analysis 98, 99, 121
third space 184
transgender 251
transnationalism 227
troll farms 226
Twitter 9, 39, 67, 68, 96, 114, 147, 184, 201,
    225, 227, 228, 240, 241, 264

Uber 175, 204
ubiquitous computing 167, 169, 178, 199
UNESCO 235
urban space 141, 165, 167, 168, 169, 176,
    216, 264
user-generated content 2, 113, 146, 150, 156,
    183, 225

video sharing services 67

virtuality 10, 135, 262
virtuality reality 1, 82, 177
visual economy 237
visually impaired 75–86
VR (see virtual reality)

warfare 21, 50, 203
Waze 163, 166
Weather Channel 189
WeChat 44, 63, 67, 68, 70
Weibo 63, 67, 68, 69, 70
WhatsApp 40, 150, 201, 203
white nationalism 226
white supremacy 210, 213, 214, 215, 259
WikiMapas 165
wireless networks 4, 82, 176

Xinhua 62, 63

Youku Tudou 150
YouTube 10, 11, 133, 145–156, 201, 225, 240
YouTube influencers 146, 147, 155

Zoom 8, 172
Zuboff, Shoshana 189–190

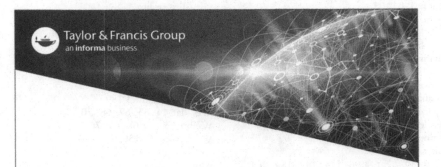

Printed in the United States
by Baker & Taylor Publisher Services